**Unsupervised and Transfer Learning**
Challenges in Machine Learning, Volume 7

# Unsupervised and Transfer Learning
## Challenges in Machine Learning, Volume 7

Isabelle Guyon, Gideon Dror, Vincent Lemaire, Graham Taylor, and Daniel Silver, editors

Nicola Talbot, production editor

Microtome Publishing
Brookline, Massachusetts
www.mtome.com

**Unsupervised and Transfer Learning**
Challenges in Machine Learning, Volume 7

Isabelle Guyon, Gideon Dror, Vincent Lemaire, Graham Taylor, and Daniel Silver, editors

Nicola Talbot, production editor



ISBN-13: 978-0-9719777-7-8

# Foreword

Over the years, machine learning has grown to be a major computer science discipline with wide applications in science and engineering. Despite its phenomenal success, a fundamental challenge to machine learning is the lack of sufficient training data to build accurate and reliable models in many practical situations. We may make the best choice of learning algorithms, but when quality data are in short supply, the resulting models can perform very poorly on a new domain.

The lack of training data problem can be taken as an opportunity to developing new theories of machine learning for developing models of complex phenonmena. As humans, we observe that we often have the ability to adapt what we have learned in one area to a new area, provided that these areas are somewhat related. We often learn in a continual fashion by standing on the shoulders of earlier models, rather than learning new models from scratch each time. We also come to recognize features that are commonly used across a set of related models. And finally, in concept learning, we often learn via not millions of examples, but through only a selected few. These phenomena call for new insights into unsupervised and transfer learning.

In the past, researchers in different subfields of machine learning have been making advances in their separate ways in the areas listed above. Recently, many of these advances have started to overlap and suggest synergistic opportunities for impact on the field. Holding a joint workshop in 2011 to bring together researchers in unsupervised and transfer learning researchers was timely. Exploring the intersection of these two fields holds strong promises for us to gain new insights, especially in their respective abilities to discover 'deep' features for domain representation. It is also an innovative idea to hold an international machine-learning contest together with the workshop, which inspires many new approaches to the difficult problem.

This volume represents a significant effort of the editors and workshop organizers, who not only put in much time in organizing the proceedings and papers, but also the challenge itself. The authors have made excellent effort in bringing us a high-quality collection with a wide coverage of topics. This collection starts with a survey paper by the editors on the state of the art of the field of unsupervised and transfer learning. It then presents papers related to theoretical advances in deep learning, model selection and clustering. The next part consists of articles by the Challenge winners on their approaches in solving the unsupervised and transfer learning contest problems. Finally, the last part consists of articles that cover various applications and specific approaches to unsupervised and transfer learning. All these articles give a complete picture of researchers' efforts for this important and challenging problem.

Looking forward, we can see several strands of emerging themes in unsupervised and transfer learning. As we march into the era of Big Data, questions on how to separate quality patterns from noise will become more pressing. The recent Google experiments on deep learning have given have show that it is possible to train a very large unsupervised neural network (16,000 computer processors with one billion connections) to automatically develop features for recognizing cat faces. The data sparsity problem associated with extremely large-scale recommendation systems provides us with strong motivation for finding new ways to transfer knowledge from auxiliary data sources. New questions about the scalability, reliability and adaptability of the unsupervised and transfer learning models will take central stage in much of ML research in the coming decade. Indeed, for unsupervised and transfer learning research, we live in very interesting times!

Qiang Yang, Huawei Noah's Ark Research Lab and Hong Kong University of Science and Technology

# Preface

Machine learning is a subarea of artificial intelligence that is concerned with systems that improve with experience. It is the science of building hardware or software that can achieve tasks by learning from examples. While much of the efforts in this field over the past twenty years have been devoted to "supervised learning", that is learning under the supervision of a "teacher" providing guidance in the form of labeled data (input output pairs), recent advances in unsupervised and transfer learning have seen a complete paradigm shift in machine learning. Unsupervised learning considers the problem of discovering regularities, features or structure in unlabeled data. Transfer learning considers the use of prior knowledge, such as learned features, from one or more source tasks when developing a hypothesis for a new target task. While human beings are adept at transfer learning using mixtures of labeled and unlabeled examples, even across widely disparate domains, we have only begun to develop machine learning systems that exhibit the combined use of unsupervised learning and knowledge transfer.

This book is a result of an international challenge on Unsupervised and Transfer Learning (UTL) that culminated in a workshop of the same name at the ICML-2011 conference in Bellevue, Washington, on July 2, 2011; it captures the best of the challenge findings and the most recent research presented at the workshop.

The book is targeted for machine learning researchers and data mining practitioners interested in "lifelong machine learning systems" that retain the knowledge from prior learning to create more accurate models for new learning problems. Such systems will be of fundamental importance to intelligent software agents and robotics in the 21st century. The articles include new theories and new theoretically grounded algorithms applied to practical problems. It addressed an audience of experienced researchers in the field as well as Masters and Doctoral students undertaking research in machine learning.

The book is organized in three major sections that can be read independently of each other. The introductory chapter is a survey on the state of the art of the field of unsupervised and transfer learning providing an overview of the book articles. The first section includes papers related to theoretical advances in deep learning, model selection and clustering. The second section presents articles by the challenge winners. The final section consists of the best articles from the ICML-2011 workshop; covering various approaches to and applications of unsupervised and transfer learning.

This project was sponsored in part by the DARPA Deep Learning program and is an activity of the Causality Workbench supported by the Pascal network of excellence funded by the European Commission and by the U.S. National Science Foundation under Grant N0. ECCS-0725746. The UTL Challenge was organized by ChaLearn (http://www.chalearn.org/); a society that works to stimulate research in the field of machine learning through international challenges. Any opinions, findings, and conclusions or recommendations expressed in this book are those of the authors and do not necessarily reflect the views of the sponsors.

This volume reprints papers from JMLR W&CP volume 27.

*November 2012*

The Editorial Team:

Daniel Silver
Acadia University
`danny.silver@acadiau.ca`

Isabelle Guyon
Clopinet
`isabelle@clopinet.com`

Gideon Dror
Academic College of Tel-Aviv-Yaffo
`gideon@mta.ac.il`

Vincent Lemaire
Orange Labs
`vincent.lemaire@orange-ftgroup.com`

Graham Taylor
University of Guelph
`gwtaylor@uoguelph.ca`

# Table of Contents

## Appendices

# ICML2011 Unsupervised and Transfer Learning Workshop


**Daniel L. Silver**                                              DANNY.SILVER@ACADIAU.CA
*Acadia University, Canada*

**Isabelle Guyon**                                                ISABELLE@CLOPINET.COM
*Clopinet, California, USA*

**Graham Taylor**                                                 GWTAYLOR@CS.NYU.EDU
*New York University, USA*

**Gideon Dror**                                                   GIDEON@MTA.AC.IL
*Academic College of Tel-Aviv-Yaffo, Israel*

**Vincent Lemaire**                                VINCENT.LEMAIRE@ORANGE-FTGROUP.COM
*Orange Labs, France*

**Editor:** Neil Lawrence


## Abstract


We organized a data mining challenge in "unsupervised and transfer learning" (the
UTL challenge) followed by a workshop of the same name at the ICML 2011 conference in Bellevue, Washington[1]. This introduction presents the highlights of the
outstanding contributions that were made, which are regrouped in this issue of JMLR
W&CP. Novel methodologies emerged to capitalize on large volumes of unlabeled
data from tasks related (but different) from a target task, including a method to learn
data kernels (similarity measures) and new deep architectures for feature learning.

**Keywords:** transfer learning, unsupervised learning, metric learning, kernel learning,
unlabeled data, challenges


## 1. Introduction

Unsupervised learning considers the problem of discovering regularities or structure in
unlabeled data (*e.g.*, finding sub-manifolds or clustering examples) based on a representation of the domain. Transfer learning considers the use of prior knowledge (such
as labeled training examples, or shared features) from one or more source tasks when
developing a hypothesis for a new target task. While human beings are adept at transfer
learning using mixtures of labeled and unlabeled examples, even across widely disparate domains, we have only begun to develop machine learning systems that exhibit
the combined use of unsupervised learning and knowledge transfer.

To foster greater research in this area we organized a international challenge on Unsupervised and Transfer Learning that culminated in a workshop of the same name at
the ICML-2011 conference in Bellevue, Washington, on July 2, 2011. This workshop

---

1. http://clopinet.com/isabelle/Projects/ICML2011/.

addressed a question of fundamental and practical interest in machine learning: the development and assessment of methods that can generate data representations (features) that can be reused across domains of tasks.

This edition of JMLR W&CP presents the challenge results and a collection of outstanding contributed articles on the subject of transfer learning and unsupervised learning. This paper and the edition focuses on unsupervised and transfer learning for classification problems based on real-valued feature representations that are related more closely to data mining tasks. Methods of transfer learning have also been investigated for reinforcement learning (Ramon et al., 2007; Taylor and Stone, 2007), however these are outside the scope of this edition.

## 2. Overview of Transfer Learning and Unsupervised Learning

### 2.1. Transfer Learning

Transfer learning refers to use of knowledge for one or more source tasks to develop efficiently a more accurate hypothesis for a new target task. Transfer learning has most frequently been applied to sets of labeled data that have a supervised target value for each example. For instance, there would be significant benefit in using an accurate diagnostic model of one disease to develop a diagnostic model for a second related disease for which you have few training examples. While all learning involves generalization across problem instances, transfer learning emphasizes the transfer of knowledge across domains, tasks, and distributions that are similar but not the same. Inductive transfer has gone by a variety of names: bias learning, learning to learn, machine life-long learning, knowledge transfer, transfer learning, meta-learning, and incremental, cumulative, and continual learning.

Research in inductive transfer began in the early 1980s with discussions on inductive bias, generalization and the necessity of heuristics for developing accurate hypotheses from small numbers of training examples (Mitchell, 1980; Utgoff, 1986). This early research suggested that the accumulation of prior knowledge for the purposes of selecting inductive bias is a useful characteristic for any learning system. Following the first major workshop on inductive transfer (NIPS1995 Workshop, 1995) a series of articles were published in special issues of Connection Science (Lorien Pratt (Editor), 1996) and Machine Learning (Pratt and Sebastian Thrun (Editors), 1997), and a book entitled "Learning to Learn" (Thrun and Lorien Y. Pratt (Editors), 1997) .

Since that time, research on inductive transfer has occurred using traditional machine learning methods (Caruana, 1997; Baxter, 1997; Silver and Mercer, 1996; Heskes, 2000; Thrun and Lorien Y. Pratt (Editors), 1997; Bakker and Heskes, 2003; Ben-David and Schuller, 2003), statistical regression methods (Greene, 2002; Zellner, 1962; Breiman and Friedman, 1998), Bayesian methods involving constraints such as hyper priors (Allenby and Rossi, 1999; Arora et al., 1998; Bakker and Heskes, 2003), and more recently kernel methods such as support vector machines (SVMs) (Jebara, 2004; Allenby and Rossi, 2005). All of these approaches rely upon the development of a hypothesis for a target task under a constraint or regularization that characterizes a similarity or *relatedness* to one or more source tasks. In 2005, a second major workshop on inductive transfer occurred at NIPS. Papers from this workshop can be found in (Silver and Bennett, 2008) as well as at (NIPS2005 Workshop, 2005).

More recently, there has been work on inductive transfer in the areas of self-taught learning (Raina et al., 2007b), transductive learning (Arnold et al., 2007), *context-sensitive* multiple task learning (Silver et al., 2008), the learning of model structure (Niculescu-Mizil and Caruana, 2007), unsupervised transfer learning [Yu,Wang], and a variety of methods that mix unsupervised and supervised learning to be discussed in greater detail below.

## 2.2. Unsupervised Learning

Unsupervised learning refers to the process of finding structure in unlabeled data resulting in new data representations (including feature representations) and/or clustering data into categories of similar examples, based on such representations (Hinton and Sejnowski, 1999). The unlabeled data distinguishes unsupervised learning from supervised learning and reinforcement learning. Important recent progress has been made in purely unsupervised learning (Smola et al., 2001; Bengio et al., 2003; Globerson and Tishby, 2003; Ghahramani, 2004; Luxburg, 2007). However, these advances tend to be ignored by practitioners who continue using a handful of popular algorithms like PCA and ICA (for feature extraction and dimensionality reduction), and K-means, and various hierarchical clustering methods for clustering (Jain et al., 1999).

## 2.3. Combining Unsupervised and Transfer Learning

It is often easier to obtain large quantities of unlabeled data from databases and sources on the web, for example images of unlabeled objects. For this reason the idea of using unsupervised learning in combination with supervised learning has attracted interest for some time. Semi-supervised learning is a machine learning approach that is halfway between supervised and unsupervised learning. In addition to the labeled data for a given task of interest, the algorithm is provided with unlabeled data for the *same* task - typically a small amount of labeled data and a large amount of unlabeled data (Blum and Mitchell, 1998). Note that these approaches usually assume that the categories of the unlabeled data, even though unknown to the learning machine, are the same as the categories of the labeled data, *i.e.,* that the "tasks" are the same.

In contrast, in the transfer learning setting, the unlabeled data does not need to come from the same task. There has been considerable progress in the past decade in developing cross-task transfer using both discriminative and generative approaches in a wide variety of settings (Pan and Yang, 2010). These approaches include multi-layer structured learning machines from the "Deep Learning" family such as convolutional neural networks, Deep Belief Networks, and Deep Boltzmann Machines (Bengio, 2009; Gutstein, 2010; Erhan et al., 2010), sparse coding (Lee et al., 2007; Raina et al., 2007a), and metric or kernel learning methods (Bromley et al., 1994; WU et al., 2009; Kulis, 2010). The "Learning to learn" and "Lifelong Learning" veins of research have continued to provide interesting results in both machine learning and cognitive science in terms of short-term learning with transfer and long-term retention of learned knowledge (Silver et al., 2008). These references include recent evidence of the value of combining unsupervised generative learning with transfer learning to generate a rich set of representation (features) upon which to build related supervised discriminative tasks. The goal of the challenge we organized was to perform an evaluation of unsupervised and transfer learning algorithms free of inventor bias to help to identify and popularize algorithms that have advanced the state of the art.

## 3. Overview of the UTL Challenge

Part of the ICML workshop was devoted to the presentation of the results of the Unsupervised and Transfer Learning challenge (UTL challenge Guyon et al., 2011a,b). The challenge, which started in December 2010 and ended in April 2011, was organized in 2 phases. The aim of **Phase 1** was to benchmark **unsupervised learning** algorithms used as preprocessors for supervised learning, in the context of transfer learning problems. The aim of **phase 2** was to encourage researchers to exploit the possibilities offered by new cutting-edge cross-task transfer learning algorithms, which **transfer supervised learning knowledge from task to task**.

To that end, the competitors were presented with five datasets illustrating classification problems from different domains: handwriting recognition, video processing, text processing, object recognition, and ecology. Each dataset was split into 3 subsets: development, validation, and final evaluation sets. In phase 1, all subsets were provided without labels to the participants. The labels remained known only to the organizers throughout the challenge. The goal of the participants was to produce the best possible data representation for the final evaluation data. This representation was then evaluated by the organizers on supervised learning classification tasks by training and testing a linear classifier on subsets of the final evaluation data, such than a learning curve would be produced. The evaluation metric was the area under the learning curve, which is a means of aggregating performance results over a range of number of training examples considered.

To avoid the possibility of participants selecting their model based on final evaluation set performance, the final results remained secret until the end of the challenge. Rather, feed-back was provided on-line during the challenge on the performance obtained on validation data, and the final evaluation set data was used only for the final ranking. For both phases, the participants could either submit a data representation (for validation data and final evaluation data) or a matrix of similarity between examples (a kernel). Hence, the competition was equivalently a data representation learning challenge and a kernel learning challenge.

In contrast with a classical evaluation of unsupervised learning as a preprocessing, the three subsets (development, validation, and final evaluation sets) were **not drawn from the same distribution**. In fact, they all had different sets of class labels. Picture for instance a problem of optical character recognition (OCR), the development set could contain only lowercase alphabetical letters, the validation set could contain uppercase letters, and the final evaluation set, digits and symbols. This setting is typical of real world problems in which there is an abundance of data available for training from a source domain, which is distinct from the target domain of interest. For instance, in face recognition, there is an abundance of pictures from unknown strangers that are available on the Internet, compared to the few images of your close family members that you care to classify. The development set represents a source domain whereas the validation and final evaluation sets represent alternative target domains on which different sets of tasks can be defined[2].

In the second phase of the challenge, a few labels of the development set were provided, offering to the participants the possibility of using supervised learning in some way to produce better data representations for the validation and final evaluation sets. The setting remained otherwise unchanged.

One of the main findings of this challenge is the power of unsupervised learning as a preprocessing tool. For all the datasets of the challenge, unsupervised learning produced results significantly better than the baseline methods (raw data or simple normalizations). The participants exploited effectively the feed-back received on the validation set to select the best data representations. The skepticism around the effectiveness of unsupervised learning is justified when no performance on a supervised task is available. However, unsupervised learning can be the object of model selection using a supervised task, similarly to preprocessing, feature selection, and hyperparameter selection. An interesting new outcome of this challenge is that the supervised tasks used for model selection can be distinct from the tasks used for the final evaluation. So, even though the learning algorithms are unsupervised, transfer learning is happening at the model selection level. This setting is related to the "self-taught learning" setting proposed in (Raina et al., 2007a). Another interesting finding is that, perhaps the development set is not useful at all. The winners of phase 1 did not use it. They devised a method to select a cascade of preprocessing steps to be used to produce a new kernel. The same cascade was then applied to produce the kernel of the final evaluation set(Aiolli, 2012). The importance of the degree of resemblance of the validation task and final task remains to be determined.

In phase 1, there was a danger of overfitting by trying too many methods and relying too heavily on the performance on the validation set. One team for instance overfitted in phase 1, ranking 1st on the validation set, but only 4th on the final evaluation set. Possibly, criteria involving both the reconstruction error and the classification accuracy on the validation tasks may be more effective for model selection. This should be the object of further research. In phase 2, the participants had available "transfer labels" for a subset of the development data (for classification tasks distinct from the classification tasks of the validation set and the final evaluation set). Therefore, they had the

---

2. In this paper, we call "domain" the input space (*e.g.*, a feature vector space) and we call "task" the output space (represented by labels for classification problems). We use the adjective "source" for an auxiliary problem, for which we have an abundance of data (*e.g.*, pictures of strangers in the Internet), and "target" for the problem of interest (*e.g.*, pictures of family members).

opportunity to use such labels to devise transfer learning strategies. The most effective strategy seems to have been to use the transfer labels for model selection again. None of the participants used those labels for learning.

Overall, an array of algorithms were used (Aiolli, 2012; Le Borgne, 2011; Liu et al., 2012; Mesnil et al, 2012; Saeed, 2011; Xu et al, 2011), including linear methods like Principal Component Analysis (PCA), and non-linear methods like clustering (K-means and hierarchical clustering being the most popular), Kernel-PCA (KPCA), non-linear auto-encoders and restricted Bolzmann machines (RBMs). A general methodology seems to have emerged. Most top ranking participants used simple normalizations (like variable standardization and/or data sphering using PCA) as a first step, followed by one or several layers of non-linear processing (stacks of auto-encoders, RBMs, KPCA, and/or clustering). Finally, "transduction" played a key role in winning first place: either the whole preprocessing chain was applied directly to the final evaluation data (this is the strategy of Fabio Aiolli who won first place in phase 1, Aiolli, 2012); or alternatively, the final evaluation data, preprocessed with a preprocessor trained on development+validation data, was post-processed with PCA (so-called "transductive PCA" used by the LISA team, who won the second phase, Mesnil et al, 2012).

## 4. Overview of Proceedings

The following provides an overview of the workshop proceedings including the tutorials, invited presentations, challenge winner articles and other refereed articles submitted to the workshop.

### 4.1. Tutorials

The workshop provided two foundational tutorials included in this proceeding. The morning tutorial covered *Deep Learning of Representations for Unsupervised and Transfer Learning* with Yoshua Bengio from the Université de Montréal (Bengio, 2012). Deep learning algorithms seek to exploit the unknown structure in the input distribution in order to discover good representations, often at multiple levels, with higher-level learned features defined in terms of lower-level features. The paper focusses on why unsupervised pre-training of representations using autoencoders and Restricted Boltzmann Machines can be useful, and how it can be exploited in the transfer learning scenario, where we care about predictions on examples that are not from the same distribution as the training distribution.

The afternoon tutorial entitled *Towards Heterogeneous Transfer Learning* was presented by Qiang Yang, Hong Kong University of Science, co-author of an authoritative review of transfer learning (Pan and Yang, 2010). Transfer learning has focused on knowledge transfer between domains with the same or similar input spaces. The heterogeneous transfer approach considers the ability to use knowledge from very different task domains and input spaces. The authors demonstrated heterogeneous transfer learning between text classification and image classification domains even when there are no explicit feature mappings provided. They explained that the key is to identify and maximize the commonalities among the internal structures (features) of the different domains.

## 4.2. Challenge Winner Articles

Three teams were presented awards at the workshop for their winning performances on the UTL Challenge and their authorship. This section will summarize the papers describing these winning entries.

The first place award for phase 1 of the UTL challenge (unsupervised learning) as well as the Pascal2 best challenge paper award for phase 1 went to Fabio Aiolli and his paper *Transfer Learning by Kernel Meta-Learning* (Aiolli, 2012). Recently, there have been a number of researchers who have investigated the problem of finding a good kernel matrix for a task. This is known as kernel learning. Kernel learning can be transformed into a semi-supervised learning problem by using a large set of unlabeled data and a smaller set of labeled data. The paper presents a novel approach to transfer learning based on kernel learning with both labeled and unlabeled data. Starting from a basic kernel, the method attempts to learn chains of kernel transforms that produce good kernel matrices for a set of source tasks. The same sequence of transformations are then applied to learn the kernel matrix for a target task. The application of this method to the five datasets of the Unsupervised and Transfer Learning (UTL) challenge produced the best results for the first phase of the competition.

The LISA team of the Université de Montréal, Canada, ranked first in the second phase of the UTL challenge (transfer learning), and their paper entitled *Unsupervised and Transfer Learning Challenge: A Deep Learning Approach* (Mesnil et al, 2012) won the Pascal2 best challenge paper award for phase 2. The LISA team demonstrated the usefulness of Deep Learning architectures to extract internal representations from a large set of unlabeled training examples. This is accomplished by introducing gradually network layers trained in an unsupervised way using the feature representation of lower layers. The final representation is then used to train a simple linear classifier with a small number of labeled training examples.

The team "1055a" of Chuanren Liu, Jianjun Xie, Hui Xiong, and Yong Ge of Core-Logic and Rutgers University won the second place award for phase 1 of the UTL challenge (unsupervised learning) and came in at third place in phase 2 (transfer learning) (Liu et al., 2012). Their paper entitled *Stochastic Unsupervised Learning on Unlabeled Data* was also selected for inclusion in these proceedings. The paper introduces a stochastic unsupervised learning method that performs as a preprocessing K-means clustering on principal components extracted from the raw unlabeled data. This removes the effect of noise and less-relevant features improving the methods robustness. The approach utilizes a stochastic process to combine multiple clustering assignments on each data point to alleviate over-fitting.

We also include in the supplemental material poster presentations and technical reports of work, which was not yet ready for publication, but shows interesting new directions of research:

The team of Zhixiang Xu (Airbus), Washington University in St. Louis, who took third place in phase 1 of the UTL challenge presented a poster entitled "Rapid Feature Learning with Stacked Linear Denoisers" (Xu et al, 2011). They investigated unsupervised pre-training of deep architectures as feature generators for shallow classifiers. They implemented a computationally efficient algorithm that mimics stacked denoising auto-encoders (SdAs). Their feature transformation improves the results of SVM classification, sometimes outperforming SdAs and deep neural networks.

Mehreen Saeed (Aliphlaila team), fourth place phase 2 UTL challenge), FAST, Pakistan, communicated a technical report entitled "Use of Representations in High Dimensional Spaces for Unsupervised and Transfer Learning Challenge" (Saeed, 2011). The author shows how manifold learning and simple similarity kernels can be used to get good results.

Yann-Aël Le Borgne (Tryan team, fourth place in second ranking of phase 2 UTL challenge), VUB, Belgium, showed a poster entitled "Supervised Dimensionality Reduction in the Unsupervised and Transfer Learning 2011 Competition" (Le Borgne, 2011). The author presented preliminary results on a technique making use of all three subsets provided for each dataset (development, validation, and final evaluation datasets) to assign labels to samples. The author then uses partial least square (PLS) to extract features of interest.

## 4.3. Fundamentals and Algorithms

The UTL workshop papers gathered in these proceedings follow two main axes. One axis ranges **from theory to application** of transfer learning and the other **from supervised learning, to unsupervised learning** and hybrid approaches. The following summarizes each of articles along those two dimensions.

In *Autoencoders, Unsupervised Learning, and Deep Architectures*, Pierre Baldi of UC Irvine, investigates the theoretical underbelly of autoencoders (Baldi, 2012). He presents a mathematical framework for the study of both linear and non-linear autoencoders - particularly the non-linear case of a Boolean autoencoder. He shows that learning with a Boolean autoencoder is equivalent to a clustering problem that can be solved in polynomial time when the number of clusters is small and becomes NP complete when the number of clusters is large. The framework sheds light on the connections between different kinds of autoencoders, their learning complexity and their composition in deep architectures. The paper brings together much of the theory on autoencoders, clustering, Hebbian learning, and information theory.

Joachim Buhmann et al. of ETH, Zurich, present a paper entitled, "Information Theoretic Model Selection for Pattern Analysis" (Buhmann et al, 2012). The authors propose a method of model and model-order selection for unsupervised data clustering based on information theory. Their approach ranks competing pattern cost functions according to their ability to extract context sensitive information from noisy data with respect to the hypothesis class. Sets of approximative solutions serve as a basis for an information theoretic communication protocol. Inferred models maximize the so-called "approximation capacity" that measures the mutual information between training data patterns and test data patterns, each of which have been made optimally "coarse" through the controlled addition of random noise. The approach is demonstrated using a Gaussian mixture model.

## 4.4. Supervised, Unsupervised and Transductive Approaches

### 4.4.1. SUPERVISED

The workshop provided new insights into supervised learning approaches to transfer. Ruslan Salakhutdinov et al. of MIT, USA, presented their paper on *One-Shot Learning with a Hierarchical Nonparametric Bayesian Model* (Salakhutdinov et al., 2012). One-shot learning is the ability to develop a general classification model from a single training example. The authors develop a hierarchical Bayesian model that can transfer acquired knowledge from previously learned categories to a novel category, in the form of a prior over category means and variances. The model discovers how to group categories into meaningful super-categories and infer to which super-category a novel example belongs, and thereby estimate not only the new category's mean but also an appropriate similarity metric. The method is tested using the MNIST and MSR Cambridge image datasets and shown to perform significantly better than simpler hierarchical Bayesian approaches, discovering new categories in a completely unsupervised fashion.

In *Inductive Transfer for Bayesian Network Structure Learning*, Alexandru Niculescu-Mizil (NEC Laboratories America) and Rich Caruana (Microsoft Research) consider the problem of jointly learning the structures of Bayesian network models from multiple related datasets (Niculescu-Mizil and Caruana, 2012). They present an algorithm that simultaneously learns a multi-task Bayesian network structure by transferring useful information between the different datasets. The algorithm extends the heuristic search techniques used in traditional structure learning to the multi-task case by defining a scoring function for sets of structures (one structure for each dataset) and an efficient procedure for searching for a set of structures that has a high score across all tasks. The approach assumes that the true dependency structures of related problems are similar: the presence or absence of arcs in some of the structures provides evidence for the presence or absence of those same arcs in the other structures.

Kohei Hayashi and Takashi Takenouchi of the Nara Institute of Science and Technology, Japan, in their paper *Self-measuring Similarity for Multi-task Gaussian Process* extend work by Bonilla et al. (2008) on a multi-task Gaussian process framework (Hayashi et al, 2012). Their approach incorporates similarities between tasks based on the observed responses, which allows for the representation of much more complex data structures. The proposed approach is able to construct covariance matrices via kernel functions even when additional information such as example target values are available. The authors propose an efficient conjugate-gradient-based algorithm that implements the approach. The method is shown to perform the best to date on the Movielens 100k dataset.

### 4.4.2. Unsupervised

The workshop also provided a number of new approaches to transfer using unsupervised learning or combinations of supervised and unsupervised transfer learning. In *Clustering: Science or Art?*, Ulrike von Luxburg et al. examine whether the quality of different clustering algorithms can be compared by a general, scientifically sound procedure, which is independent of particular clustering algorithms (von Luxburg et al., 2012). They conclude that clustering should not be treated as an application-independent mathematical problem, but should always be studied in the context of its end-use. Different reasons for clustering bring with it different metrics for success. They argue that research spent on developing a "taxonomy of clustering problems" will be more fruitful than efforts spent on developing a domain independent clustering algorithm.

Preferably, high dimensional data, such as pixels of an image, are better described in terms of a small number of meta-features. In their paper *Unsupervised dimensionality reduction via gradient-based matrix factorization with two learning rates and their automatic updates*, Vladimir Nikulin and Tian-Hsiang Huang, of University of Queensland, Australia prescribe three related methods that combine to reduce noise while still capturing the essential features of the original data (Nikulin and Huang, 2012). The resulting features can then be used to do supervised classification. The proposed methods are demonstrated on the classification of gene expression data from cancer research where the number of labeled samples is relatively small compared to the number of genes in each sample.

### 4.4.3. Transductive

Ayan Acharya et al. report in their paper *Transfer Learning with Cluster Ensembles* a method of transferring learned knowledge from a set of source tasks when the target task has no labeled examples (Acharya et al, 2012). They present an optimization framework that applies the outputs of a cluster ensemble on a target task to moderate posterior probability estimates provided by classifiers previously induced on a related domain of tasks, so that the posterior probabilities are better adapted to the new context. This framework is general in that it admits a wide range of loss functions and classification/clustering methods. Empirical results on both text and hyperspectral data indicate that the proposed method can yield substantially superior classification results as compared to competing transductive learning techniques (Transductive SVM, Locally Weighted Ensemble).

## 4.5. Case studies

The final five papers from the workshop can be considered applications or case studies of unsupervised and transfer learning. The first paper, entitled *Transfer Learning in Computational Biology* applies multiple task learning to several problems in computational biology where the generation of training labels is often very costly (Widmer and Rätsch, 2012). The authors, Christian Widmer and Gunnar Raetsch, of MPI in Germany, received the Pascal2 best paper award at the workshop for this work. The paper presents two problems from sequence biology and uses regularization (SVM) based transfer learning methods, with a special focus on the case of a hierarchical relationship between tasks. The authors propose strategies to learn or refine a measure of task relatedness so as to optimize the transfer from source to target task.

In *Transfer Learning in Sequential Decision Problems: A Hierarchical Bayesian Approach*, Aaron Wilson et al, of Oregon State University, show that transfer is doubly beneficial in reinforcement learning where the agent not only needs to generalize from sparse experience, but also needs to discover good opportunities to learn in the first place (Wilson et al., 2012). They show that the hierarchical Bayesian framework can be readily adapted to sequential decision problems and provides a natural formalization of transfer learning.

*Transfer Learning for Auto-gating of Flow Cytometry Data.* by Gyemin Lee et al. of the University of Michigan, Ann Arbor, apply transfer learning to flow cytometry, a technique for rapidly quantifying physical and chemical properties of large numbers of cells (Lee et al., 2012). In clinical applications, flow cytometry data must be manually "gated"(scored) to identify cell populations of interest. The authors leverage existing datasets, previously gated by experts, to automatically gate a new flow cytometry dataset while accounting for biological variation. An empirical study demonstrates the approach by automatically gating lymphocytes from peripheral blood samples. The authors received the Pascal2 best student paper award for this work.

Philemon Brakel and Benjamin Schrauwen, of Ghent University, Belgium, use a hierarchical Bayesian logistic regression model to perform a binary document classification task in the paper *Transfer Learning for Document Classification: Sampling Informative Priors* Their approach estimates the covariance matrix of a multivariate Gaussian prior over the model parameters using a set of related tasks. Inference was done using a combination of Hybrid Monte Carlo and Gibbs sampling. They demonstrate that the obtained priors contain information that is beneficial for developing a model for document classification from small training sets.

Finally, in *Divide and Transfer: an Exploration of Segmented Transfer to Detect Wikipedia Vandalism*, Si-Chi Chin and W. Nick Street, of the University of Iowa, apply knowledge transfer methods to the problem of detecting Wikipedia vandalism (Chin and Street, 2012). Transfer is used to address the problem of small amounts of labeled data by leveraging unlabeled data and previously acquired knowledge from related source tasks. Avoiding negative transfer becomes a primary concern given the diverse nature of Wikipedia modifications that can occur. The proposed two segmented transfer approaches map unlabeled data from the target task to the most related cluster from the source task, classifying the unlabeled data using the most relevant learned models.

## 5. Summary

Challenges foster progress in particular scientific domains, but their specific formulation may bias research in too narrow ways. For that reason, our workshop invited diverse contributions on the theme of transfer learning, in addition to discussing the results of the challenge we organized. As a result, it is more difficult to draw general conclusions summarizing the enormous body of work that this represents. In some sense, transfer learning covers all the aspects of machine learning, with the only particularity that training data includes "source" domains and/or tasks that do are different from the "target" domains and/or tasks of interest. Within this general setting, many types of transfer learning formulations have been made. From our point of view, the most notable contribution of these proceedings is to demonstrate the effectiveness of recently proposed methods in the context of a wide variety of real world applications, both through the results of the challenge and other contributed papers. We hope that the mix of articles collected in this proceedings issue will spark further interest and curiosity in transfer learning. There is much work to be done in this area in terms of new computational learning theory and the application of existing algorithms and techniques.

## Acknowledgements

We are very grateful to all authors for their submissions to the workshop and for their work in improving and polishing their articles for this proceedings issue. We would also like to thank the referees for their reviews and helpful comments to authors that have led to improvements in content and format. Finally, we are particularly grateful to the Editor-in-Chief, Neil Lawrence, for the opportunity to compile this proceedings issue and to the editorial staff of JMLR Workshop and Conference Proceedings for their kind assistance. Our generous sponsors and data donors and our dedicated advisors, beta testers listed on our challenge website (`http://clopinet.com/ul`) are gratefully acknowledged. This project is part of the DARPA Deep Learning program and is an activity of the Causality Workbench supported by the Pascal network of excellence funded by the European Commission and by the U.S. National Science Foundation under Grant N0. ECCS-0725746. Any opinions, findings, and conclusions or recommendations expressed in this material are those of the authors and do not necessarily reflect the views of the funding agencies.

## References

Ayan Acharya et al. Transfer learning with cluster ensembles. In *ICML 2011 Unsupervised and Transfer Learning Workshop*. JMLR W&CP, this volume, 2012.

Fabio Aiolli. Transfer learning by kernel meta-learning. In *ICML 2011 Unsupervised and Transfer Learning Workshop*. JMLR W&CP, this volume, 2012.

G. M. Allenby and P. E. Rossi. Marketing models of consumer heterogeneity. *Journal of Econometrics*, 89:57–78, 1999.

G. M. Allenby and P. E. Rossi. Learning multiple tasks with kernel methods. *Journal of Machine Learning Research*, 6:615–637, 2005.

Andrew Arnold, Ramesh Nallapati, and William W. Cohen. A comparative study of methods for transductive transfer learning. In *In ICDM Workshop on Mining and Management of Biological Data*, 2007.

N. Arora, G.M Allenby, and J. Ginter. A hierarchical bayes model of primary and secondary demand. *Marketing Science*, 17(1):29–44, 1998.

B. Bakker and T. Heskes. Task clustering and gating for bayesian multi-task learning. *Journal of Machine Learning Research*, 4:83–99, 2003.

Pierre Baldi. Autoencoders, unsupervised learning, and deep architectures. In *ICML 2011 Unsupervised and Transfer Learning Workshop*. JMLR W&CP, this volume, 2012.

Jonathan Baxter. Theoretical models of learning to learn. *Learning to Learn*, pages 71–94, 1997.

S. Ben-David and R. Schuller. Exploiting task relatedness for multiple task learning. In *Proceedings of Computational Learning Theory (COLT)*, pages 185–192, 2003.

Yoshua Bengio. Learning deep architectures for AI. *Foundations and Trends in Machine Learning*, 2(1):1–127, 2009. doi: 10.1561/2200000006. Also published as a book. Now Publishers, 2009.

Yoshua Bengio. Deep learning of representations for unsupervised and transfer learning. In *ICML 2011 Unsupervised and Transfer Learning Workshop*. JMLR W&CP, this volume, 2012.

Yoshua Bengio, Jean-François Paiement, Pascal Vincent, Olivier Delalleau, Nicolas Le Roux, and Marie Ouimet. Out-of-sample extensions for lle, isomap, mds, eigenmaps, and spectral clustering. In *NIPS03*, 2003.

Avrim Blum and Tom Mitchell. Combining labeled and unlabeled data with co-training. In *COLT' 98 Proceedings of the eleventh annual conference on Computational learning theory*, pages 92–100. Morgan Kaufmann Publishers, 1998.

L. Breiman and J.H Friedman. Predicting multivariate responses in multiple linear regression. *Royal Statistical Society Series B*, 1998.

Jane Bromley, Isabelle Guyon, Yann LeCun, Eduard Säckinger, and Roopak Shah. Signature verification using a "siamese" time delay neural network. In *In NIPS Proc*, 1994.

Joachim Buhmann et al. Information theoretic model selection for pattern analysis. In *ICML 2011 Unsupervised and Transfer Learning Workshop*. JMLR W&CP, this volume, 2012.

R. Caruana. Multitask learning. *Machine Learning*, 28(1):41–75, 1997.

Si-Chi Chin and W. Nick Street. Divide and transfer: an exploration of segmented transfer to detect wikipedia vandalism. In *ICML 2011 Unsupervised and Transfer Learning Workshop*. JMLR W&CP, this volume, 2012.

D. Erhan, Y. Bengio, A. Courville, P.-A. Manzagol, P. Vincent, and S. Bengio. Why does unsupervised pre-training help deep learning? *JMLR*, 11:625–660, 2010.

Zoubin Ghahramani. *Unsupervised Learning*, volume 3176, pages 72–112. Springer-Verlag, Berlin, 2004.

Amir Globerson and Naftali Tishby. Sufficient dimensionality reduction. *J. Mach. Learn. Res.*, 3:1307–1331, March 2003. ISSN 1532-4435. URL http://portal.acm.org/citation.cfm?id=944919.944975.

W. Greene. *Econometric Analysis, 5th Edition*. Prentice Hall, 2002.

S. M. Gutstein. *Transfer Learning Techniques for Deep Neural Nets*. PhD thesis, The University of Texas at El Paso, 2010.

Isabelle Guyon, Gideon Dror, Vincent Lemaire, Danny Silver, Graham Taylor, and Aha David W. Analysis of the ijcnn 2011 utl challenge. *Neural Networks*, In press 2011a.

Isabelle Guyon, Gideon Dror, Vincent Lemaire, Graham Taylor, and Aha David W. Unsupervised and transfer learning challenge. In *The 2011 International Join Conference on Neural Networks*, pages 793–800, July 2011b.

Kohei Hayashi et al. Self-measuring similarity for multi-task gaussian process. In *ICML 2011 Unsupervised and Transfer Learning Workshop*. JMLR W&CP, this volume, 2012.

T. Heskes. Empirical bayes for learning to learn. In P. Langley, editor, *Proceedings of the International Conference on Machine Learning (ICML'00)*, pages 367–374, 2000.

G.E. Hinton and T.J. Sejnowski. *Unsupervised learning: foundations of neural computation*. Computational neuroscience. MIT Press, 1999. ISBN 9780262581684. URL http://books.google.ca/books?id=yj04Y0lje4cC.

A K Jain, M N Murty, and P. J. Flynn. Data clustering: A review, 1999.

Tony Jebara. Multi-task feature and kernel selection for svms. In *Proceedings of the International Conference on Machine Learning (ICML 04)*, pages 185–192, 2004.

Brian Kulis. Icml tutorial on metric learning, 2010. URL http://www.cs.berkeley.edu/~kulis/icml2010_tutorial.htm.

Yann-Aël Le Borgne. Supervised dimensionality reduction in the unsupervised and transfer learning 2011 competition, 2011.

Gyemin Lee, Lloyd Stoolman, and Clayton Scott. Transfer learning for auto-gating of flow cytometry data. In *ICML 2011 Unsupervised and Transfer Learning Workshop.* JMLR W&CP, this volume, 2012.

Honglak Lee, Alexis Battle, Rajat Raina, and Andrew Y. Ng. Efficient sparse coding algorithms. In B. Schölkopf, J. Platt, and T. Hoffman, editors, *Advances in Neural Information Processing Systems 19*, pages 801–808. MIT Press, Cambridge, MA, 2007.

Chuanren Liu, Jianjun Xie, Hui Xiong, and Yong Ge. Stochastic unsupervised learning on unlabeled data. In *ICML 2011 Unsupervised and Transfer Learning Workshop.* JMLR W&CP, this volume, 2012.

Lorien Pratt (Editor). Reuse of neural networks through transfer. *Connection Science,* 8(2), 1996.

Ulrike Luxburg. A tutorial on spectral clustering. *Statistics and Computing,* 17:395–416, December 2007. ISSN 0960-3174. doi: 10.1007/s11222-007-9033-z. URL `http://portal.acm.org/citation.cfm?id=1288822.1288832`.

Grégoire Mesnil et al. Unsupervised and transfer learning challenge: a deep learning approach. In *ICML 2011 Unsupervised and Transfer Learning Workshop.* JMLR W&CP, this volume, 2012.

Tom M. Mitchell. The need for biases in learning generalizations. *Readings in Machine Learning,* pages 184–191, 1980. ed. Jude W. Shavlik and Thomas G. Dietterich.

Alexandru Niculescu-Mizil and Rich Caruana. Inductive transfer for bayesian network structure learning. *Journal of Machine Learning Research - Proceedings Track,* 2: 339–346, 2007.

Alexandru Niculescu-Mizil and Richard Caruana. Inductive transfer for bayesian network structure learning. In *ICML 2011 Unsupervised and Transfer Learning Workshop.* JMLR W&CP, this volume, 2012.

Vladimir Nikulin and Tian-Hsiang Huang. Unsupervised dimensionality reduction via gradient-based matrix factorization with two adaptive learning rates. In *ICML 2011 Unsupervised and Transfer Learning Workshop.* JMLR W&CP, this volume, 2012.

NIPS1995 Workshop. Learning to learn. `http://plato.acadiau.ca/courses/comp/dsilver/NIPS95_LTL/transfer.workshop.1995.html`, 1995.

NIPS2005 Workshop. Inductive transfer – 10 years later. `http://iitrl.acadiau.ca/itws05/`, 2005.

Sinno Jialin Pan and Qiang Yang. A survey on transfer learning. *IEEE Transactions on Knoweledge and Data Engineering,* 22(10):1345–1359, October 2010.

Lorien Pratt and Sebastian Thrun (Editors). Transfer in inductive systems. *Machine Learning*, 28(1), 1997.

Rajat Raina, Alexis Battle, Honglak Lee, Benjamin Packer, and Andrew Y. Ng. Self-taught learning: Transfer learning from unlabeled data. In *Proceedings of the Twenty-fourth International Conference on Machine Learning*, 2007a.

Rajat Raina, Alexis Battle, Honglak Lee, Benjamin Packer, and Andrew Y. Ng. Self-taught learning: transfer learning from unlabeled data. In *Proceedings of the 24th international conference on Machine learning*, ICML '07, pages 759–766, New York, NY, USA, 2007b. ACM. ISBN 978-1-59593-793-3. doi: 10.1145/1273496.1273592. URL http://doi.acm.org/10.1145/1273496.1273592.

J. Ramon, K. Driessens, and T. Croonenborghs. Transfer learning in reinforcement learning problems through partial policy recycling. *Proc. 18th European Conf. Machine Learning (ECML 2007)*, pages 699–707, 2007.

Mehreen Saeed. Use of representations in high dimensional spaces for unsupervised and transfer learning challenge, 2011.

Ruslan Salakhutdinov, Josh Tenenbaum, and Antonio Torralba. One-shot learning with a hierarchical nonparametric bayesian model. In *ICML 2011 Unsupervised and Transfer Learning Workshop*. JMLR W&CP, this volume, 2012.

Daniel L. Silver and Kristin P. Bennett. Special issue on inductive transfer. *Machine Learning*, 73, 2008.

Daniel L. Silver and Robert E. Mercer. The parallel transfer of task knowledge using dynamic learning rates based on a measure of relatedness. *Connection Science Special Issue: Transfer in Inductive Systems*, 8(2):277–294, 1996.

Daniel L. Silver, Ryan Poirier, and Duane Currie. Inductive transfer with context-sensitive neural networks. *Machine Learning*, 73:323–336, 2008.

Alexander J. Smola, Sebastian Mika, Bernhard Schölkopf, and Robert C. Williamson. Regularized principal manifolds. *JMLR*, 1:179–209, 2001.

M.E. Taylor and P. Stone. Cross-domain transfer for reinforcement learning. *Proc. 24th International Conf. Machine Learning (ICML 2007)*, pages 879–886, 2007.

Sebastian Thrun and Lorien Y. Pratt (Editors). *Learning To Learn*. Kluwer Academc Publisher, Boston, MA, 1997.

Paul E. Utgoff. *Machine Learning of Inductive Bias*. Kluwer Academc Publisher, Boston, MA, 1986.

Ulrike von Luxburg, Robert C. Williamson, and Isabelle Guyon. Clustering: Science or art? In *ICML 2011 Unsupervised and Transfer Learning Workshop*. JMLR W&CP, this volume, 2012.

Christian Widmer and Gunnar Rätsch. Multitask learning in computational biology. In *ICML 2011 Unsupervised and Transfer Learning Workshop*. JMLR W&CP, this volume, 2012.

Aaron Wilson, Alan Fern, and Prasad Tadepalli. Transfer learning in sequential decision problems: A hierarchical bayesian approach. In *ICML 2011 Unsupervised and Transfer Learning Workshop*. JMLR W&CP, this volume, 2012.

Xiao-Ming WU, Anthony Man-Cho So, Zhenguo Li, and Shuo-Yen Robert Li. Fast graph laplacian regularized kernel learning via semidefinite quadratic linear programming. In Y. Bengio, D. Schuurmans, J. Lafferty, C. K. I. Williams, and A. Culotta, editors, *Advances in Neural Information Processing Systems 22*, pages 1964–1972. Curran Associates, Inc., 2009.

Zhixiang Xu et al. Rapid feature learning with stacked linear denoisers, 2011.

A. Zellner. An efficient method for estimating seemingly unrelated regression equations and tests for aggregation bias. *Journal of the American Statistical Association*, 57: 348–368, 1962.

# Deep Learning of Representations for Unsupervised and Transfer Learning

**Yoshua Bengio**       YOSHUA.BENGIO@UMONTREAL.CA
*Dept. IRO, Université de Montréal. Montréal (QC), H2C 3J7, Canada*

**Editor:** I. Guyon, G. Dror, V. Lemaire, G. Taylor, and D. Silver

## Abstract

Deep learning algorithms seek to exploit the unknown structure in the input distribution in order to discover good representations, often at multiple levels, with higher-level learned features defined in terms of lower-level features. The objective is to make these higher-level representations more abstract, with their individual features more invariant to most of the variations that are typically present in the training distribution, while collectively preserving as much as possible of the information in the input. Ideally, we would like these representations to disentangle the unknown factors of variation that underlie the training distribution. Such unsupervised learning of representations can be exploited usefully under the hypothesis that the input distribution $P(x)$ is structurally related to some task of interest, say predicting $P(y|x)$. This paper focuses on the context of the Unsupervised and Transfer Learning Challenge, on why unsupervised pre-training of representations can be useful, and how it can be exploited in the transfer learning scenario, where we care about predictions on examples that are not from the same distribution as the training distribution.

**Keywords:** Deep Learning, unsupervised learning, representation learning, transfer learning, multi-task learning, self-taught learning, domain adaptation, neural networks, Restricted Boltzmann Machines, Autoencoders.

## 1. Introduction

Machine learning algorithms attempt to discover structure in data. In their simpler forms, that often means discovering a predictive relationship between variables. More generally, that means discovering where probability mass concentrates in the joint distribution of all the observations. Many researchers have found that the way in which data are represented can make a huge difference in the success of a learning algorithm.

Whereas many practitioners have relied solely on hand-crafting representations, thus exploiting human insights into the problem, there is also a long tradition of learning algorithms that attempt to *discover good representations*. Representation learning is the general context where this paper is situated. What can a good representation buy us? What is a good representation? What training principles might be used to discover good representations?

Supervised machine learning tasks can be abstracted in terms of $(X, Y)$ pairs, where $X$ is an input random variable and $Y$ is a *label* that we wish to predict given $X$. This paper considers the use of representation learning in the case where *labels for the task of interest are not available at the time of learning the representation*. One wishes to learn the representation either in a purely unsupervised way, or using *labels for other tasks*. This type of setup has been called self-taught learning (Raina et al., 2007) but also falls in the areas of transfer learning, domain adaptation, and multi-task learning (where typically one also has labels for the task of interest) and is related to semi-supervised learning (where one has many unlabeled examples and a few labeled ones).

In order to address this challenge (and *the Unsupervised and Transfer Learning Challenge*[1]), the algorithms described here exploit **Deep Learning**, i.e., learning *multiple levels of representation*. The intent is to discover more *abstract* features in the higher levels of the representation, which hopefully make it easier to *separate from each other the various explanatory factors extent in the data.*

## 1.1. The Context of The Unsupervised and Transfer Learning Challenge

The challenge was organized according to the following learning setup. The test (and validation) sets have examples from classes not well represented in the training set. They have only a small number of unlabeled examples (4096) and very few labeled examples (1 to 64 per class) available to a Hebbian linear classifier (which discriminates according to the median between the centroids of two classes compared) applied separately to each class against the others. In the second phase of the competition, some labeled examples (from classes other than those in the test or validation sets) are available for the training set. Participants can use the training set (with some irrelevant labels, in the second phase) to construct a representation for test set examples. Typically this is achieved by learning a transformation of the raw input vectors into a new space, which hopefully captures the most important factors of variation present in the unknown generating distribution. That transformation can then be applied to test examples. The challenge server then trains the Hebbian linear classifier on top of that representation, on a small random subset of the test set, and evaluates generalization on the rest of the test set (many random subsets are computed to get an average score). The main difficulties are the following:

1. The input distribution is very different in the test (or validation) set, compared to the training set (for example, the set of classes to be discriminated in the test set may be absent or rare in the training set), making it unclear if anything can be transfered from the training to the test set.

2. Very few labels are available to the linear classifier on the test set, meaning that generalization of the classifier is inherently difficult and sensitive to the particulars of the representation chosen.

3. No labels for the classes of interest (of the test set) are available at all when learning the representation. The labels from the training set might in fact mislead a representation-learning algorithm, because the directions of discrimination

---

1. http://www.causality.inf.ethz.ch/unsupervised-learning.php

which are useful among the training set classes could be useless to discriminate among the test set classes.

This puts great pressure on the representation learning algorithm applied on the training set (unlabeled, in the experiments we performed) to discover really *generic* features likely to be of interest for many classification tasks on such data. Our intuition is that *more abstract features* are more likely to fit that stringent requirement, which motivates the use of Deep Learning algorithms.

## 1.2. Representations as Coordinate Systems

Representation learning is also intimately related to the research in *manifold learning* (Hinton et al., 1997; Tenenbaum et al., 2000; Saul and Roweis, 2002; Belkin and Niyogi, 2003). The objective of manifold learning algorithms is two-fold: identify low-dimensional regions of high-density (called manifold), and construct a coordinate system on these manifolds, i.e., a low-dimensional representation for input examples. Principal Components Analysis (PCA) is the linear ancestor of manifold learning algorithms: it provides a projection of each input vector to a low-dimensional coordinate vector, implicitly defining a low-dimensional hyperplane in input space near which density is hopefully concentrating. The extent of this mass concentration can be measured by the proportion of variance explained by principal eigenvectors of the data covariance matrix. Changes in the directions of the principal components are perfectly captured by the PCA, whereas changes in the orthogonal directions are completely lost. The proportion of the variance in the data captured by the PCA is a good measure of the effectiveness of a PCA dimensionality reduction (for a given choice of number of dimensions). The assumption is that directions where there is very little change in the data do not matter and can be considered noise, but this is not always true, especially when one assumes that the manifold is linear (as with PCA). Non-linear manifold learning algorithms avoid the linear assumption but retain the notion of a drastic dimensionality reduction.

As we argue more at the end of this paper, although cutting the low-variance directions out (i.e., considering those directions as noise) is often very effective, *it is not always clear what is signal and what is noise*: although the extent of variability is a good hint, it is not perfect. As an example, consider images of faces, and two factors: person identity and pose. Most of the variation in pixel space can be explained by pose (especially the 2-D translation, scaling, and rotation), and two persons of the same sex, age, and hair type will be distinguishable only by looking at low variance components. That is why one often starts by preprocessing such images to align them as much as possible or focus only on images of faces in the same pose, e.g. frontal view.

It is a good thing to test, for one's data, if one can get better classification by simply removing the low-variance components, and in that case one should definitely do it[2].

---

2. and in fact, removing some of the low-variance directions with a preliminary PCA has worked well in the challenge.

However, we believe that a more encompassing and more conservative but more ambitious approach is to use a learning algorithm that separates the explanatory factors from each other as much as possible, and let a discriminant classifier pick out those that are relevant to a particular task.

In this context, *overcomplete*[3] *sparse*[4] *representations* have often (Ranzato et al., 2007b, 2008; Goodfellow et al., 2009) been found to work better than dense undercomplete representations (such as produced by PCA). Consider a sparse overcomplete representation in the neighborhood of an input example $x$. Most local changes around $x$ involve a continuous change of the "active" (non-zero) elements of the representation. Hence the set of active elements of the representation defines a local *chart*, a local coordinate system. Those charts are stitched together through the zero/non-zero transitions that occur when crossing some boundaries in input space. Goodfellow et al. (2009) have found that sparse autoencoders gave rise to more invariant representations (compared to non-sparse ones), in the sense that a subset of the representation elements (also called features) were more insensitive to input transformations such as translation or rotation of the camera. One advantage of such as an overcomplete representation is that it is not "cramped" in a small-dimensional space. The effective dimensionality (number of non-zeros) can vary depending on where we look. It is very plausible that in some regions of space it may be more appropriate to have more dimensions than in others.

Let $h(x)$ denote the mapping from an input $x$ to its representation $h(x)$. Overcomplete representations which are not necessarily sparse but where $h$ is non-linear are characterized at a particular input point $x$ by a "soft" dimensionality and a "soft" subset of the representation coordinates that are active. The degree to which $h_i(x)$ is active basically depends on $\|\frac{\partial h_i(x)}{\partial x}\|$. When $\|\frac{\partial h_i(x)}{\partial x}\|$ is close to zero, coordinate $i$ is inactive and unresponsive to changes in $x$, while the active coordinates encode (i.e., respond to) local changes around $x$. This is what happens with the Contracting Autoencoder described a bit more in section 3.5.

## 1.3. Depth

Depth is a notion borrowed from complexity theory, and that is defined for *circuits*. A circuit is a directed acyclic graph where each node is associated with a computation. The results of the computation of a node are used as input by the successors of that node. In the circuit, *input nodes* have no predecessor and *output nodes* have no successor. The depth of a circuit is the longest path from an input to an output node. A longstanding question in complexity theory is the extent to which depth-limited circuits can efficiently represent functions that can otherwise be efficiently represented. A depth-2 circuit (with appropriate choice of computational elements, e.g. logic gates or formal neurons) can compute or approximate any function, but it may require an exponentially large number of nodes. This is a relevant question for machine learning, because many learning algorithms learn "shallow architectures" (Bengio and LeCun, 2007), typically

---

3. overcomplete representation: with more dimensions than the raw input
4. sparse representation: with many zeros or near-zeros

of depth 1 (linear predictors) or 2 (most non-parametric predictors). If AI-tasks require deeper circuits (and human brains certainly appear deep), then we should find ways to incorporate depth into our learning algorithms. The consequences of using a too shallow predictor would be that it may not generalize well, unless given huge numbers of examples and capacity (i.e., computational resources and statistical resources).

The early results on the limitations of shallow circuits regard functions such as the parity function (Yao, 1985), showing that logic gates circuits of depth-2 require exponential size to implement $d$-bit parity where a deep circuit of depth $O(\log(d))$ could implement it with $O(d)$ nodes. Håstad (1986) then showed that there are functions computable with a polynomial-size logic gate circuit of depth $k$ that require exponential size when restricted to depth $k - 1$ (Håstad, 1986). Interestingly, a similar result was proven for the case of circuits made of linear threshold units (formal neurons) (Håstad and Goldmann, 1991), when trying to represent a particular family of functions. A more recent result brings an example of a very large class of functions that cannot be efficiently represented with a small-depth circuit (Braverman, 2011). It is particularly striking that the main theorem regards the representation of functions that capture dependencies in joint distributions. Basically, dependencies that involve more than $r$ variables are difficult to capture by shallow circuits. An $r$-independent distribution is one that cannot be distinguished from the uniform distribution when looking only at $r$ variables at a time. The proof of the main theorem (which concerns distribution over bit vectors) relies on the fact that order-$r$ polynomials over the reals cannot capture $r$-independent distributions. The main result is that bounded-depth circuits cannot distinguish data from $r$-independent distributions from independent noisy bits. We have also recently shown (Bengio and Delalleau, 2011) results for *sum-product networks* (where nodes either compute sums or products, over the reals). We found two families of polynomials that can be efficiently represented with deep circuits, but require exponential size with depth-2 circuits. Interestingly, sum-product networks were recently proposed to efficiently represent high-dimensional joint distributions (Poon and Domingos, 2010).

Besides the complexity-theory hints at their representational advantages, there are other motivations for studying learning algorithms which build a deep architecture. The earliest one is simply inspiration from brains. By putting together anatomical knowledge and measures of the time taken for signals to travel from the retina to the frontal cortex and then to motor neurons (about 100 to 200 ms), one can gather that at least 5 to 10 feedforward levels are involved for some of the simplest visual object recognition tasks. Slightly more complex vision tasks require iteration and feedback top-down signals, multiplying the overall depth by an extra factor of 2 to 4 (to about half a second).

Another motivation derives from what we know of cognition and abstractions: as argued in Bengio (2009), it is natural for humans to represent concepts at one level of abstraction as the *composition* of concepts at lower levels. Engineers often craft representations at multiple levels, with higher levels obtained by transformation of lower levels. Instead of a flat `main` program, software engineers structure their code to obtain plenty of *re-use*, with functions and modules re-using other functions and modules. This inspiration is directly linked to machine learning: deep architectures appear well suited to *represent higher-level abstractions* because they lend themselves to *re-use*. For example, some of the features that are useful for one task may be useful for another, making Deep Learning particularly well suited for *transfer learning* and *multi-task learning* (Caruana, 1995; Collobert and Weston, 2008). Here one is exploiting the existence of underlying common explanatory factors that are useful for both tasks. This is also true of *semi-supervised learning*, which exploits connections between the input distribution $P(X)$ and a target conditional distribution $P(Y|X)$. In general these two distributions, seen as functions of $x$, may be unrelated to each other. But in the world around us, it is often the case that some of the factors that shape the distribution of input variables $X$ are also predictive of the output variables $Y$. Deep Learning relies heavily on unsupervised or semi-supervised learning, and assumes that *representations of X that are useful to capture $P(X)$ are also in part useful to capture $P(Y|X)$*.

In the context of the Unsupervised and Transfer Learning Challenge, the assumption exploited by Deep Learning algorithms goes even further, and is related to the Self-Taught Learning setup (Raina et al., 2007). In the unsupervised representation-learning phase, one may have access to examples of only some of the classes, and the representation learned should be useful for other classes. One therefore assumes that some of the factors that explain $P(X|Y)$ for $Y$ in the training classes, and that will be captured by the learned representation, will be useful to predict different classes, from the test set. In phase 1 of the competition, only $X$'s from the training classes are observed, while in phase 2 some corresponding labels are observed as well, but no labeled examples from the test set are ever revealed. In our team (LISA), we only used the phase 2 training set labels to help perform model selection, since selecting and fine-tuning features based on their dicriminatory ability on training classes greatly increased the risk of removing important information for test classes. See Mesnil et al. (2011) for more details.

## 2. Greedy Layer-Wise Learning of Representations

The following basic recipe was introduced in 2006 (Hinton and Salakhutdinov, 2006; Hinton et al., 2006; Ranzato et al., 2007a; Bengio et al., 2007):

1. Let $h_0(x) = x$ be the lowest-level representation of the data, given by the observed raw input $x$.

2. For $\ell = 1$ to $L$

   Train an unsupervised learning model taking as observed data the training examples $h_{\ell-1}(x)$ represented at level $\ell - 1$, and after training, producing representations $h_\ell(x) = R_\ell(h_{\ell-1}(x))$ at the next level.

From this point on, several variants have been explored in the literature. For supervised learning with fine-tuning, which is the most common variant (Hinton et al., 2006; Ranzato et al., 2007b; Bengio et al., 2007):

3. Initialize a supervised predictor whose first stage is the parametrized representation function $h_L(x)$, followed by a linear or non-linear predictor as the second stage (i.e., taking $h_L(x)$ as input).

4. Fine-tune the supervised predictor with respect to a supervised training criterion, based on a labeled training set of $(x, y)$ pairs, and optimizing the parameters in both the representation stage and the predictor stage.

A supervised variant involves using all the levels of representation as input to the predictor, keeping the representation stage fixed, and optimizing only the predictor parameters (Lee et al., 2009a,b):

3. Train a supervised learner taking as input $(h_k(x), h_{k+1}(x), \ldots, h_L(x))$ for some choice of $0 \leq k \leq L$, using a labeled training set of $(x, y)$ pairs.

A special case of the above is to have $k = L$, i.e., we keep only the top level as input to the classifier without supervised fine-tuning of the representation. Since labels for the test classes are not available (for fine-tuning) in the Unsupervised and Transfer Learning Challenge, the latter approach makes more sense, but in other settings (especially when the number of labeled examples is large) we have often found fine-tuning to be helpful (Lamblin and Bengio, 2010).

Finally, there is a common unsupervised variant, e.g. for training deep autoencoders (Hinton and Salakhutdinov, 2006) or a Deep Boltzmann Machine (Salakhutdinov and Hinton, 2009):

3. Initialize an unsupervised model of $x$ based on the parameters of all the stages.

4. Fine-tune the unsupervised model with respect to a global (all-levels) training criterion, based on the training set of examples $x$.

As detailed in Mesnil et al. (2011), it turned out for the challenge to always work better to have a low-dimensional $h_L$ (i.e. the input to the classifier), e.g., a handful of dimensions. This top-level representation was typically obtained by choosing PCA as the last stage of the hierarchy. In experiments with other kinds of data, with many more labeled examples, we had obtained better results with *high-dimensional* top-level representations (thousands of dimensions). We found higher dimensional top-level representations to be most hurtful for the cases where there are very few labeled examples. Note that the challenge criterion is an average over 1, 2, 4, 8, 16, 32 and 64 labeled examples per class. Because the cases with very few labeled examples were those on which there was most room for improvement (compared to other competitors), it makes sense that the low-dimensional solutions were the most successful.

Another remark is important here. On datasets with a larger labeled training set, we found that the supervised fine-tuning variant (where all the levels are finally tuned with respect to the supervised training criterion) can perform *substantially better* than without supervised fine-tuning (Lamblin and Bengio, 2010).

## 2.1. Transductive Specialization to Transfer Learning and Domain Adaptation

On the other hand, in the context of the challenge, there were no training labels for the task of interest, i.e., the classes of the test set, so it would not even have been possible to perform meaningful supervised fine-tuning. Worse than that in fact, the *input distribution as well was very different* between the training set and the test set. The large majority of examples from the training set were from classes other than those in the test set. This is a particularly extreme **transfer learning** or **domain adaptation** setup.

How could one hope to generalize in this context? If the training set input distribution had nothing to do with the test set input distribution, then even unsupervised representation-learning on the training set might not be helpful as a learned preprocessing for the test set. The only hope is that a representation-learning algorithm would discover features that capture the generic factors of variation present in all the classes, and that the classifier trained on the test set would then just need to pick up those factors relevant to the discrimination among test set classes. Unfortunately, because of the very small number of labeled examples available to the test set classifier, we found that we could not obtain good results (on the validation set) with high-dimensional representations. This implied that some selection of the relevant features had to be performed even before seeing any label from the test set. We believe that we achieved some of that by using a **transductive strategy**. *The top levels(s) of the unsupervised feature-learning hierarchy were trained purely or mostly on the test set examples.* Since the training set was much larger, we used it to extract a large set of general-purpose features that covered the variations across many classes. The unlabeled test set was then used transductively to *select among the non-linear factors extracted from the training set those few factors varying most in the test set.* Typically this was simply achieved by a simple PCA applied at the last level, trained only on test examples, and with very few leading eigenvectors selected.

## 3. A Zoo of Possible Layer-Wise Unsupervised Learning Algorithms

### 3.1. PCA, ICA, Normalization

Existing linear models such as PCA or ICA can be useful as one or more of the levels of a deep hierarchy. In fact, on several of the challenge datasets, we found that using a PCA as **the first and the last level** often worked very well. PCA preserves the global linear directions of maximum variance, and separates them into orthogonal components. The representation learned is the projection on the principal eigenvectors of the input covariance matrix. This corresponds to a coordinate system associated with a linear manifold spanned by these eigenvectors (centered at the center of mass of the data). For the first PCA layer, we typically kept a fairly large number of directions, so the main effect of that step is to smooth the input distribution by eliminating some of the variations involved in the least globally varying directions. Optionally, the PCA transformation can include a whitening step which means that the projections are normalized to variance 1 (by dividing each projection by the square root of the corresponding eigenvalue, i.e., component variance).

Whereas PCA can already perform a kind of normalization across examples (by subtracting the mean over examples and dividing by the standard deviation over examples, in the chosen directions), there is a complementary form of normalization which we have found useful. It is a simple variant of the *contrast normalization* commonly employed as an intermediate step in deep convolutional neural networks (LeCun et al., 1989, 1998a). The idea is to normalize across the elements of each input vector, by subtracting the mean and dividing by the standard deviation across elements of the input vector.

### 3.2. Autoencoders

An autoencoder defines a reconstruction $r(x) = g(h(x))$ of the input $x$ from the composition of an *encoder* $h(\cdot)$ and a *decoder* $g(\cdot)$. In general both are parametrized and the most common parametrization corresponds to $r(x)$ being the output of a one-hidden layer neural network (taking $x$ as input) and $h(x)$ being the non-linear output of the hidden layer. Training proceeds by minimizing the average of reconstruction errors, $L(r(x), x)$. If the encoder and decoder are linear and $L(r(x), x) = \|r(x) - x\|^2$ is the square error, then $h(x)$ learns to span the principal eigenvectors of the input, i.e., being equivalent (up to a rotation) to a PCA (Bourlard and Kamp, 1988). However, with a non-linear encoder, one obtains a representation that can be greedily stacked and often yields better representations with deeper encoders (Bengio et al., 2007; Goodfellow et al., 2009). A probabilistic interpretation of reconstruction error is simply as a particular form of *energy function* (Ranzato et al., 2008) (the logarithm of an unnormalized probability density function). It means that examples with low reconstruction error have higher probability according to the model. A sparsity term in the energy function has been used to allow overcomplete representations (Ranzato et al., 2007b, 2008) and shown to yield features that are (for some of them) more invariant to geometric transformations of images (Goodfellow et al., 2009). A successful alternative (Bengio et al., 2007; Vincent et al., 2008) to the square reconstruction error in the case of inputs that are binary or in the $(0,1)$ interval (like pixel intensities) is the sum of KL divergences between the binomial probabilities associated with each input $x_i$ and with each reconstruction $r_i(x)$ (both seen as binary probabilities).

### 3.3. RBMs

As shown in Bengio and Delalleau (2009), the reconstruction error gradient for autoencoders can be seen as an approximation of the Contrastive Divergence (Hinton, 1999, 2002) update rule for Restricted Boltzmann Machines (Hinton et al., 2006). Boltzmann Machines are undirected graphical models, defined by an energy function which is related to the joint probability of inputs (visible) $x$ and hidden (latent) variables $h$ through

$$P(x,h) = e^{-\text{energy}(x,h)}/Z$$

where the normalization constant $Z$ is called the partition function, and the marginal probability of the observed data (which is what we want to maximize) is simply $P(x) = \sum_h P(x,h)$ (summing or integrating over all possible configurations of $h$). A Boltzmann Machine is the equivalent for binary random variables of the multivariate Gaussian distribution for continuous random variables, in the sense that it is defined by an energy function that is a second-order polynomial in the random bit values. Both are particular kinds of Markov Random Fields (undirected graphical models), but the partition function of the Boltzmann Machine is intractable, which means that approximations of the log-likelihood gradient $\frac{\partial \log P(x)}{\partial \theta}$ must be employed to train it (where $\theta$ is the set of parameters).

Restricted Boltzmann Machines (RBMs) are Boltzmann Machines with a restriction in the connection pattern of the graphical model between variables, forming a bipartite graph with two groups of variables: the input (or visible) and latent (or hidden) variables. Whereas the original RBM employs binomial hidden and visible units, which worked well on data such as MNIST (where grey levels actually correspond to probabilities of turning on a pixel), the original extension of RBMs to continuous data (the Gaussian RBM) has not been as successful as more recent continuous-data RBMs such as the mcRBM (Ranzato and Hinton, 2010), the mPoT model (Ranzato et al., 2010) and the **spike-and-slab RBM** (Courville et al., 2011), which was used in the challenge. The spike-and-slab RBM energy function allows hidden units to either push variance up or down in different directions, and it can be efficiently trained thanks to a 3-way block Gibbs sampling procedure.

RBMs are defined by their energy function, and when it is tractable (which is usually the case), their *free energy* function is:

$$\mathrm{FreeEnergy}(x) = -\log \sum_h e^{-\mathrm{energy}(x,h)}.$$

The log-likelihood gradient can be defined in terms of the gradient of the free energy on observed (so-called positive) data samples $x$ and on (so-called negative) model samples $\tilde{x} \sim P(\tilde{x})$:

$$-\frac{\partial \log P(x)}{\partial \theta} = \frac{\partial \mathrm{FreeEnergy}(x)}{\partial \theta} - E[\frac{\partial \mathrm{FreeEnergy}(\tilde{x})}{\partial \theta}]$$

where the expectation is over $\tilde{x} \sim P(\tilde{x})$. When the free energy is tractable, the first term can be computed readily, whereas the second term involves sampling from the model. Various kinds of RBMs can be trained by approximate maximum likelihood stochastic gradient descent, often involving a Monte-Carlo Markov Chain to obtain those model samples. See Bengio (2009) for a much more complete tutorial on this subject, along with Hinton (2010) for tips and tricks.

### 3.4. Denoising Autoencoders

Denoising autoencoders training is a simple variation on autoencoder training: try to reconstruct the clean original input from an artificially and stochastically corrupted version of it, by minimizing the denoising reconstruction error. Denoising autoencoders are simply trained by stochastic gradient descent, typically using mini-batches (of 20 to 200 examples) in order to take advantage of faster matrix-matrix operations on CPUs or GPUs. Denoising autoencoders solve one of the thorny limitations of ordinary autoencoders: the representation can be overcomplete without causing any problem. More hidden units just means that the model can more finely represent the input distribution. Denoising autoencoders have recently been shown to be directly related to score matching (Vincent, 2011), an induction principle that can replace maximum likelihood when it is not tractable (and the inputs are continuous-valued). The score matching criterion is the squared norm of the difference between the model's *score* (gradient $\frac{\partial \log P(x)}{\partial x}$ of the log-likelihood with respect to the input $x$) and the score of the true data generating density (which is unknown, but from which we have samples). A simple way to understand the connection between denoising autoencoders and score matching is the following. Considering that reconstruction error is an energy function, the reconstruction from an autoencoder normally goes from a lower-probability (higher energy) input configuration to a nearby higher-probability (lower energy) one, so the difference $r(\tilde{x}) - \tilde{x}$ between reconstruction and input is the model's view of a direction of maximum increase in probability (i.e., the model's score). On the other hand, when one takes a training sample $x$ and one randomly corrupts it into $\tilde{x}$, one typically obtains a lower probability neighbor, i.e., the vector $x - \tilde{x}$ is nature's hint about a direction of rapid increase in probability (when starting at $\tilde{x}$). The squared difference of these two differences is just the denoising reconstruction error $(r(\tilde{x}) - x)^2$, in the case of the squared error reconstruction loss.

In the challenge, we used (Mesnil et al., 2011) particular denoising autoencoders that are well suited for data with sparse high-dimensional inputs. Instead of the usual sigmoid or tanh non-linear hidden unit activations, these autoencoders are based on *rectifier* units ($\max(x,0)$ instead of tanh) with L1 penalty in the training criterion, which tends to make the hidden representation sparse. Stochastic rectifier units had been introduced in the context of RBMs earlier (Nair and Hinton, 2010) and we have found them to be extremely useful for deterministic deep networks (Glorot et al., 2011a) and denoising autoencoders (Glorot et al., 2011b). A recent extension of denoising autoencoders is particularly useful for two of the challenge datasets in which the input vectors are very large and sparse. It addresses a particularly troubling issue when training autoencoders on large sparse vectors: whereas the encoder can take advantage of the numerous zeros in the input vector (it does not need to do any computation for them), the decoder needs to make reconstruction predictions and compute reconstruction error for all the inputs, including the zeros. With the sampled reconstruction algorithm (Dauphin et al., 2011), one only needs to compute reconstructions and reconstruction error for a small stochastically selected subset of the zeros, yielding very substantial speed-ups (20-fold in the experiments of Dauphin et al. (2011)), the more so as the fraction of non-zeros decreases.

### 3.5. Contractive Autoencoders

Contractive autoencoders (Rifai et al., 2011) minimize a training criterion that is the sum of a reconstruction error and a "contraction penalty", which encourages the learnt representation $h(x)$ to be as invariant as possible to the input $x$, while still allowing to distinguish the training examples from each other (i.e., to reconstruct them). As a consequence, the representation is faithful to changes in input space in the directions of the manifold near which examples concentrate, but it is highly contractive in the orthogonal directions. This is similar in spirit to a PCA (which only keeps the leading directions of variation and completely ignores the others), but is softer (no hard cutting at a particular dimension), is non-linear and can contract in different directions depending on where one looks in the input space (hence can capture non-linear manifolds). To prevent a trivial solution in which the encoder weights go to zero and the decoder weights to infinity, the contractive autoencoder uses tied weights (the decoder weights are forced to be the transpose of the encoder weights). Because of the contractive criterion, what we find empirically is that for any particular input example, many of the hidden units saturate while a few remain sensitive to changes in the input (corresponding to changes in the directions of changes expected under the data distribution). That subset of active units changes as we move around in input space, and defines a kind of *local chart*, or local coordinate system, in the neighborhood of each input point. This can be visualized to some extent by looking at the singular values and singular vectors of the Jacobian matrix $\frac{\partial h(x)}{\partial x}$ (containing the derivatives of each hidden unit output with respect to each input unit). Contrary to other autoencoders, one tends to find only few dominant eigenvalues, and their number corresponds to a local rank or local dimension (which can change as we move in input space). This is unlike other dimensionality reduction algorithms in which the number of dimensions is fixed by hand (rather than learnt) and fixed across the input domain. In fact the learnt representation can be overcomplete (larger than the input): it is only in the sense of its Jacobian that it has an effective small dimensionality for any particular input point. The large number of hidden units can be exploited to model complicated non-linear manifolds.

## 4. Tricks and Tips

A good starting point for tricks and tips relevant to training deep architectures, and in particular Restricted Boltzmann Machines (RBMs), is Hinton (2010). An older guide which is also useful to some extent is Orr and Muller (1998), and in particular LeCun et al. (1998b), since many of the ideas from neural networks training can be exploited here.

### 4.1. Monitoring Performance During Training

RBMs are tricky because although there are good estimators of the log-likelihood *gradient*, there are no known cheap ways of estimating the log-likelihood itself (Annealed Importance Sampling (Murray and Salakhutdinov, 2009) is an expensive way of doing it). A poor man's option is to measure reconstruction error (as if the parameters were those of an autoencoder), which works well for the beginning of training but does not help to choose a stopping point (e.g. to avoid overfitting). The practical solution is to save the model weights at different numbers of epochs (e.g., 5, 10, 20, 50) and plug the learned representation into a supervised classifier (for each of these training durations) in order to decide what training duration to select.

In the case of denoising autoencoders, on the other hand, the denoising reconstruction error is a good measure of the model's progress (since it corresponds to the training criterion) and it can be used for early stopping. However, the best generative model or the one with the best denoising is not always the one that works best in terms of providing a good representation for a classifier. This is especially true in the transfer setting of the competition, where the training distribution is different from the test and validation distributions. In that case, the expensive solution of evaluating validation classification error at different training durations is the approach we have chosen. The training criterion of contractive autoencoders can also be used as a good monitoring device (and its value on a validation set used to perform early stopping). Note that we did not really "stop" training, we only recorded representations at different points in the training trajectory and estimated Area under the Learning Curve (ALC[5]) or other criteria associated with each. The advantage is that we do not need to retrain a separate model from scratch for each of the possible durations tested.

## 4.2. Random Search and Greedy Layer-wise Strategy

Because one can have a different type of representation-learning model at each layer, and because each of these learning algorithms has several hyper-parameters, there is a huge number of possible configurations and choices one can make in exploring the kind of deep architectures that led to the winning entry of the challenge. There are two approaches that practitioners of machine learning typically employ to deal with hyper-parameters. One is manual trial and error, i.e., a human-guided search. The other is a grid search, i.e., choosing a set of values for each hyper-parameter and training and evaluating a model for each combination of values for all the hyper-parameters. Both work well when the number of hyper-parameters is small (e.g. 2 or 3) but break down when there are many more[7]. More systematic approaches are needed. An approach that we have found to scale better is based on random search and greedy exploration. The idea of random search (Bergstra and Bengio, 2011, 2012) is simple and can advantageously replace grid search. Instead of forming a regular grid by choosing a small set of values for each hyper-parameter, one defines a distribution from which to sample values for each hyper-parameter, e.g., the log of the learning rate could be taken as uniform between $\log(0.1)$ and $\log(10^{-6})$, or the log of the number of hidden units or principal components could be taken as uniform between $\log(2)$ and $\log(5000)$. The main advantage of random (or quasi-random) search over a grid is that when some

---

5. The ALC is the sum of the accuracy measure for different number of labeled training examples. The accuracy measure used here is the AUC or Area Under the Curve. See [6]

7. Experts can handle many hyper-parameters, but results become less reproducible and algorithms less accessible to non-experts.

hyper-parameters have little or no influence, random search does not waste any computation, whereas grid search will redo experiments that are equivalent and do not bring any new information (because many of them have the same value for hyper-parameters that matter and different values for hyper-parameters that do not). Instead, with random search, every experiment is different, thus bringing more information. In addition, random search is convenient because even if some jobs are not finished, one can draw conclusions from the jobs that are finished. In fact, one can use the results on subsets of the experiments to establish confidence intervals (the experiments are now all iid), and draw a curve (with confidence interval) showing how performance improves as we do more exploration. Of course, it would be even better to perform a sequential optimization (Bergstra et al., 2011) (such as Bayesian Optimization (Brochu et al., 2009)) in order to take advantage of results of training experiments as they are obtained and sample in more promising regions of configuration space, but more research needs to be done towards this. On the other hand, random search is very easy and does not introduce hyper-hyper-parameters.

Another trick that we have used successfully in the past and in the challenge is the idea of a greedy search. Since the deep architecture is obtained by stacking layers and each layer comes with its own choices and hyper-parameters, the general strategy is the following. First optimize the choices for the first layer (e.g., try to find which single-layer learning algorithm and its hyper-parameters give best results according to some criterion such as validation set classification error or ALC). Then keep that best choice (or a few of the best choices found) and explore choices for the second layer, keeping only the best overall choice (or a few of the best choices found) among the 2-layer systems tested. This procedure can then be continued to add more layers, without the computational cost exploding with the number of layers (it just grows linearly).

## 4.3. Hyper-Parameters

The single most important hyper-parameter for most of the algorithms described here is the learning rate. A too small learning rate means slow convergence, or convergence to a poor performance given a finite budget of computation time. A too large learning rate gives poor results because the training criterion may increase or oscillate. The optimal learning rate for one data set and architecture may be too large or too small for another one, so it is worth optimizing the learning rate. Like most numerical hyper-parameters, the learning rates should be explored in the log-domain, and there is not much to be gained by refining it more than a factor of 2, whereas the dynamic range explored could be around $10^6$ learning rates are typically below 1. To efficiently search for a good learning rate, a greedy heuristic that we used is based on the following strategy. Start with a large learning rate and reduce it (by a factor 3) until training does not diverge. The largest learning rate which does not give divergent training (increasing training error) is usually a very good choice of learning rate.

For the challenge, another very sensitive hyper-parameter is the number of dimensions of the top-level representation fed to the classifier. It should probably be close to or related to the true number of classes (more classes would require more dimensions to be separated easily by a linear classifier).

Early stopping is another handy trick to speed-up model search, since it can be used to detect overfitting (even in the unsupervised learning sense) for a low computational cost. After each training iteration one can compute an indicator of generalization error (either from the application of the unsupervised learning criterion on the validation set or even by training a linear classifier on a pseudo-validation set, as described below, sec. 4.5).

## 4.4. Visualization

Since the validation set ALC was an unreliable indicator of test set ALC, we used several strategies in the second phase of the competition to help guide the model selection. One of them is simply visualization of the representations as cloud points. One can visualize 3 dimensions at a time, either the leading 3 or a subset of the leading ones. To order dimensions we used PCA or t-SNE dimensionality reduction (van der Maaten and Hinton, 2008).

## 4.5. Simulating the Final Evaluation Scenario

Another strategy that often comes handy is to simulate (as much as possible) the final evaluation scenario, even in the absence of the test labels. In the case of the second phase of the competition, some of the training classes labels are available. Thus we could simulate the final evaluation scenario by choosing a subset of the training classes as "pseudo training set" and the rest as "pseudo test set", doing unsupervised training on the pseudo training set (or the union of pseudo training and pseudo test sets) and training the linear classifier using the pseudo test set. We then considered hyper-parameter settings that led not only to high accuracy in average across different choices of class subsets (for the pseudo train/test split), but also to high robustness (low variance) across these splits.

## 5. Examples in Transfer Learning

Deep Learning seems well suited to transfer learning because it focuses on learning representations and in particular "abstract" representations, representations that ideally disentangle the factors of variation present in the input.

This has been demonstrated already not only in this competition (see Mesnil et al. (2011)) for more details), but also in several other instances. For example, in Bengio et al. (2011), it has been shown that a deep learner can take more advantage of out-of-distribution training examples (whose distribution differs from the test distribution) than a shallow learner. Two settings were explored, with both results illustrated in Figure 1. In the first one (left hand side of figure), training examples (character images) are distorted by adding all kinds of noises and random transformations (coherent with character images), but the target distribution contains clean examples. Training on the distorted examples was found to help the deep learners (SDA{1,2}) more than the shallow ones (MLP{1,2}), when the goal is to test on the clean examples. In the second setting (right hand side of figure), i.e., the multi-task setting, training is on all 62 classes available, but the target distribution of interest is a restriction to a subset of the classes.

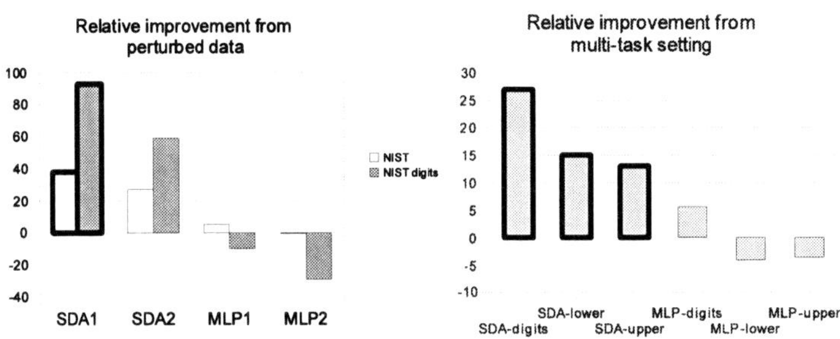



Figure 1: Relative improvement in character classification error rate due to out-of-distribution examples. Left: Improvement (or loss, when negative) induced by out-of-distribution examples (perturbed data). Right: Improvement (or loss, when negative) induced by multi-task learning (training on all character classes and testing only on either digits, upper case, or lower-case). The deep learner (stacked denoising autoencoder) benefits more from out-of-distribution examples, compared to a shallow MLP. {SDA,MLP}1 and {SDA,MLP}2 are trained on different types of distortions. The NIST set includes all 62 classes while NIST digits include only the 10 digits. Reproduced from Bengio et al. (2011).

Another successful transfer example also using stacked denoising autoencoders arises in the context of *domain adaptation*, i.e., where one trains an unsupervised representation based on examples from a set of domains but a classifier is then trained from few examples of only one domain. In that case, unlike in the challenge, the output variable always has the same semantics, but the input distribution (and to a lesser extent the relation between input and output) changes from domain to domain. Glorot et al. (2011b) applied stacked denoising autoencoders with sparse rectifiers (the same as used for the challenge) to domain adapation in *sentiment analysis* (predicting whether a user liked a disliked a product based on a short review). Some of the results are summarized in Figure 2, comparing transfer ratio, which indicates relative error when testing in-domain vs out-of-domain, i.e., how well transfer works (see Glorot et al. (2011b) for more details). The stacked denoising autoencoders (SDA) are compared with the state of the art methods: SCL (Blitzer et al., 2006) or Structural Correspondence Learning, MCT (Li and Zong, 2008) or Multi-label Consensus Training, SFA (Pan et al., 2010) or Spectral Feature Alignment, and T-SVM (Sindhwani and Keerthi, 2006) or Transductive SVM. The SDA$_{sh}$ (Stacked Denoising Autoencoder trained on all domains) clearly beats the state-of-the-art.

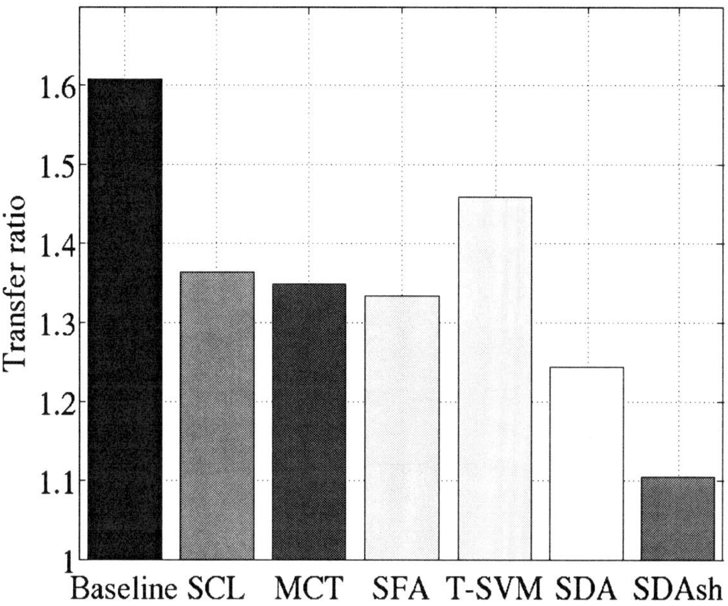

Figure 2: **Transfer ratios** on the Amazon benchmark. Both SDA-based systems outperforms the rest, and SDA$_{sh}$ (unsupervised training on all domains) is best. Reproduced from Glorot et al. (2011b).

## 6. Moving Forward: Disentangling Factors of Variation

In spite of all the nice results described here, and in spite of winning the final evaluation of the challenge, it is clear to the author that research in Deep Learning of representations is only in its infancy, and that much more should be done to improve the learning algorithms. In particular, it is the author's belief that these algorithms would be much more useful in transfer learning if they could better *disentangle the underlying factors of variation*. In a sense it is obvious that if we had algorithms that could do that really well, than most learning tasks (supervised learning, transfer learning, reinforcement learning, etc.) would become much easier, and the effect would be most felt when only very few labeled examples for the final task of interest are present. The question is whether we can actually improve in this direction, and the hypothesis we propose is that by explicitly designing the models and training criterion towards that objective, there is much to be gained. Ultimately, one can view the problem of learning from very few labeled examples of the task of interest almost as a an inference problem ("given that I define a new class based on this particular example, what is the probability that this other example also belongs to it?"), where the parameters of the model (i.e., the repre-

BENGIO

sentation) have already been established through prior training on many more related examples (labeled or not) which help to capture the underlying factors of variation, some of which are relevant in the target task.

## Acknowledgments

The author wants to thank NSERC, the CRC, mPrime, Compute Canada and FQRNT for their support, and the other authors of the companion paper (Mesnil et al., 2011) for their contributions to many of the results and ideas illustrated here.

## References

Mikhail Belkin and Partha Niyogi. Using manifold structure for partially labeled classification. In S. Becker, S. Thrun, and K. Obermayer, editors, *Advances in Neural Information Processing Systems 15 (NIPS'02)*, Cambridge, MA, 2003. MIT Press.

Yoshua Bengio. Learning deep architectures for AI. *Foundations and Trends in Machine Learning*, 2(1):1–127, 2009. Also published as a book. Now Publishers, 2009.

Yoshua Bengio and Olivier Delalleau. Justifying and generalizing contrastive divergence. *Neural Computation*, 21(6):1601–1621, June 2009.

Yoshua Bengio and Olivier Delalleau. Shallow versus deep sum-product networks, 2011. *The Learning Workshop*, Fort Lauderdale, Florida.

Yoshua Bengio and Yann LeCun. Scaling learning algorithms towards AI. In L. Bottou, O. Chapelle, D. DeCoste, and J. Weston, editors, *Large Scale Kernel Machines*. MIT Press, 2007.

Yoshua Bengio, Pascal Lamblin, Dan Popovici, and Hugo Larochelle. Greedy layerwise training of deep networks. In Bernhard Schölkopf, John Platt, and Thomas Hoffman, editors, *Advances in Neural Information Processing Systems 19 (NIPS'06)*, pages 153–160. MIT Press, 2007.

Yoshua Bengio, Frédéric Bastien, Arnaud Bergeron, Nicolas Boulanger-Lewandowski, Thomas Breuel, Youssouf Chherawala, Moustapha Cisse, Myriam Côté, Dumitru Erhan, Jeremy Eustache, Xavier Glorot, Xavier Muller, Sylvain Pannetier Lebeuf, Razvan Pascanu, Salah Rifai, François Savard, and Guillaume Sicard. Deep learners benefit more from out-of-distribution examples. In *JMLR W&CP: Proceedings of the Fourteenth International Conference on Artificial Intelligence and Statistics (AISTATS 2011)*, April 2011.

James Bergstra and Yoshua Bengio. Random search for hyper-parameter optimization, 2011. *The Learning Workshop*, Fort Lauderdale, Florida.

James Bergstra and Yoshua Bengio. Random search for hyper-parameter optimization. *Journal of Machine Learning Research*, 2012.

James Bergstra, Rémy Bardenet, Yoshua Bengio, and Balázs Kégl. Algorithms for hyper-parameter optimization. In *NIPS'2011*, 2011.

John Blitzer, Ryan McDonald, and Fernando Pereira. Domain adaptation with structural correspondence learning. In *Proc. of EMNLP '06*, 2006.

Hervé Bourlard and Yves Kamp. Auto-association by multilayer perceptrons and singular value decomposition. *Biological Cybernetics*, 59:291–294, 1988.

Mark Braverman. Poly-logarithmic independence fools bounded-depth boolean circuits. *Communications of the ACM*, 54(4):108–115, April 2011.

Eric Brochu, Vlad M. Cora, and Nando de Freitas. A tutorial on bayesian optimization of expensive cost functions, with application to active user modeling and hierarchical reinforcement learning. Technical Report TR-2009-23, Department of Computer Science, University of British Columbia, November 2009.

Rich Caruana. Learning many related tasks at the same time with backpropagation. In G. Tesauro, D.S. Touretzky, and T.K. Leen, editors, *Advances in Neural Information Processing Systems 7 (NIPS'94)*, pages 657–664, Cambridge, MA, 1995. MIT Press.

Ronan Collobert and Jason Weston. A unified architecture for natural language processing: Deep neural networks with multitask learning. In William W. Cohen, Andrew McCallum, and Sam T. Roweis, editors, *Proceedings of the Twenty-fifth International Conference on Machine Learning (ICML'08)*, pages 160–167. ACM, 2008.

Aaron Courville, James Bergstra, and Yoshua Bengio. Unsupervised models of images by spike-and-slab RBMs. In *Proceedings of the Twenty-eight International Conference on Machine Learning (ICML'11)*, June 2011.

Yann Dauphin, Xavier Glorot, and Yoshua Bengio. Sampled reconstruction for large-scale learning of embeddings. In *Proceedings of the Twenty-eight International Conference on Machine Learning (ICML'11)*, June 2011.

Xavier Glorot, Antoire Bordes, and Yoshua Bengio. Deep sparse rectifier neural networks. In *JMLR W&CP: Proceedings of the Fourteenth International Conference on Artificial Intelligence and Statistics (AISTATS 2011)*, April 2011a.

Xavier Glorot, Antoire Bordes, and Yoshua Bengio. Domain adaptation for large-scale sentiment classification: A deep learning approach. In *Proceedings of the Twenty-eight International Conference on Machine Learning (ICML'11)*, June 2011b.

Ian Goodfellow, Quoc Le, Andrew Saxe, and Andrew Ng. Measuring invariances in deep networks. In Yoshua Bengio, Dale Schuurmans, Christopher Williams, John Lafferty, and Aron Culotta, editors, *Advances in Neural Information Processing Systems 22 (NIPS'09)*, pages 646–654, 2009.

Johan Håstad. Almost optimal lower bounds for small depth circuits. In *Proceedings of the 18th annual ACM Symposium on Theory of Computing*, pages 6–20, Berkeley, California, 1986. ACM Press.

Johan Håstad and Mikael Goldmann. On the power of small-depth threshold circuits. *Computational Complexity*, 1:113–129, 1991.

G. E. Hinton, P. Dayan, and M. Revow. Modelling the manifolds of images of handwritten digits. *IEEE Transactions on Neural Networks*, 8:65–74, 1997.

Geoffrey E. Hinton. Products of experts. In *Proceedings of the Ninth International Conference on Artificial Neural Networks (ICANN)*, volume 1, pages 1–6, Edinburgh, Scotland, 1999. IEE.

Geoffrey E. Hinton. Training products of experts by minimizing contrastive divergence. *Neural Computation*, 14:1771–1800, 2002.

Geoffrey. E. Hinton. A practical guide to training restricted Boltzmann machines. Technical Report UTML TR 2010-003, Department of Computer Science, University of Toronto, 2010.

Geoffrey E. Hinton and Ruslan Salakhutdinov. Reducing the dimensionality of data with neural networks. *Science*, 313(5786):504–507, July 2006.

Geoffrey E. Hinton, Simon Osindero, and Yee Whye Teh. A fast learning algorithm for deep belief nets. *Neural Computation*, 18:1527–1554, 2006.

Pascal Lamblin and Yoshua Bengio. Important gains from supervised fine-tuning of deep architectures on large labeled sets. NIPS*2010 Deep Learning and Unsupervised Feature Learning Workshop, 2010.

Yann LeCun, Bernhard Boser, John S. Denker, Donnie Henderson, Richard E. Howard, Wayne Hubbard, and Lawrence D. Jackel. Backpropagation applied to handwritten zip code recognition. *Neural Computation*, 1(4):541–551, 1989.

Yann LeCun, Leon Bottou, Yoshua Bengio, and Patrick Haffner. Gradient-based learning applied to document recognition. *Proceedings of the IEEE*, 86(11):2278–2324, November 1998a.

Yann LeCun, Léon Bottou, Genevieve B. Orr, and Klaus-Robert Müller. Efficient backprop. In *Neural Networks, Tricks of the Trade*, Lecture Notes in Computer Science LNCS 1524. Springer Verlag, 1998b.

Honglak Lee, Roger Grosse, Rajesh Ranganath, and Andrew Y. Ng. Convolutional deep belief networks for scalable unsupervised learning of hierarchical representations. In Léon Bottou and Michael Littman, editors, *Proceedings of the Twenty-sixth International Conference on Machine Learning (ICML'09)*. ACM, Montreal (Qc), Canada, 2009a.

Honglak Lee, Peter Pham, Yan Largman, and Andrew Ng. Unsupervised feature learning for audio classification using convolutional deep belief networks. In Yoshua Bengio, Dale Schuurmans, Christopher Williams, John Lafferty, and Aron Culotta, editors, *Advances in Neural Information Processing Systems 22 (NIPS'09)*, pages 1096–1104, 2009b.

Shoushan Li and Chengqing Zong. Multi-domain adaptation for sentiment classification: Using multiple classifier combining methods. In *Proc. of NLP-KE '08*, 2008.

Grégoire Mesnil, Yann Dauphin, Xavier Glorot, Salah Rifai, Yoshua Bengio, Ian Goodfellow, Erick Lavoie, Xavier Muller, Guillaume Desjardins, David Warde-Farley, Pascal Vincent, Aaron Courville, and James Bergstra. Unsupervised and transfer learning challenge: a deep learning approach. In Isabelle Guyon, G. Dror, V Lemaire, G. Taylor, and D. Silver, editors, *JMLR W& CP: Proceedings of the Unsupervised and Transfer Learning challenge and workshop*, volume 7, 2011.

Iain Murray and Ruslan Salakhutdinov. Evaluating probabilities under high-dimensional latent variable models. In Daphne Koller, Dale Schuurmans, Yoshua Bengio, and Leon Bottou, editors, *Advances in Neural Information Processing Systems 21 (NIPS'08)*, volume 21, pages 1137–1144, 2009.

V. Nair and G. E Hinton. Rectified linear units improve restricted boltzmann machines. In *Proc. of ICML '10*, 2010.

Genevieve Orr and Klaus-Robert Muller, editors. *Neural networks: tricks of the trade*, volume 1524 of *Lecture Notes in Computer Science*. Springer-Verlag Inc., New York, NY, USA, 1998. ISBN 3-540-65311-2 (paperback).

Sinno Jialin Pan, Xiaochuan Ni, Jian-Tao Sun, Qiang Yang, and Zheng Chen. Cross-domain sentiment classification via spectral feature alignment. In *Proc. of WWW '10*, 2010.

H. Poon and P. Domingos. Sum-product networks: A new deep architecture. In *NIPS 2010 Workshop on Deep Learning and Unsupervised Feature Learning*, Whistler, Canada, 2010.

Rajat Raina, Alexis Battle, Honglak Lee, Benjamin Packer, and Andrew Y. Ng. Self-taught learning: transfer learning from unlabeled data. In Zoubin Ghahramani, editor, *Proceedings of the Twenty-fourth International Conference on Machine Learning (ICML'07)*, pages 759–766. ACM, 2007.

M. Ranzato, C. Poultney, S. Chopra, and Y. LeCun. Efficient learning of sparse representations with an energy-based model. In *NIPS'06*, 2007a.

M. Ranzato, V. Mnih, and G. Hinton. Generating more realistic images using gated MRF's. In J. Lafferty, C. K. I. Williams, J. Shawe-Taylor, R.S. Zemel, and A. Culotta, editors, *Advances in Neural Information Processing Systems 23 (NIPS'10)*, pages 2002–2010, 2010.

M.A. Ranzato and G.E. Hinton. Modeling pixel means and covariances using factorized third-order Boltzmann machines. In *Computer Vision and Pattern Recognition (CVPR), 2010 IEEE Conference on*, pages 2551–2558. IEEE, 2010.

Marc'Aurelio Ranzato, Christopher Poultney, Sumit Chopra, and Yann LeCun. Efficient learning of sparse representations with an energy-based model. In B. Schölkopf, J. Platt, and T. Hoffman, editors, *Advances in Neural Information Processing Systems 19 (NIPS'06)*, pages 1137–1144. MIT Press, 2007b.

Marc'Aurelio Ranzato, Y-Lan Boureau, and Yann LeCun. Sparse feature learning for deep belief networks. In J.C. Platt, D. Koller, Y. Singer, and S. Roweis, editors, *Advances in Neural Information Processing Systems 20 (NIPS'07)*, pages 1185–1192, Cambridge, MA, 2008. MIT Press.

Salah Rifai, Pascal Vincent, Xavier Muller, Xavier Glorot, and Yoshua Bengio. Contracting auto-encoders: Explicit invariance during feature extraction. In *Proceedings of the Twenty-eight International Conference on Machine Learning (ICML'11)*, June 2011.

Ruslan Salakhutdinov and Geoffrey E. Hinton. Deep Boltzmann machines. In *Proceedings of The Twelfth International Conference on Artificial Intelligence and Statistics (AISTATS'09)*, volume 5, pages 448–455, 2009.

L. Saul and S. Roweis. Think globally, fit locally: unsupervised learning of low dimensional manifolds. *Journal of Machine Learning Research*, 4:119–155, 2002.

Vikas Sindhwani and S. Sathiya Keerthi. Large scale semi-supervised linear svms. In *Proc. of SIGIR '06*, 2006.

Joshua Tenenbaum, Vin de Silva, and John C. Langford. A global geometric framework for nonlinear dimensionality reduction. *Science*, 290(5500):2319–2323, December 2000.

Laurens van der Maaten and Geoffrey E. Hinton. Visualizing data using t-sne. *Journal of Machine Learning Research*, 9:2579–2605, November 2008. URL `http://www.jmlr.org/papers/volume9/vandermaaten08a/vandermaaten08a.pdf`.

Pascal Vincent. A connection between score matching and denoising autoencoders. *Neural Computation*, to appear, 2011.

Pascal Vincent, Hugo Larochelle, Yoshua Bengio, and Pierre-Antoine Manzagol. Extracting and composing robust features with denoising autoencoders. In Andrew McCallum and Sam Roweis, editors, *Proceedings of the 25th Annual International Conference on Machine Learning (ICML 2008)*, pages 1096–1103. Omnipress, 2008.

Andrew Yao. Separating the polynomial-time hierarchy by oracles. In *Proceedings of the 26th Annual IEEE Symposium on Foundations of Computer Science*, pages 1–10, 1985.

# Autoencoders, Unsupervised Learning, and Deep Architectures

**Pierre Baldi**                                               PFBALDI@ICS.UCI.EDU
*Department of Computer Science*
*University of California, Irvine*
*Irvine, CA 92697-3435*

**Editor:** I. Guyon, G. Dror, V. Lemaire, G. Taylor and D. Silver

## Abstract

Autoencoders play a fundamental role in unsupervised learning and in deep architectures for transfer learning and other tasks. In spite of their fundamental role, only linear autoencoders over the real numbers have been solved analytically. Here we present a general mathematical framework for the study of both linear and non-linear autoencoders. The framework allows one to derive an analytical treatment for the most non-linear autoencoder, the Boolean autoencoder. Learning in the Boolean autoencoder is equivalent to a clustering problem that can be solved in polynomial time when the number of clusters is small and becomes NP complete when the number of clusters is large. The framework sheds light on the different kinds of autoencoders, their learning complexity, their horizontal and vertical composability in deep architectures, their critical points, and their fundamental connections to clustering, Hebbian learning, and information theory.

**Keywords:** autoencoders, unsupervised learning, compression, clustering, principal component analysis, boolean, complexity, deep architectures, hebbian learning, information theory

## 1. Introduction

Autoencoders are simple learning circuits which aim to transform inputs into outputs with the least possible amount of distortion. While conceptually simple, they play an important role in machine learning. Autoencoders were first introduced in the 1980s by Hinton and the PDP group (Rumelhart et al., 1986) to address the problem of "back-propagation without a teacher", by using the input data as the teacher. Together with Hebbian learning rules (Hebb, 1949; Oja, 1982), autoencoders provide one of the fundamental paradigms for unsupervised learning and for beginning to address the mystery of how synaptic changes induced by local biochemical events can be coordinated in a self-organized manner to produce global learning and intelligent behavior.

More recently, autoencoders have taken center stage again in the "deep architecture" approach (Hinton et al., 2006; Hinton and Salakhutdinov, 2006; Bengio and LeCun, 2007; Erhan et al., 2010) where autoencoders, particularly in the form of Restricted Boltzmann Machines (RBMS), are stacked and trained bottom up in unsupervised fashion, followed by a supervised learning phase to train the top layer and fine-tune the entire architecture. The bottom up phase is agnostic with respect to the final task and thus can obviously be used in transfer learning approaches. These deep architectures have been shown to lead to state-of-the-art results on a number of challenging classification and regression problems.

In spite of the interest they have generated, and with a few exceptions (Baldi and Hornik, 1988; Sutskever and Hinton, 2008; Montufar and Ay, 2011), little theoretical understanding of autoencoders and deep architectures has been obtained to this date. Additional confusion may have been created by the use of the term "deep". A deep architecture from a computer science perspective should have $n^\alpha$ polynomial-size layers, for some small $\alpha > 0$, where $n$ is the size of the input vectors (see Clote and Kranakis (2002) and references therein). But that is not the case in the architectures described in Hinton et al. (2006) and Hinton and Salakhutdinov (2006), which seem to have constant or at best logarithmic depth, the distinction between finite and logarithmic depth being almost impossible for the typical values of $n$ used in computer vision, speech recognition, and other typical problems. Thus the main motivation behind this work is to derive a better theoretical understanding of autoencoders, with the hope of gaining better insights into the nature of unsupervised learning and deep architectures.

If general theoretical results about deep architectures exist, these are unlikely to depend on a particular hardware realization, such as RBMs. Similar results ought to be true for alternative, or more general, forms of computation. Thus the strategy proposed here is to introduce a general framework and study different kinds of autoencoder circuits, in particular Boolean autoencoders which can be viewed as the most extreme form of non-linear autoencoders. The expectation is that certain properties of autoencoders and deep architectures may be easier to identify and understand mathematically in simpler hardware embodiments, and that the study of different kinds of autoencoders may facilitate abstraction and generalization through the identifications of common properties.

For this purpose, we begin in Section 2 by describing a fairly general framework for studying autoencoders. In Section 3, we review and extend the known results on linear autoencoders. In the light of deep architectures, we look at novel properties such a vertical composition (stacking) and connection of critical points to stability under recycling (feeding outputs back to the input layer). In Section 4, we study Boolean autoencoders, and prove several properties including their fundamental connection to clustering. In Section 5, we address the complexity of Boolean autoencoder learning. In Section 6, we study autoencoders with large hidden layers, and introduce the notion of horizontal composition of autoencoders. In Section 7, we address other classes of autoencoders and generalizations. Finally, in Section 8, we summarize the results and their possible consequences for the theory of deep architectures.

## 2. A General Autoencoder Framework

To derive a fairly general framework, an $n/p/n$ autoencoder (Figure 1) is defined by a t-uple $n, p, m, \mathbb{F}, \mathbb{G}, \mathcal{A}, \mathcal{B}, \mathcal{X}, \Delta$ where:

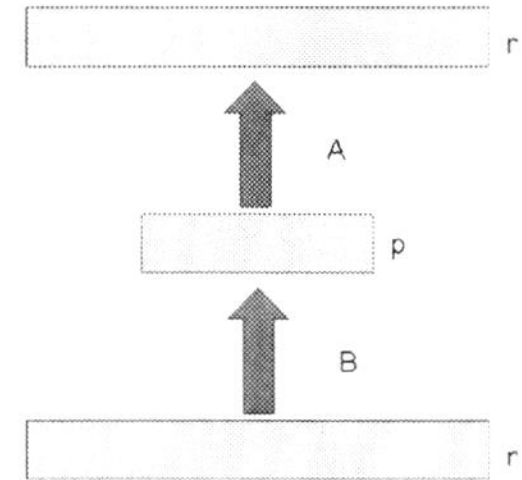

Figure 1: An n/p/n Autoencoder Architecture.

1. $\mathbb{F}$ and $\mathbb{G}$ are sets.

2. $n$ and $p$ are positive integers. Here we consider primarily the case where $0 < p < n$.

3. $\mathcal{A}$ is a class of functions from $\mathbb{G}^p$ to $\mathbb{F}^n$.

4. $\mathcal{B}$ is a class of functions from $\mathbb{F}^n$ to $\mathbb{G}^p$.

5. $\mathcal{X} = \{x_1, \ldots, x_m\}$ is a set of $m$ (training) vectors in $\mathbb{F}^n$. When external targets are present, we let $\mathcal{Y} = \{y_1, \ldots, y_m\}$ denote the corresponding set of target vectors in $\mathbb{F}^n$.

6. $\Delta$ is a dissimilarity or distortion function (e.g. $L_p$ norm, Hamming distance) defined over $\mathbb{F}^n$.

For any $A \in \mathcal{A}$ and $B \in \mathcal{B}$, the autoencoder transforms an input vector $x \in \mathbb{F}^n$ into an output vector $A \circ B(x) \in \mathbb{F}^n$ (Figure 1). The corresponding *autoencoder problem* is to find $A \in \mathcal{A}$ and $B \in \mathcal{B}$ that minimize the overall distortion function:

$$\min E(A,B) = \min_{A,B} \sum_{t=1}^{m} E(x_t) = \min_{A,B} \sum_{t=1}^{m} \Delta(A \circ B(x_t), x_t) \tag{1}$$

In the non auto-associative case, when external targets $y_t$ are provided, the minimization problem becomes:

$$\min E(A,B) = \min_{A,B} \sum_{t=1}^{m} E(x_t, y_t) = \min_{A,B} \sum_{t=1}^{m} \Delta(A \circ B(x_t), y_t) \tag{2}$$

Note that $p < n$ corresponds to the regime where the autoencoder tries to implement some form of compression or feature extraction. The case $p \geq n$ is discussed towards the end of the paper.

Obviously, from this general framework, different kinds of autoencoders can be derived depending, for instance, on the choice of sets $\mathbb{F}$ and $\mathbb{G}$, transformation classes $\mathcal{A}$ and $\mathcal{B}$, distortion function $\Delta$, as well as the presence of additional constraints,such as regularization. To the best of our knowledge, neural network autoencoders were first introduced by the PDP group as a special case of this definition, with all vectors components in $\mathbb{F} = \mathbb{G} = \mathbb{R}$ and $A$ and $B$ corresponding to matrix multiplications followed by non-linear sigmoidal transformations with an $L_2^2$ error function. [For regression problems, the non-linear sigmoidal transformation is typically used only in the hidden layer]. As an approximation to this case, in the next section, we study the linear case with $\mathbb{F} = \mathbb{G} = \mathbb{R}$. More generally, linear autoencoders correspond to the case where $\mathbb{F}$ and $\mathbb{G}$ are fields and $\mathcal{A}$ and $\mathcal{B}$ are the classes of linear transformations, hence $A$ and $B$ are matrices of size $p \times n$ and $n \times p$ respectively. The linear real case where $\mathbb{F} = \mathbb{G} = \mathbb{R}$ and $\Delta$ is the squared Euclidean distance was addressed in Baldi and Hornik (1988) (see also Bourlard and Kamp (1988)). More recently the theory has been extended also to complex-valued linear autoencoders (Baldi et al., 2011)

## 3. The Linear Autoencoder

We partly restate without proof the results derived in Baldi and Hornik (1988) for the linear real case with $\Delta = L_2^2$, but organize them in a way meant to highlight the connections to other kinds of autoencoders, and extend their results from a deep architecture perspective. We use $A^t$ to denote the transpose of any matrix $A$.

**1) Group Invariance.** Every solution is defined up to multiplication by an invertible $p \times p$ matrix $C$, or equivalently up to a change of coordinates in the hidden layer. This is obvious since since $AC^{-1}CB = AB$.

**2) Problem Complexity.** While the cost function is quadratic and all the operations are linear, the overall problem is not convex because the hidden layer limits the rank of the overall transformation to be at most $p$, and the set of matrices of rank $p$ or less is *not* convex. However the linear autoencoder problem over $\mathbb{R}$ can be solved analytically. Note that in this case one is interested in finding a low rank approximation to the identity function.

**3) Fixed Layer Solution.** The problem becomes convex if $A$ is fixed, or if $B$ is fixed. When $A$ is fixed, assuming $A$ has rank $p$ and that the data covariance matrix $\Sigma_{XX} = \sum_i x_i x_i^t$ is invertible, then at the optimum $B^* = B(A) = (A^t A)^{-1} A^t$. When $B$ is fixed, assuming $B$ has rank $p$ and that $\Sigma_{XX}$ is invertible, then at the optimum $A^* = A(B) = \Sigma_{XX} B^t (B \Sigma_{XX} B^t)^{-1}$.

**4) The Landscape of E.** The overall landscape of $E$ has no local minima. All the critical points where the gradient of $E$ is zero, correspond to projections onto subspaces associated with subsets of eigenvectors of the covariance matrix $\Sigma_{XX}$. Projections onto the subspace associated with the $p$ largest eigenvalues correspond to the global minimum and Principal Component Analysis. All other critical point, corresponding to projections onto subspaces associated with other set of eigenvalues, are saddle points. More precisely, if $\mathcal{I} = i_1, \ldots, i_p$ ($1 \le i_i < \ldots < i_p \le n$) is any ordered list of indices, let $U_\mathcal{I} = [u_1, \ldots, u_p]$ denote the matrix formed by the orthonormal eigenvectors of $\Sigma_{XX}$ associated with the eigenvalues $\lambda_{i_1}, \ldots, \lambda_{i_p}$. Then two matrices $A$ and $B$ of rank $p$ define a critical point if and only if there is a set $\mathcal{I}$ and an invertible $p \times p$ matrix $C$ such that $A = U_\mathcal{I} C$, $B = C^{-1} U_\mathcal{I}^t$, and $W = AB = P_{U_\mathcal{I}}$, where $P_{U_\mathcal{I}}$ is the orthogonal projection onto the subspace spanned by the columns of $U_\mathcal{I}$. At the global minimum, assuming that

$C = I$, the activities in the hidden layer are given by the dot products $u_1^t x \ldots u_p^t x$ and correspond to the coordinates of $x$ along the first $p$ eigenvectors of $\Sigma_{XX}$.

**5) Clustering.** The global minimum performs a form of clustering by hyperplane, with respect to $KerB$, the kernel of $B$. For any given vector $x$, all the vectors of the form $x + Ker(B)$ are mapped onto the same vector $y = AB(x) = AB(x + KerB)$.

**6) Recycling Stability.** At any critical point, $AB$ is a projection operator and thus recycling outputs is stable at the first pass: $(AB)^n(x) = AB(x) = U_I U_I^t(x)$ for any $n \geq 1$.

**7) Generalization.** At any critical point, for any $x$, $AB(x)$ is equal to the projection of $x$ onto the corresponding subspace and the corresponding error can be expressed easily as the squared distance of $x$ to the projection space.

**8) Vertical Composition.** The global minimum of $E$ remains the same if additional matrices of rank greater or equal to $p$ are introduced between the input layer and the hidden layer or the hidden layer and the output layer. Thus there is no reduction in overall distortion by introducing such matrices. However, if such matrices are introduced for other reasons, there is a composition law so that the optimum solution for a deep autoencoder with a stack of matrices, can be obtained by combining the optimal solutions of shallow autoencoders. More precisely, consider an autoencoder network with layers of size $n/p_1/p/p_1/n$ (Figure 2) with $n > p_1 > p$. Then the optimal solution of this network can be obtained by first computing the optimal solution for an $n/p_1/n$ autoencoder network, and combining it with the optimal solution of an $p_1/p/p_1$ autoencoder network using the activities in the hidden layer of the first network as the training set for the second network, exactly as in the case of stacked RBMs (Hinton et al., 2006; Hinton and Salakhutdinov, 2006). This is because the projection onto the subspace spanned by the top $p$ eigenvectors can be composed by a projection onto the subspace spanned by the top $p_1$ eigenvectors, followed by a projection onto the subspace spanned by the top $p$ eigenvectors.

**9) External Targets.** With the proper adjustments, the results above remain essentially the same if a set of target output vectors $y_1, \ldots, y_m$ is provided, instead of $x_1, \ldots, x_m$ serving as the targets (see (Baldi and Hornik, 1988)).

**10) Symmetries and Hebbian Rules.** At the global minimum, for $C = I$, $A = B^t$. The constraint $A = B^t$ can be imposed during learning by "weight sharing" and is consistent with a Hebbian rule that is symmetric between the pre- and post synaptic neurons and is applied to the network by clamping the output units to be equal to the input units (or having a folded autoencoder).

## 4. The Boolean Autoencoder

The Boolean autoencoder is the most extreme form of non-linear autoencoder. In the purely Boolean case, we have $\mathbb{F} = \mathbb{G} = \{0, 1\}$, $A$ and $B$ are unrestricted Boolean functions, and $\Delta$ is the Hamming distance. Many variants of this problem can be obtained by restricting the classes $\mathcal{A}$ and $\mathcal{B}$ of Boolean functions, for instance by bounding the connectivity of the hidden units. The linear case with $\mathbb{F} = \mathbb{G} = \{0, 1\} = \mathbb{F}_2$, where $\mathbb{F}_2$ is the Galois field with two elements, is a special case of the Boolean case and will be discussed later. For lack of space, proofs are only briefly sketched.

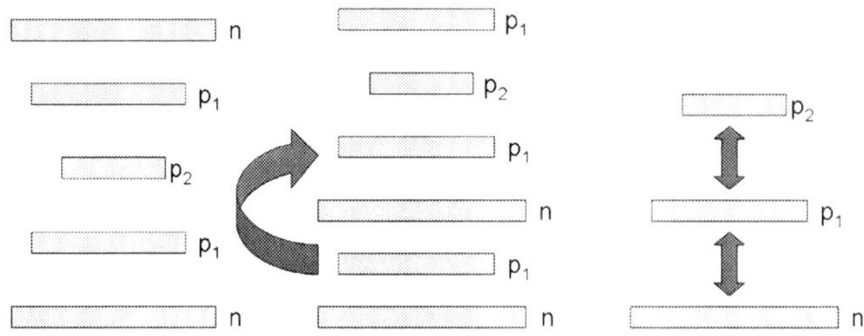

Figure 2: Vertical Composition of Autoencoders.

**1) Group Invariance.** Every solution is defined up to a permutation of the $2^p$ points of the hypercube $\mathbb{H}^p$. This is because the Boolean function are unrestricted and therefore their lookup tables can accommodate any such permutation, or relabeling of the hidden states.

**2) Problem Complexity.** In general, the overall optimization problem is NP-hard. To be more precise, one must specify the regime of interest characterized by which variables among $n$, $m$, and $p$ are going to infinity. Obviously one must have $n \to \infty$. If $p$ does not go to infinity, then the problem can be polynomial, for instance when the centroids must belong to the training set. If $p \to \infty$ and $m$ is a polynomial in $n$, which is the case of interest in machine learning where typically $m$ is a low degree polynomial in $n$, then the problem of finding the best boolean mapping (i.e. the Boolean mapping that minimizes the distortion $E$ associated with the Hamming distance on the training set) is NP hard, or the corresponding decision problem is NP-complete. More precisely the optimisation problem is NP-hard in the regime where $p \sim \epsilon \log_2 m$ with $\epsilon > 0$. A proof of this result is given in the next section.

**3) Fixed Layer Solution.** If the $A$ mapping is fixed, then it is easy to find the optimal $B$ mapping. Conversely if the $B$ mapping is fixed, it is easy to find the optimal $A$ mapping. To see this, assume first that $A$ is fixed. Then for each of the $2^p$ possible Boolean vectors $h_1, \ldots, h_{2^p}$ of the hidden layer, $A(h_1) \ldots, A(h_{2^p})$ provide $2^p$ points (centroids) in the hypercube $\mathbb{H}^n$. One can build the corresponding Voronoi partition by assigning each point of $\mathbb{H}^n$ to its closest centroid, breaking ties arbitrarily, thus forming a partition of $\mathbb{H}^n$ into $2^p$ corresponding clusters $C_1, \ldots, C_{2^p}$, with $C_i = C^{Vor}(A(h_i))$. The optimal mapping $B^*$ is then easily defined by setting $B^*(x) = h_i$ for any $x$ in $C_i = C^{Vor}(A(h_i))$. Conversely, assume that $B$ is fixed. Then for each of the $2^p$ possible Boolean vectors $h_1, \ldots, h_{2^p}$ of the hidden layer, let $C^B(h_i) = \{x \in \mathbb{H}^n : B(x) = h_i\}$. To minimize the reconstruction error, $A^*$ must map $h_i$ onto a point $y$ of $\mathbb{H}^n$ minimizing the sum of Hamming distances to points in $X \cap C^B(h_i)$. It is easy to see that the minimum is realized by the component-wise majority vector $A^*(h_i) = Majority[X \cap C^B(h_i)]$, breaking ties arbitrarily (e.g. by a coin flip).

**4) The Landscape of E.** In general $E$ has many local minima (e.g with respect to the Hamming distance applied to the lookup tables of $A$ and $B$). Critical points are defined to be the points satisfying simultaneously the equations above for $A^*$ and $B^*$.

**5) Clustering.** The overall optimization problem is a problem of optimal clustering. The clustering is defined by the transformation $B$. Approximate solutions can be sought by many algorithms, such as k-means, belief propagation (Frey and Dueck, 2007), minimum spanning paths and trees (Slagle et al., 1975), and hierarchical clustering.

**6) Recycling Stability.** At any critical point, recycling outputs is stable at the first pass so that for any $x$ $(AB)^n(x) = AB(x)$ (and is equal to the majority vector of the corresponding Voronoi cluster).

**7) Generalization.** At any critical point, for any $x$, $AB(x)$ is equal to the centroid of the corresponding Voronoi cluster and the corresponding error can be expressed easily.

**8) Vertical Composition.** The global optimum remains the same if additional Boolean layers of size equal or greater to $p$ are introduced between the input layer and the hidden layer and/or the hidden layer and the output layer. Thus there is no reduction in overall distortion $E$ by adding such layers. Consider a Boolean autoencoder network with layers of size $n, p_1, p, p_1, n$ (Figure 2) with $n > p_1 > p$. Then the optimal solution of this network can be obtained by first computing the optimal solution for an $n, p_1, n$ autoencoder network, and combining it with the optimal solution of an $p_1, p, p_1$ autoencoder network using the activity in the hidden layer of the first network as the training set, exactly as in the case of stacked RBMs. The reason for this is that the global optimum correspond to clustering into $2^p$ clusters, and this can be obtained by first clustering into $2^{p_1}$ clusters, and then clustering these clusters into $2^p$ clusters. The stack of Boolean functions performs hierarchical clustering with respect to the input space.

**9) External Targets.** With the proper adjustments, the results above remain essentially the same if a set of target output vectors $y_1, \ldots, y_m$ is provided, instead of $x_1, \ldots, x_m$ serving as the targets. To see this, consider a deep architecture consisting of a stack of autoencoders along the lines of Hinton et al. (2006). For any activity vector $h$ in the last hidden layer before the output layer, compute the set of points $C(h)$ in the training set that are mapped to $h$ by the stacked architecture. Assume, without any loss of generality, that $C(h) = \{x_1, \ldots, x_k\}$ with corresponding targets $\{y_1, \ldots, y_k\}$. Then it is easy to see that the final output for $h$ produced by the top layer ought to be the centroid of the targets given by $Majority(y_1, \ldots, y_k)$

## 5.  Clustering Complexity on the Hypercube

In this section, we briefly review some results on clustering complexity and then prove that the hypercube clustering decision problem is in general NP-complete. The complexity of various clustering problems, in different spaces, or with different objective functions, has been studied in the literature. There are primarily two kind of results: (1) graphical results derived on graphs $G = (V, E, \Delta)$ where the dissimilarity $\Delta$ is not necessarily a distance; and (2) geometric results derived in the Euclidean space $\mathbb{R}^d$ where $\Delta = L_2^2$, $L_2$, or $L_1$. In general, the clustering decision problem is NP-complete and the clustering optimization problem is NP-hard, except in some simple cases involving either a constant number $k$ of clusters or clustering in the 1-dimensional Euclidean space. In general, the results in Euclidean spaces are harder to derive than the results on graphs. When polynomial time algorithms exist, geometric problems tend to have faster solutions taking advantage of the geometric properties. However, none of the existing complexity theorems directly addresses the problem of clustering on the hypercube with respect to the Hamming distance.

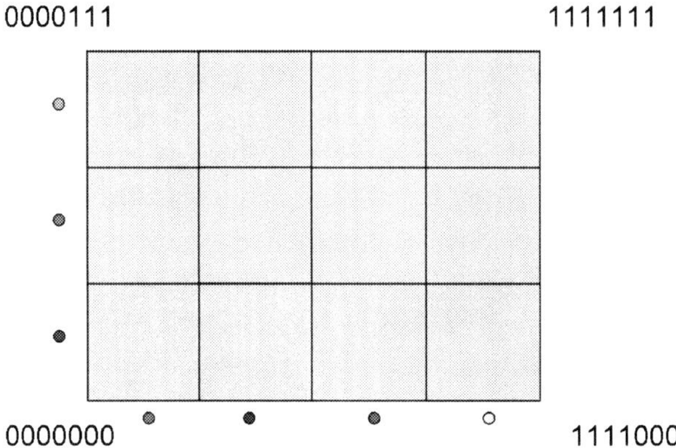


Figure 3: Embedding of a $3 \times 4$ Square Lattice onto $\mathbb{H}^7$ by Edge Coloring. All edges in the same row or column are given the same color. Each color corresponds to one of the dimensions of the 7-dimensional hypercube. For any pair of points, their Manhattan distance on the lattice is equal to the Hamming distance between their images in the 7-dimensional hypercube.

To deal with the hypercube clustering problem one must first understand which quantities are allowed to go to infinity. If $n$ is not allowed to go to infinity, then the number $m$ of training examples is also bounded by $2^n$ and, since we are assuming $p < n$, there is no quantity that can scale. Thus by necessity we must have $n \to \infty$. We must also have $m \to \infty$. The case of interest for machine learning is when $m$ is a low degree polynomial of $n$. Obviously the hypercube clustering problem is in NP, and it is a special case of clustering in $\mathbb{R}^n$. Thus the only important problem to be addressed is the reduction of a known NP-complete problem to a hypercube clustering problem.

For the reduction, it is natural to start from a known NP-complete graphical or geometric clustering problem. In both case, one must find ways to embed the original problem with its original metric into the hypercube with the Hamming distance. There are theorems for homeomorphic or squashed-embedding of graphs into the hypercube (Hartman, 1976; Winkler, 1983), however these emebeddings do not map the original dissimilarity function onto the the Hamming metric. Thus here we prefer to start from some of the known geometric results and use a strict cubical graph embedding. A graph is cubical if it is the subgraph of some hypercube $\mathbb{H}^d$ for some $d$ (Harary, 1988; Livingston and Stout, 1988). Although deciding whether a graph is cubical is NP-complete (Afrati et al., 1985), there is a theorem (Havel and Morávek, 1972) providing a necessary and sufficient condition for a graph to be cubical. A graph $G(V, E)$ is cubical and embeddable in $\mathbb{H}^d$ if and only if it is possible to color the edges of $G$ with $d$ colors such that: (1) All edges incident with a common vertex are of different color; (2) In each path of $G$, there is some color that appears an odd number of times; and (3) In each cycle of $G$, no color appears an odd number of times. We can now state and prove the following theorem.

**Theorem.** Consider the following hypercube clustering problem:

Input: $m$ binary vectors $x_1, \ldots, x_m$ of length $n$ and an integer $k$.

Output: $k$ binary vectors $c_1, \ldots, c_k$ of length $n$ (the centroids) and a function $f$ from $\{x_1, \ldots, x_m\}$ to $\{c_1, \ldots, c_k\}$ that minimizes the distortion $E = \sum_{t=1}^{m} \Delta(x_t, f(x_t))$ where $\Delta$ is the Hamming distance. The hypercube clustering problem is NP hard when $k \sim m^{\epsilon}$ $(\epsilon > 0)$.

**Proof.** To sketch the reduction, we start from the problem of clustering $m$ points in the plane $\mathbb{R}^2$ using cluster centroids and the $L_1$ distance, which is NP-complete (Megiddo and Supowit, 1984) by reduction from 3-SAT (Garey and Johnson, 1979) when $k \sim m^{\epsilon}$ $(\epsilon > 0)$ (see, also related results in Mahajan et al. (2009) and Vattani (2010)). Without any loss of generality, we can assume that the points in these problems lie on the vertices of a square lattice. Using the theorem in Havel and Morávek (1972), one can show that a $n \times m$ square lattice in the plane can be embedded into $\mathbb{H}^{n+m}$. In fact, an explicit embedding is given in Figure 3. It is easy to check that the $L_1$ or Manhattan distance between any two points on the square lattice is equal to the corresponding Hamming distance in $\mathbb{H}^{n+m}$. This polynomial reduction completes the proof that if the number of cluster satisfies $k = 2^p \sim m^{\epsilon}$, or equivalently $p \sim \epsilon \log_2 m \sim C \log n$, then the hypercube clustering problem associated with the Boolean autoencoder is NP-hard, and the corresponding decision problem NP-complete. If the numbers $k$ of clusters is fixed and the centroids must belong to the training set, there are only $\binom{m}{k} \sim m^k$ possible choices for the centroids inducing the corresponding Voronoi clusters. This yields a trivial, albeit not efficient, polynomial time algorithm. When the centroids are not required to be in the training set, we conjecture also the existence of polynomial time algorithms by adapting the corresponding theorems in Euclidean space.

## 6. The Case $p \geq n$

When the hidden layer is larger than the input layer and $\mathbb{F} = \mathbb{G}$, there is an optimal 0-distortion solution involving the identity function. Thus this case is interesting only if additional constraints are added to the problem. These can come in the form of regularization, for instance to ensure sparsity of the hidden-layer representation, or restrictions on the classes of functions $\mathcal{A}$ and $\mathcal{B}$, or noise in the hidden layer (see next section). When these constraints force the hidden layer to assume only $k$ different values and $k < m$, for instance in the case of a sparse Boolean hidden layer, then the previous analyses hold and the problem reduces to a $k$ clustering problem.

In this context of large hidden layers, in addition to vertical composition, there is also a natural horizontal composition for autoencoders that can be used to create large hidden layer representations (Figure 4) simply by horizontally combining autoencoders. Two (or more) autoencoders with architectures $n/p_1/n$ and $n/p_2/n]$ can be trained and the hidden layers can be combined to yield an expanded hidden representation of size

$p_1 + p_2$ that can then be fed to the subsequent layers of the overall architecture. Differences in the $p_1$ and $p_2$ hidden representations could be introduced by many different mechanisms, for instance using different learning algorithms, different initializations, different training samples, different learning rates, or different distortion measures. It is also possible to envision algorithms that incrementally add (or remove) hidden units to the hidden layer (Reed, 1993; Kwok and Yeung, 1997). In the linear case over $\mathbb{R}$, for instance, a first hidden unit can be trained to extract the first principal component, a second hidden unit can then be added to extract the second principal component, and so forth.

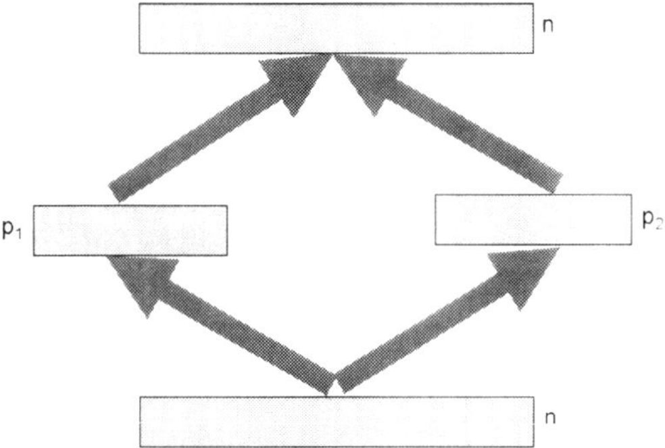


Figure 4: Horizontal Composition of Autoencoders to Expand the Hidden Layer Representation. The hidden layers of two separate autoencoders can be combined to yield a larger hidden representation of size $p_1 + p_2$ (see text).

## 7. Other Generalizations

Within the general framework introduced here, other kinds of autoencoders can be considered. First, one can consider mixed autoencoders with different constraints on $\mathbb{F}$ and $\mathbb{G}$, or different constraints on $\mathcal{A}$ and $\mathcal{B}$. A simple example is when the input and output layers are real $\mathbb{F} = \mathbb{R}$ and the hidden layer is binary $\mathbb{G} = \{0, 1\}$ (and $\Delta = L_2^2$). It is easy to check that in this case, as long as $2^p = k < m$, the autoencoder aims at clustering the real data into $k$ clusters and all the results obtained in the Boolean case are applicable with the proper adjustments. For instance, the centroid associated with a hidden state $h$ should be the center of mass of the input vectors mapped onto $h$. In general, this mixed autoencoder is also NP hard when $k \sim m^\epsilon$ and, from a probabilistic view point, it corresponds to a mixture of $k$ Gaussians model.

A second natural direction is to consider autoencoders that are linear but over fields other than the real numbers, for instance over the field $\mathbb{C}$ of complex numbers (Baldi et al., 2011), or over finite fields. For all these linear autoencoders, the Kernel of $B$ plays an important role since inputs vectors are basically clustered modulo this kernel. These autoencoders are not without theoretical and practical interests. Consider the linear autoencoder over the Galois field with two elements $GF(2) = \mathbb{F}_2$. It is easy to see that this is a special case of the Boolean autoencoder, where the Boolean functions are restricted to parity functions. This autoencoder can also be seen as implementing a linear code (McEliece, 1977). When there is noise in the 'transmission" of the hidden layer and $p > n$, one can consider solutions where $n$ units in the hidden layer correspond to the identity function and the remaining $p - n$ units implement additional parity check bits that are linearly computed from the input and used for error correction. Thus all well known linear codes, such as Hamming or Reed-Solomon codes, can be viewed within this linear autoencoder framework. While the linear autoencoder over $\mathbb{F}_2$ will be discussed elsewhere, it is worth noting that it is likely to yield an NP-hard problem, as for the case of the unrestricted Boolean autoencoder. This can be seen by considering that finding the minimum (non-zero) weight vector in the kernel of a binary matrix, or the radius of a code, are NP-complete problems (McEliece and van Tilborg, 1978; Frances and Litman, 1997). A simple overall classification of autoencoders is given in Figure 5.

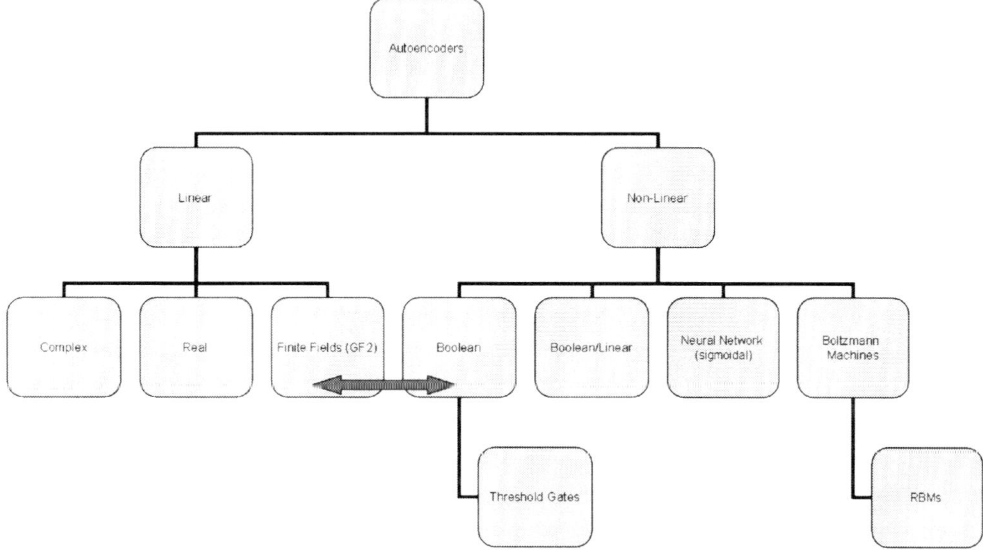


Figure 5: Simple Autoencoder Classification.

## 8. Discussion

Studying the linear and Boolean autoencoder in detail enables one to gain a general perspective on autoencoders, define key properties that are shared by different autoencoders and that ought to be checked systematically in any new kind of autoencoder (e.g. group invariances, clustering, recycling stability). The general perspective also shows the intimate connections between autoencoders and information and coding theory: (1) autoencoders with $n < p$ and a noisy hidden layer correspond to the classical noisy channel transmission and coding problem, with linear code over finite fields as a special case; (2) autoencoders with $n > p$ correspond to compression which can be lossy when the number of states in the hidden layer is less than the number of training examples ($m > k$) and lossless otherwise ($m < k$).

When $n > p$ the general emerging picture is that autoencoders learning is in general NP-complete [1] except in simple but important cases (e.g. linear over $\mathbb{R}$, Boolean with fixed $k$) and that in essence all autoencoders are performing some form of clustering suggesting a unified view of different forms of unsupervised learning, where Hebbian learning, autoencoders, and clustering are three faces of the same die. While autoencoders and Hebbian rules provide unsupervised learning implementations, it is clustering that provides the basic conceptual operation that underlies them. In addition, it is important to note that in general clustering objects without providing a label for each cluster is useless for subsequent higher levels of processing. In addition to clustering, autoencoders provide a label for each cluster through the activity in the hidden layer and thus elegantly address both the clustering and labeling problems at the same time.

RBMs and their efficient contrastive learning algorithm may provide an efficient form of autoencoder and autoencoder learning, but it is doubtful that there is anything special about RBMs at a deeper conceptual level. Thus it ought to be possible to derive results comparable to those described in Hinton et al. (2006) and Hinton and Salakhutdinov (2006) by stacking other kinds of autoencoders, and more generally by hierarchically stacking a series of clustering algorithms using vertical composition, perhaps also in combination with horizontal composition. As pointed out in the previous sections, it is easy to add a top layer for supervised regression or classification tasks on top of the hierarchical clustering stack. In aggregate, these results suggest that: (1) the so-called deep architectures may in fact have a non-trivial but constant (or logarithmic) depth, which is also consistent with what is observed in sensory neuronal circuits; (2) the fundamental unsupervised operation behind deep architectures, in one form or the other, is clustering, which is composable both horizontally and vertically–in this view, clustering may emerge as the central and almost sufficient ingredient for building intelligent systems; and (3) the generalization properties of deep architectures may be easier to understand when ignoring many of the hardware details, in terms of the most simple forms of autoencoders (e.g Boolean), or in terms of the more fundamental underlying clustering operations.

## Acknowledgments

Work in part supported by grants NSF IIS-0513376, NIH LM010235, and NIH-NLM T15 LM07443 to PB.

---

1. RBM learning is NP-complete by similarity with minimizing a quadratic form over the hypercube.

# References

F. Afrati, C.H. Papadimitriou, and G. Papageorgiou. The complexity of cubical graphs. *Information and control*, 66(1-2):53–60, 1985.

P. Baldi and K. Hornik. Neural networks and principal component analysis: Learning from examples without local minima. *Neural Networks*, 2(1):53–58, 1988.

P. Baldi, S. Forouzan, and Z. Lu. Complex-Valued Autoencoders. *Neural Networks*, 2011. Submitted.

Yoshua Bengio and Yann LeCun. Scaling learning algorithms towards AI. In L. Bottou, O. Chapelle, D. DeCoste, and J. Weston, editors, *Large-Scale Kernel Machines*. MIT Press, 2007.

H. Bourlard and Y. Kamp. Auto-association by multilayer perceptrons and singular value decomposition. *Biological cybernetics*, 59(4):291–294, 1988. ISSN 0340-1200.

P. Clote and E. Kranakis. *Boolean functions and computation models*. Springer Verlag, 2002.

Dumitru Erhan, Yoshua Bengio, Aaron Courville, Pierre-Antoine Manzagol, Pascal Vincent, and Samy Bengio. Why does unsupervised pre-training help deep learning? *Journal of Machine Learning Research*, 11:625–660, February 2010.

M. Frances and A. Litman. On covering problems of codes. *Theory of Computing Systems*, 30(2):113–119, 1997.

B.J. Frey and D. Dueck. Clustering by passing messages between data points. *Science*, 315(5814):972, 2007.

M.R. Garey and D.S. Johnson. *Computers and Intractability*. Freeman San Francisco, 1979.

F. Harary. Cubical graphs and cubical dimensions. *Computers & Mathematics with Applications*, 15(4):271–275, 1988.

J. Hartman. The homeomorphic embedding of Kn in the m-cube* 1. *Discrete Mathematics*, 16(2):157–160, 1976.

I. Havel and J. Morávek. *B*-valuations of graphs. *Czechoslovak Mathematical Journal*, 22(2):338–351, 1972.

D.O. Hebb. The organization of behavior: A neurophychological study. *WileyInterscience, New York*, 1949.

G.E. Hinton and R.R. Salakhutdinov. Reducing the dimensionality of data with neural networks. *Science*, 313(5786):504, 2006.

G.E. Hinton, S. Osindero, and Y.W. Teh. A fast learning algorithm for deep belief nets. *Neural Computation*, 18(7):1527–1554, 2006.

T. Kwok and D. Yeung. Constructive Algorithms for Structure Learning in Feedforward Neural Networks for Regression Problems. *IEEE Transactions on Neural Networks*, 8:630–645, 1997.

M. Livingston and Q.F. Stout. Embeddings in hypercubes. *Mathematical and Computer Modelling*, 11:222–227, 1988.

M. Mahajan, P. Nimbhorkar, and K. Varadarajan. The planar k-means problem is NP-hard. *WALCOM: Algorithms and Computation*, pages 274–285, 2009.

R. McEliece and H. van Tilborg. On the inherent intractability of certain coding problems(Corresp.). *IEEE Transactions on Information Theory*, 24(3):384–386, 1978.

R. J. McEliece. *The Theory of Information and Coding*. Addison-Wesley Publishing Company, Reading, MA, 1977.

N. Megiddo and K.J. Supowit. On the complexity of some common geometric location problems. *SIAM J. COMPUT.*, 13(1):182–196, 1984.

G. Montufar and N. Ay. Refinements of Universal Approximation Results for Deep Belief Networks and Restricted Boltzmann Machines. *Neural Computation*, pages 1–14, 2011. ISSN 0899-7667.

E. Oja. Simplified neuron model as a principal component analyzer. *Journal of mathematical biology*, 15(3):267–273, 1982.

R. Reed. Pruning algorithms-a survey. *Neural Networks, IEEE Transactions on*, 4(5): 740–747, 1993.

D.E. Rumelhart, G.E. Hinton, and R.J. Williams. Learning internal representations by error propagation. In *Parallel Distributed Processing. Vol 1: Foundations*. MIT Press, Cambridge, MA, 1986.

JL Slagle, CL Chang, and SR Heller. A clustering and data reorganization algorithm. *IEEE Transactions on Systems, Man and Cybernetics*, 5:121–128, 1975.

I. Sutskever and G.E. Hinton. Deep, narrow sigmoid belief networks are universal approximators. *Neural Computation*, 20(11):2629–2636, 2008.

A. Vattani. A simpler proof of the hardness of k-means clustering in the plane. *UCSD Technical Report*, 2010.

P.M. Winkler. Proof of the squashed cube conjecture. *Combinatorica*, 3(1):135–139, 1983.

# Information Theoretic Model Selection for Pattern Analysis


**Joachim M. Buhmann**    JBUHMANN@INF.ETHZ.CH
**Morteza Haghir Chehreghani**    MORTEZA.CHEHREGHANI@INF.ETHZ.CH
**Mario Frank**    MARIO.FRANK@INF.ETHZ.CH
**Andreas P. Streich**    ANDREAS.STREICH@ALUMNI.ETHZ.CH
*Department of Computer Science, ETH Zurich, Switzerland*


**Editor:** I. Guyon, G. Dror, V. Lemaire, G. Taylor, and D. Silver

## Abstract


Exploratory data analysis requires (i) to define a set of patterns hypothesized to exist in the data, (ii) to specify a suitable quantification principle or cost function to rank these patterns and (iii) to validate the inferred patterns. For data clustering, the patterns are object partitionings into $k$ groups; for PCA or truncated SVD, the patterns are orthogonal transformations with projections to a low-dimensional space. We propose an information theoretic principle for model selection and model-order selection. Our principle ranks competing pattern cost functions according to their ability to extract context sensitive information from noisy data with respect to the chosen hypothesis class. Sets of approximative solutions serve as a basis for a communication protocol. Analogous to Buhmann (2010), inferred models maximize the so-called approximation capacity that is the mutual information between coarsened training data patterns and coarsened test data patterns. We demonstrate how to apply our validation framework by the well-known Gaussian mixture model and by a multi-label clustering approach for role mining in binary user privilege assignments.

**Keywords:** Unsupervised learning, data clustering, model selection, information theory, maximum entropy, approximation capacity


## 1. Model Selection via Coding

Model selection and model order selection [Burnham and Anderson (2002)] are fundamental problems in pattern analysis. A variety of models and algorithms has been proposed to extract patterns from data, i.e., for clustering, but a comprehensive theory on how to choose the "right" pattern model given the data is still missing. Statistical learning theory as in Vapnik (1998) advocates to measure the generalization ability of models and to employ the prediction error as a measure of model quality. In particular, stability analysis of clustering solutions has shown very promising results for model order selection in clustering [Dudoit and Fridlyand (2002); Lange et al. (2004)], although discussed controversially by Ben-David et al. (2006). Stability is, however, only one aspect of statistical modeling, e.g., for unsupervised learning. The other aspect of the modeling tradeoff is characterized by the informativeness of the extracted patterns. A tolerable decrease in the stability of inferred patterns in the data (data model) might be compensated by a substantial increase of their information content (see also the discussion in Tishby et al. (1999)).

We formulate a principle that balances these two antagonistic objectives by transforming the model selection problem into a coding problem for a generic communication scenario. Thereby, the objective function or cost function that maps patterns to quality scores is considered as a noisy channel. Different objectives are ranked according to their transmission properties and the cost function with the highest channel capacity is then selected as the most informative model for a given data set. Thereby, we generalize the set-based coding scheme proposed by Buhmann (2010) to sets of weighted hypotheses in order to simplify the embedding of pattern inference problems in a communication framework. Learning, in general, resembles communication from a conceptual viewpoint: For *communication*, one demands a high rate (a large amount of information transferred per channel use) together with a decoding rule that is stable under the perturbations of the messages by the noise in the channel. For *learning* patterns in data, the data analyst favors a rich model with high complexity (e.g., a large number of clusters in grouping), while the generalization error of test patterns is expected to remain stable and low. We require that solutions of pattern analysis problems are reliably inferred from noisy data.

This article first summarizes the information theoretic framework of weighted approximation set coding (wASC) for validating statistical models. We then demonstrate, for the first time, how to practically transform a pattern recognition task into a communication setting and how to compute the capacity of clustering solutions. The feasibility of wASC is demonstrated for mixture models and real world data.

## 2. Brief Introduction to Approximation Set Coding

In this section, we briefly describe the theory of weighted Approximation Set Coding (wASC) for pattern analysis as proposed by Buhmann (2010).

Let $\mathbf{X} = \{X_1, \ldots, X_n\} \in \mathcal{X}$ be a set of $n$ objects $\mathbf{O}$ and $n$ measurements in a data space $\mathcal{X}$, where the measurements characterize the objects. Throughout the paper, we assume the special case of a bijective map between objects and measurements, i.e., the $i^{\text{th}}$ object is isomorphic to the vector $\mathbf{x}_i \in \mathbb{R}^D$. In general, the (object, measurement) relation might be more complex than an object-specific feature vector.

A **hypothesis**, i.e. a solution of a pattern analysis problem, is a function $c$ that assigns objects (e.g. data) to patterns of a pattern space $\mathcal{P}$:

$$c : \mathcal{X} \to \mathcal{P}, \qquad \mathbf{X} \mapsto c(\mathbf{X}). \tag{1}$$

Accordingly, the **hypothesis class** is the set of all such functions, i.e. $C(\mathbf{X}) := \{c(\mathbf{X}) : \mathbf{X} \in \mathcal{X}\}$. For clustering, the patterns are object partitionings $\mathcal{P} = \{1, \ldots, k\}^n$. A model for pattern analysis is characterized by a **cost** or **objective function** $R(c, \mathbf{X})$ that assigns a real value to a pattern $c(\mathbf{X})$. To simplify the notation, model parameters $\theta$ (e.g., centroids) are not explicitly listed as arguments of the objective function. Let $c^{\perp}(\mathbf{X})$ be the pattern that minimizes the cost function, i.e. $c^{\perp}(\mathbf{X}) \in \arg\min_c R(c, \mathbf{X})$. As the measurements $\mathbf{X}$ are random variables, the global minimum $c^{\perp}(\mathbf{X})$ of the empirical costs is a random variable as well. In order to rank all solutions of the pattern analysis problem, we introduce *approximation weights*

$$w : C \times \mathcal{X} \times \mathbb{R}_+ \to [0,1], \qquad (c,\mathbf{X},\beta) \mapsto w_\beta(c,\mathbf{X}). \tag{2}$$

The weights are chosen to be non-negative $w_\beta(c,\mathbf{X}) \geq 0$, the maximal weight is allocated to the global minimizer $c^{\perp}$ and it is normalized to one ($w_\beta(c^{\perp},\mathbf{X}) = 1$). Semantically, the weights $w_\beta(c,\mathbf{X})$ quantify the quality of a solution w.r.t. the global minimizer of $R(.,\mathbf{X})$. The scaling parameter $\beta$ controls the size of the solution set. Large $\beta$ yields a small solution set and small $\beta$ renders many solutions as good approximations of the minimizer $c^{\perp}(\mathbf{X})$ in terms of costs, i.e., $w_\beta(c,\mathbf{X}) > 1 - \epsilon$ denotes that $c$ is regarded as an $\epsilon/\beta$-good approximation of the minimal costs $R(c^{\perp},\mathbf{X})$. Therefore, we also require that weights fulfil the inverse order constraints compared to costs, i.e., a solution $c$ with lower or equal costs than $\tilde{c}$ should have a larger or equal weight, i.e.,

$$R(c,\mathbf{X}) \leq R(\tilde{c},\mathbf{X}) \quad \Longleftrightarrow \quad w_\beta(c,\mathbf{X}) \geq w_\beta(\tilde{c},\mathbf{X}). \tag{3}$$

Given a cost function $R(c,\mathbf{X})$) these order constraints determine the weights up to a monotonic (possibly nonlinear) transformation $f(.)$ which effectively rescales the costs ($\tilde{R}(c,\mathbf{X}) = f(R(c,\mathbf{X}))$). The family of (Boltzmann) weights

$$w_\beta(c,\mathbf{X}) := \exp(-\beta \Delta R(c,\mathbf{X})), \text{ with } \Delta R(c,\mathbf{X}) := R(c,\mathbf{X}) - R(c^{\perp},\mathbf{X})) \tag{4}$$

parameterized by the inverse computational temperature $\beta$, fulfils these requirements. Although the Boltzmann weights are a special choice, all other weighting schemes can be explained by a monotonic rescaling of the costs.

Conceptually, wASC assumes a "two sample set scenario" as in Tishby et al. (1999). Let $\mathbf{X}^{(q)}, q \in \{1,2\}$, be two datasets with the same inherent structure but different noise instances. In most cases, their sets of global minima differ, i.e. $\{c^{\perp}(\mathbf{X}^{(1)})\} \cap \{c^{\perp}(\mathbf{X}^{(2)})\} = \emptyset$, demonstrating that the global minimizers often lack robustness to fluctuations. The approximation weights (2) have been introduced to cure this instability. Solutions with large approximation weights $w_\beta(c,\mathbf{X}) \geq 1 - \epsilon$, $\epsilon \ll 1$ can be accepted as substitutes of the global minimizers. Adopting a learning theoretic viewpoint, the set of solutions with large weights generalizes significantly better than the set of global minimizers, provided that $\beta$ is suitably chosen. The concept wASC serves the purpose to determine such an appropriate scale $\beta$. The two data sets $\mathbf{X}^{(q)}, q \in \{1,2\}$, define two weight sets $w_\beta(c,\mathbf{X}^{(q)})$. These weights give rise to the two *weight sums* $\mathcal{Z}_q$ and the *joint weight sum* $\mathcal{Z}_{12}$

$$\mathcal{Z}_q := \mathcal{Z}(\mathbf{X}^{(q)}) = \sum_{c \in C(\mathbf{X}^{(q)})} \exp(-\beta \Delta R(c,\mathbf{X}^{(q)})), q = 1,2 \tag{5}$$

$$\mathcal{Z}_{12} := \mathcal{Z}(\mathbf{X}^{(1)},\mathbf{X}^{(2)}) = \sum_{c \in C(\mathbf{X}^{(2)})} \exp(-\beta(\Delta R(c,\mathbf{X}^{(1)}) + \Delta R(c,\mathbf{X}^{(2)}))), \tag{6}$$

where $\exp(-\beta(\Delta R(c,\mathbf{X}^{(1)}) + \Delta R(c,\mathbf{X}^{(2)})))$ measures how well a solution $c$ minimizes costs on *both* datasets. The sums (5,6) play a central role in our framework. If $\beta = 0$, all weights $w_\beta(c, \mathbf{X}) = 1$ are independent of the costs. In this case, $\mathcal{Z}_q = |C(\mathbf{X}^{(q)})|$ indicates the size of the hypothesis space, and $\mathcal{Z}_{12} = \mathcal{Z}_1 = \mathcal{Z}_2$. For high $\beta$, all weights are small compared to the weight $w_\beta(c^\perp, \mathbf{X}^{(q)})$ of the global optimum and the weight sum essentially counts the number of globally optimal solutions. For intermediate $\beta$, $\mathcal{Z}(\cdot)$ takes a value between 0 and $|C(\mathbf{X}^{(q)})|$, giving rise to the interpretation of $\mathcal{Z}(\cdot)$ as the effective number of patterns that approximately fit the dataset $\mathbf{X}^{(q)}$, where $\beta$ defines the precision of this approximation. Essentially, $\mathcal{Z}_q$ counts all statistically indistinguishable data patterns that approximate the minimum of the objective function. The global optimum $c^\perp(\mathbf{X})$ can change whenever we optimize on another random subset of the data, whereas, for a well-tuned $\beta$, the set of weights $\{w_\beta(c, \mathbf{X})\}$ remains approximately invariant. Therefore, $\beta$ defines the resolution of the hypothesis class that is relevant for inference. Noise in the measurements $\mathbf{X}$ reduces this resolution and thus coarsens the hypothesis class. As a consequence, the key problem of learning is to control the resolution optimally: How high can $\beta$ be chosen to still ensure identifiability of $\{w_\beta : w_\beta(c, \mathbf{X}) \geq 1 - \epsilon\}$ in the presence of data fluctuations? Conversely, choosing $\beta$ too low yields a too coarse resolution of solutions and does not capture the maximal amount of information in the data.

We answer this key question by means of a communication scenario. The communication architecture includes a *sender* $\mathfrak{S}$ and a *receiver* $\mathfrak{R}$ with a *problem generator* $\mathfrak{P}\mathfrak{G}$ between the two terminals $\mathfrak{S}, \mathfrak{R}$ (see Fig. 1). The communication protocol is organized in two stages: (i) design of a communication code and (ii) the communication process.

For the **communication code**, we adapt Shannon's random coding scenario, where a codebook of random bit strings covers the space of all bit strings. In random coding, the sender sends a bit string and the receiver observes a perturbed version of this bit string. For decoding, the receiver has to find the most similar codebook vector in the codebook which is the decoded message. In the same spirit, for our scenario, the sender must communicate patterns to the receiver via noisy datasets. Since we are interested in patterns with low costs, the optimal pattern $c^\perp(\mathbf{X}^{(1)})$ can serve as a message. The other patterns in the codebook are generated by transforming the training data $\tau \circ \mathbf{X}^{(1)}$ with the transformation $\tau \in \mathbb{T} := \{\tau_1, ..., \tau_{2^{n_\rho}}\}$. The number of codewords is $2^{n\rho}$ and $\rho$ is the rate of the protocol. The choice of such transformations depends on the hypothesis class and they have to be equivariant, i.e., the transformed optimal pattern equals the optimal pattern of the transformed data $\tau \circ c(\mathbf{X}^{(1)}) = c(\tau \circ \mathbf{X}^{(1)})$. In data clustering, *permuting* the indices of the objects defines the group of transformations to cover the pattern space. Each clustering solution $c \in C(\mathbf{X}^{(1)})$ can be transformed into another solution by a permutation $\tau$ on the indices of $c$.

To **communicate**, $\mathfrak{S}$ selects a transformation $\tau_s \in \mathbb{T}$ and sends it to a *problem generator* $\mathfrak{P}\mathfrak{G}$ as depicted in Fig. 1. $\mathfrak{P}\mathfrak{G}$ then generates a new dataset $\mathbf{X}^{(2)}$, applies the transformation $\tau_s$, and sends the resulting data $\tilde{\mathbf{X}} := \tau_s \circ \mathbf{X}^{(2)}$ to $\mathfrak{R}$. On the receiver side, the lack of knowledge on the transformation $\tau_s$ is mixed with the stochastic variability of the source generating the data $\mathbf{X}$. $\mathfrak{R}$ has to estimate the transformation $\hat{\tau}$ based on $\tilde{\mathbf{X}}$. The decoding rule of $\mathfrak{R}$ selects the pattern transformation $\hat{\tau}$ that yields the highest joint

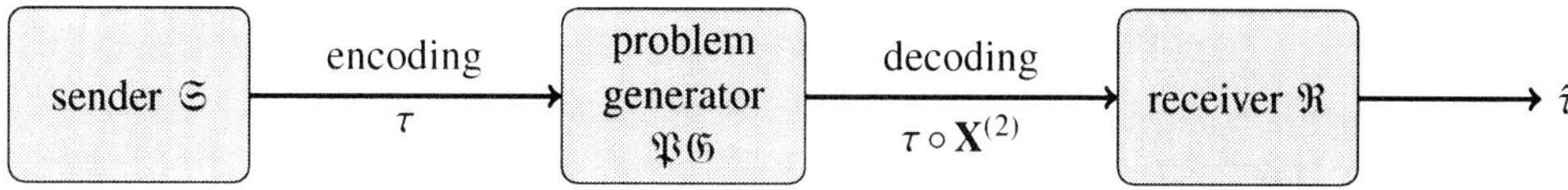


Figure 1: Communication process: (1) the sender selects transformation $\tau$, (2) the problem generator draws $\mathbf{X}^{(2)} \sim \mathbb{P}(\mathbf{X})$ and applies $\tau$ to it, and (3) the receiver estimates $\hat{\tau}$ based on $\tilde{\mathbf{X}} = \tau \circ \mathbf{X}^{(2)}$.

weight sum of $\hat{\tau} \circ \mathbf{X}^{(1)}$ and $\tilde{\mathbf{X}}$, i.e.,

$$\hat{\tau} \in \arg\max_{\tau \in \mathbb{T}} \sum_{c \in C(\mathbf{X}^{(1)})} \exp(-\beta(R(c, \tau \circ \mathbf{X}^{(1)}) + R(c, \tilde{\mathbf{X}}))) . \tag{7}$$

In the absence of noise in the data, we have $\mathbf{X}^{(1)} = \mathbf{X}^{(2)}$, and error-free communication works even for $\beta \to \infty$. The higher the noise level, the lower we have to choose $\beta$ in order to obtain weight sums that are approximately invariant under the stochastic fluctuations in the measurements thus preventing decoding errors. The error analysis of this protocol investigates the probability of decoding error $\mathbb{P}(\hat{\tau} \neq \tau_s | \tau_s)$. As derived for an equivalent channel in Buhmann (2011), an asymptotically vanishing error rate is achievable for rates

$$\rho \leq \mathcal{I}_\beta(\tau_s, \hat{\tau}) = \frac{1}{n} \log\left(\frac{|\{\tau_s\}| \mathcal{Z}_{12}}{\mathcal{Z}_1 \cdot \mathcal{Z}_2}\right) = \frac{1}{n}\left(\log \frac{|\{\tau_s\}|}{\mathcal{Z}_1} + \log \frac{|C^{(2)}|}{\mathcal{Z}_2} - \log \frac{|C^{(2)}|}{\mathcal{Z}_{12}}\right) \tag{8}$$

The three logarithmic terms in eq.(8) denote the mutual information between the coarsening of the pattern space on the sender side and the coarsening of the pattern space on the receiver side.

The cardinality $|\{\tau_s\}|$ is determined by the number of realizations of the random transformation $\tau$, i.e. by the entropy of the type (in an information theoretic sense) of the empirical minimizer $c^\perp(\mathbf{X})$. As the entropy increases for a large number of patterns, $|\{\tau_s\}|$ accounts for the model complexity or informativeness of the solutions. For noisy data, the communication rate is reduced as otherwise the solutions can not be resolved by the receiver. The relative weights are determined by the term $\mathcal{Z}_{12}/(\mathcal{Z}_1 \cdot \mathcal{Z}_2) \in [0, 1]$ which accounts for the stability of the model under noise fluctuations.

In analogy to information theory, we define the *approximation capacity* as

$$\mathcal{CAP}(\tau_s, \hat{\tau}) = \max_\beta \mathcal{I}_\beta(\tau_s, \hat{\tau}) . \tag{9}$$

Using these entities, we can describe how to apply the wASC principle for model selection from a set of cost functions $\mathcal{R}$: Randomly split the given dataset $\mathbf{X}$ into two subsets $\mathbf{X}^{(1)}$ and $\mathbf{X}^{(2)}$. For each candidate cost function $R(c, \mathbf{X}) \in \mathcal{R}$, compute the mutual information (eq. 8) and maximize it with respect to $\beta$. Then select the cost function that achieves highest capacity at the best resolution $\beta^\star$.

There exists a long history of information theoretic approaches to model selection, which traces back at least to Akaike's extension of the Maximum Likelihood principle. AIC penalizes fitted models by twice the number of free parameters. The Bayesian Information Criterion (BIC) suggests a stronger penalty than AIC, i.e., number of model parameters times logarithm of the number of samples. Rissanen's minimum description length principles is closely related to BIC (see e.g. Hastie et al. (2008) for model selection penalties). Tishby et al. (1999) proposed to select the number of clusters according to a difference of mutual informations which they called the imformation bottleneck. This asymptotic concept is closely related to rate distortion theory with side information (see Cover and Thomas (2006)). Finite sample size corrections of the information bottleneck allowed Still and Bialek (2004) to determine an optimal temperature with a prefered number of clusters.

## 3. Approximation Capacity for Parametric Clustering Models

Let $\mathbf{X}^{(q)}, q \in \{1, 2\}$ be two datasets drawn from the same source. We consider a parametric clustering model with $K$ clusters. Then the cost function can be written as

$$R(c, \mathbf{X}) = \sum_{i=1}^{n} \epsilon_{i,c(i)} \text{ with } \forall i, c(i) \in \{1, .., K\}. \tag{10}$$

$\epsilon_{i,c(i)}$ indicates the costs of assigning object $i$ to cluster $c(i)$. These costs $\epsilon_{i,c(i)}$ also contains all relevant parameters to identify a clustering solution, e.g. centroids. In the well-known case of k-means clustering we derive $\epsilon_{i,c(i)} = \|x_i - y_{c(i)}\|^2$.

Calculating the approximation capacity requires the following steps:

1. Identify the hypothesis space of the models and compute the cardinality of the set of possible transformations $|\{\tau_s\}|$.

2. Calculate the weight sums $\mathcal{Z}_q$, $q = 1, 2$, and the joint weight sum $\mathcal{Z}_{12}$.

3. Maximize $\mathcal{I}_\beta$ in Eq. (8) with respect to $\beta$.

In clustering problems, the hypothesis space is spanned by all possible assignments of objects to sources. The appropriate transformation in clustering problems is the permutation of objects. Albeit a solution contains the cluster assignments *and* cluster parameters like centroids, the centroid parameters contribute almost no entropy to the solution. With given cluster assignments the solution is fully determined as the objects of each cluster pinpoint the centroids to a particular vector. With the permutation transformations one can construct all clusterings starting from a single clustering. However, as the mutual information in Eq. (8) is estimated solely based on the identity transformation, one can ignore the specific kind of transformations when computing this estimate. The cardinality $|\{\tau_s\}|$ is then the number of all distinct clusterings on $\mathbf{X}^{(1)}$.

We obtain the individual weight sums and the joint weight sum by summing over all possible clustering solutions

$$
\mathcal{Z}_q \; = \; \sum_{c \in C(\mathbf{X}^{(q)})} \exp\left(-\beta \sum_{i=1}^{n} \epsilon_{i,c(i)}^{(q)}\right) = \prod_{i=1}^{n} \sum_{k=1}^{K} \exp\left(-\beta \epsilon_{i,k}^{(q)}\right), q = 1, 2,
\tag{11}
$$

$$
\mathcal{Z}_{12} \; = \; \sum_{c \in C(\mathbf{X}^{(2)})} \exp\left(-\beta \sum_{i=1}^{n} (\epsilon_{i,c(i)}^{(1)} + \epsilon_{i,c(i)}^{(2)})\right) = \prod_{i=1}^{n} \sum_{k=1}^{K} \exp\left(-\beta (\epsilon_{i,k}^{(1)} + \epsilon_{i,k}^{(2)})\right).
\tag{12}
$$

By substituting these weight sums to Eq. (8), the mutual information amounts to

$$
\mathcal{I}_\beta = \frac{1}{n} \log |\{\tau_s\}| + \frac{1}{n} \sum_{i=1}^{n} \left( \log \sum_{k=1}^{K} e^{-\beta(\epsilon_{i,k}^{(1)} + \epsilon_{i,k}^{(2)})} - \log \sum_{k=1}^{K} e^{-\beta \epsilon_{i,k}^{(1)}} \sum_{k'=1}^{K} e^{-\beta \epsilon_{i,k'}^{(2)}} \right).
\tag{13}
$$

The approximation capacity is numerically determined as the maximum of $\mathcal{I}_\beta$ over $\beta$.

## 4. Approximation Capacity for Mixtures of Gaussians

In this section, we demonstrate the principle of maximum approximation capacity on the well known Gaussian mixture model (GMM). We first study the approximation set coding for GMMs and then we experimentally compare it against other model selection principles.

### 4.1. Experimental Evaluation of Approximation Capacity

A GMM with $K$ components is defined as $p(\mathbf{x}) = \sum_{k=1}^{K} \pi_k \, \mathcal{N}(\mathbf{x} \,|\, \mu_k, \Sigma)$, with non-negative $\pi_k$ and $\sum_k \pi_k = 1$. For didactical reasons, we do not optimize the covariance matrix $\Sigma$ and simply fix it to $\Sigma - 0.5 \cdot \mathbf{I}$. Then, maximizing the GMM likelihood essentially reduces to centroid-based clustering. Therefore, $\epsilon_{i,k} := \|\mathbf{x}_i - \mu_k\|^2$ indicates the costs of assigning object $i$ to cluster $k$.

For experimental evaluation, we define $K = 5$ Gaussians with parameters $\pi_k = 1/K$, $\mu \in \{(1,0), (0,1.5), (-2,0), (0,-3), (4.25,-4)\}$, and with covariance $\Sigma = 0.5 \cdot \mathbf{I}$. Let $\mathbf{X}^{(q)}, q \in \{1,2\}$ be two datasets of identical size $n = 10{,}000$ drawn from these Gaussians. We optimize the assignment variables and the centroid parameters of our GMM model via annealed Gibbs sampling [Geman and Geman (1984)]. The computational temperature in Gibbs sampling is equivalent to the assumed width of the distributions. Thereby, we provide twice as many clusters to the model in order to enable overfitting. Starting from a high temperature, we successively cool down while optimizing the model parameters. In Figure 2($a$), we illustrate the positions of the centroids with respect to the center of mass. At high temperature, all centroids coincide, indicating that the optimizer favors one cluster. As the temperature is lowered further, the centroids separate into increasingly many clusters until, finally, the sampler uses all 10 clusters to fit the data.

Figure 2($b$) shows the numerical analysis of the mutual information in Eq. (13). When the stopping temperature of the Gibbs sampler coincides with the temperature

$\beta^{-1}$ that maximizes mutual information, we expect the best tradeoff between robustness and informativeness. And indeed, as illustrated in Figure 2(a), the correct model-order $\hat{K} = 5$ is found at this temperature. At lower stopping temperatures, the clusters split into many instable clusters which increases the decoding error, while at higher temperatures informativeness of the clustering solutions decreases.

## 4.2. Comparison with other principles

We compare approximation capacity against two other model order selection principles: i) generalization ability, and ii) BIC score.

**Relation to generalization ability:**  A properly regularized clustering model explains not only the dataset at hand, but also new datasets from the same source. The inferred model parameters and assignment probabilities from the first dataset $\mathbf{X}^{(1)}$ can be used to compute the costs for the second dataset $\mathbf{X}^{(2)}$. The appropriate clustering model yields low costs on $\mathbf{X}^{(2)}$, while very informative but unstable structures and also very stable but little informative structures have high costs due to overfitting or underfitting, respectively.

We measure this generalization ability by computing the "transfer costs" $R(c^{(1)}, \mathbf{X}^{(2)})$ [Frank et al. (2011)]: At each stopping temperature of the Gibbs sampler, the current parameters $\mu^{(1)}$ and assignment probabilities $\mathbf{P}^{(1)}$ inferred from $\mathbf{X}^{(1)}$ are transfered to $\mathbf{X}^{(2)}$. The assignment probabilities $\mathbf{P}^{(1)}$ assume the form of a Gibbs distribution

$$p(\mu_k^{(1)}|\mathbf{x}_i^{(1)}) = Z_x^{-1} \exp\left(-\beta\|\mathbf{x}_i^{(1)} - \mu_k^{(1)}\|^2\right), \tag{14}$$

with $Z_x$ as the normalization constant. The expected transfer costs with respect to these probabilities are then

$$\left\langle R(c^{(1)}, \mathbf{X}^{(2)})\right\rangle = \sum_{i=1}^{n}\sum_{k=1}^{K} p(\mathbf{x}_i^{(1)}, \mu_k^{(1)})\|\mathbf{x}_i^{(2)} - \mu_k^{(1)}\|^2 \approx \frac{1}{n}\sum_{n=1}^{n}\sum_{k=1}^{K} p(\mu_k^{(1)}|\mathbf{x}_i^{(1)})\|\mathbf{x}_i^{(2)} - \mu_k^{(1)}\|^2 ,$$
$$\tag{15}$$

Figure 2(b) illustrates the transfer costs as a function of $\beta$ and compares it with the approximation capacity. The optimal transfer costs are obtained at the stopping temperature that corresponds to the approximation capacity.

**Relation to BIC**  Arguably the most popular criterion for model-order selection is BIC as proposed by Schwarz (1978). It is, like wASC, an asymptotic principle, i.e. for sufficiently many observations, the fitted model preferred by BIC ideally corresponds to the candidate which is a posteriori most probable. However, the application of BIC is limited to models where one can determine the number of free parameters as here with GMM. Figure 2(c) confirms the consistency of wASC with BIC in finding the correct model order in our experiment.

## 5. Model Selection for Boolean matrix factorization

To proceed with studying different applicability aspects of wASC, we now consider the task to select one out of four models for factorizing a Boolean matrix with the

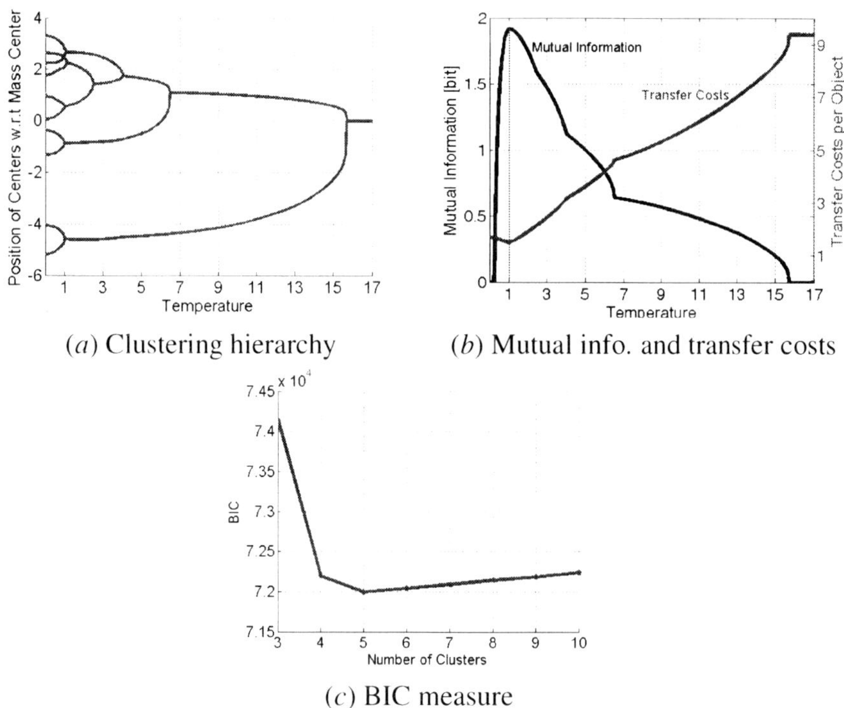

(*a*) Clustering hierarchy  (*b*) Mutual info. and transfer costs

(*c*) BIC measure

Figure 2: Annealed Gibbs sampling for GMM: Influence of the stopping temperature for annealed optimization on the mutual information, on the transfer costs and on the positions of the cluster centroids. The lowest transfer cost is achieved at the temperature with highest mutual information. This is the lowest temperature at which the correct number of clusters $\hat{K} = 5$ is found. The hierarchy in Fig. 2(*a*) is obtained by projecting the two-dimensional centroids at each stopping temperature to the optimal one-dimensional subspace using multidimensional scaling. BIC verifies correctness of $\hat{K} = 5$.

clustering method proposed in Streich et al. (2009). Experiments with known ground truth allow us to rank these models according to their parameter estimation accuracy. We investigate whether wASC reproduces this ranking.

## 5.1. Data and Models

Consider binary data $\mathbf{X} \in \{0, 1\}^{n \times D}$ in $D$ dimensions, where a row $\mathbf{x}_i$ describes a single data item. Each data item $i$ is assigned to a set of sources $\mathcal{L}_i$, and these sources generate the measurements $\mathbf{x}_i$ of the data item. The probabilities $v_{k,d}$ of a source $k$ to emit a zero in dimension $d$ parameterize the sources. To generate a data item $i$, one sample is drawn from each source in $\mathcal{L}_i$. In each dimension $d$, the individual samples are then combined via the Boolean OR to obtain the structure part of the data item $\tilde{\mathbf{x}}_i$. Finally, a noise process generates the $\mathbf{x}_i$ by randomly selecting a fraction of $\epsilon$ elements and replacing them with random values. Following this generative process, the negative log-likelihood is $R = \sum_i R_{i,\mathcal{L}_i}$, where the individual costs of assigning data item $i$ to source set $\mathcal{L}_i$ are

$$R_{i,\mathcal{L}_i} = -\sum_{d=1}^{D} \log\left((1-\epsilon)(1-v_{\mathcal{L}_i,d})^{x_{id}} v_{\mathcal{L}_i,d}^{1-x_{id}} + \epsilon\, r^{x_{id}}(1-r)^{1-x_{id}}\right). \tag{16}$$

Multi-Assignment Clustering (MAC) supports the simultaneous assignment of one data item to more than one source, i.e. the source sets can contain more than one element ($|\mathcal{L}_i| \geq 1$), while Single-Assignment Clustering (SAC) has the constraint $|\mathcal{L}_i| = 1$ for all $i$. Hence, when MAC has $K$ sources and $L$ different source combinations, the SAC model needs $L$ independent sources for an equivalent model complexity. For MAC, $v_{\mathcal{L}_i,d} := \prod_{\lambda \in \mathcal{L}_i} v_{\lambda,d}$ is the product of all source parameters in assignment set $\mathcal{L}_i$, while for SAC, $v_{\mathcal{L}_i,d}$ is an independent parameter of the cluster indexed by $\mathcal{L}_i$. SAC thus has to estimate $L \cdot D$ parameters, while MAC uses the data more efficiently to only learn $K \cdot D$ parameters of the individual modes.

The model parameter $\epsilon$ is the mixture weight of the noise process, and $r$ is the probability for a noisy bit to be 1. Fixing $\epsilon = 0$ corresponds to a generative model without noise process.

In summary, there are four model variants, each one defined by the constraints of its parameters: MAC models are characterized by $|\mathcal{L}_i| \geq 1$, $\mathbf{v} \in [0,1]^{K \cdot D}$, SAC models by $|\mathcal{L}_i| = 1$, $\mathbf{v} \in [0,1]^{L \cdot D}$; generative models without noise are described by $\epsilon = 0$ and its noisy version by $\epsilon \in [0,1[$.


## 5.2. Computation of the approximation capacity.

For the cost function in Eq. (16) the models. solutions in the hypothesis space. the hypothesis space is spanned by all possible assignments of objects to source combinations. A solution (a point in this hypothesis space) is encoded by the $n$ source-sets $\mathcal{L}_i, i \in \{1,..,n\}$ with $|\mathcal{L}_i| \in \{1,..,K\}$. We explained in the last section that $L$ has the same magnitude for all four model variants. Therefore, the hypothesis space of all four models equals in cardinality. In the following, we use the running index $\mathcal{L}$ to sum over all $L$ possible assignment sets.

As the probabilistic model factorizes over the objects (and therefore the costs are a sum over object-wise costs $R(v_{\mathcal{L}_i*}, \mathbf{x}_{i*}^{(q)})$ in Eq. (16)) we can conveniently sum over the entire hypothesis space by summing over all possible assignment sets for each object, similar as described in Section 3. The weight sums are then

$$\mathcal{Z}^{(q)} = \prod_{i=1}^{n} \sum_{\mathcal{L}=1}^{L} \exp\left(-\beta R(v_{\mathcal{L}*}, \mathbf{x}_{i*}^{(q)})\right), \quad q = 1,2\,, \tag{17}$$

$$\mathcal{Z}_{12} = \prod_{i=1}^{n} \sum_{\mathcal{L}=1}^{L} \exp\left(-\beta(R(v_{\mathcal{L}*}, \mathbf{x}_{i*}^{(1)}) + R(v_{\mathcal{L}*}, \mathbf{x}_{i*}^{(2)}))\right). \tag{18}$$

where the two datasets must be aligned before computing $R(v_{\mathcal{L}*}, \mathbf{x}_{i*}^{(2)})$ such that $\mathbf{x}_{i*}^{(1)}$ and $\mathbf{x}_{i*}^{(2)}$ have a high probability to be generated by the same sources. In this particular experiment we guaranteed alignment by generation of the data. With real-world data one must use a mapping function as, for instance, in Frank et al. (2011).

The weight sums of the four model variants differ only in the combined source estimates $v_{\mathcal{L}*}, \forall \mathcal{L}$. We train these estimates on the first dataset $\mathbf{x}^{(1)}$ prior to computing the mutual information Eq. (8). Having the formulas for the weight sums, one can readily evaluate the mutual information as a function of the inverse computational temperature $\beta$. We maximize this function numerically for each of the model variants.

### 5.3. Experiments

We investigate the dependency between the accuracy of the source parameter estimation and the approximation capacity. We choose a setting with 2 sources and we draw 100 samples from each source as well as from the combination of the two sources. The sources have 150 dimensions and a Hamming distance of 40 bits. To control the difficulty of the inference problem, the fraction $\epsilon$ of random bits varies between 0 and 0.99. The parameter of the Bernoulli-noise process is set to $r = 0.75$. The model parameters are then estimated by MAC and SAC both with and without a noise model. We use the true model order, i.e. $K = 2$ and $L = 3$ and infer the parameters by deterministic annealing Rose (1998).

The mismatch of the estimates to the true sources and the approximation capacity are displayed in Figures 3($a$) and 3($b$), both as a function of the noise fraction $\epsilon$. Each method has very precise estimates up to a model-dependent critical noise level. For higher noise values, the accuracy breaks down. For both MAC and SAC, adding a noise model shifts this performance decay to a higher noise level. Moreover, MACmix estimates the source parameters more accurately than SACmix and shows its performance decay at a significantly increased noise levels. The approximation capacity (Fig. 3($b$)) confirms this ranking. For noise-free data ($\epsilon = 0$), all four models attain the theoretical maximum of the approximation capacity, $\log_2(3)$ bits. As $\epsilon$ increases, the approximation capacity decreases for all models, but we observe vast differences in the sensitivity

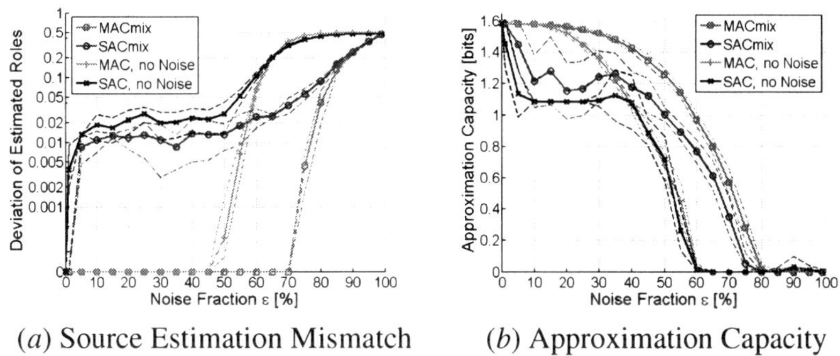

($a$) Source Estimation Mismatch      ($b$) Approximation Capacity

Figure 3: Error of source parameter estimation versus approximation capacity.

of the capacity to the noise level. Two effects decrease the capacity: inaccurately estimated parameters and (even with perfect estimates) the noise in the data that favors the assignment probabilities of an object to clusters to be more uniform (for $\epsilon = 1$ all clusters are equally probable). In conclusion, the wASC agrees with the ranking by parameter accuracy. We emphasize that parameter accuracy requires knowledge of the true source parameters while wASC requires only the data at hand.

## 6. Phase Transition in Inference

This section discusses phase transitions and learnability limits. We review theoretical results and show how the wASC principle can be employed to verify them.

### 6.1. Phase Transition of Learnability

Let the two centroids $\mu_k$, $k = 1, 2$, be orthogonal to each other and let them have equal magnitudes: $|\mu_1| = |\mu_2|$. The normalized separation $u$ is defined as $u := |\mu_1 - \mu_2| / \sqrt{2\sigma_0}$, where $\sigma_0$ indicates the variance of the underlying Gaussian probability distributions with $\Sigma = \sigma_0 \cdot \mathbf{I}$. We consider the asymptotic limit $D \to \infty$ while $\alpha := n/D$, $\sigma_0$ and $u$ are kept finite as described in Barkai et al. (1993). In this setting, the complexity of the problem, measured by the Bayes error, is proportional to $\sqrt{1/D}$. Therefore, we decrease the distance between the centroids by a factor of $\sqrt{1/D}$ when going to higher dimensions in order to keep the problem complexity constant. Similar to the two dimensional study, we use annealed Gibbs sampling to estimate the centroids $\mu_1$, $\mu_2$ at different temperatures. The theory of this problem is studied in Barkai and Sompolinsky (1994) and Witoelar and Biehl (2009). The study shows the presence of different phases depending on the values of stopping temperature and $\alpha$. We introduce the same parameters as in Barkai and Sompolinsky (1994): The separation vector $\Delta\hat{\mu} = (\hat{\mu}_1 - \hat{\mu}_2)/2$, as well as the order parameters $s = \sigma_0|\Delta\hat{\mu}|^2$ (the separation between the two estimated centers) and $r = \Delta\hat{\mu} \cdot \Delta\mu/u$ (the projection of the distance vector between the estimated

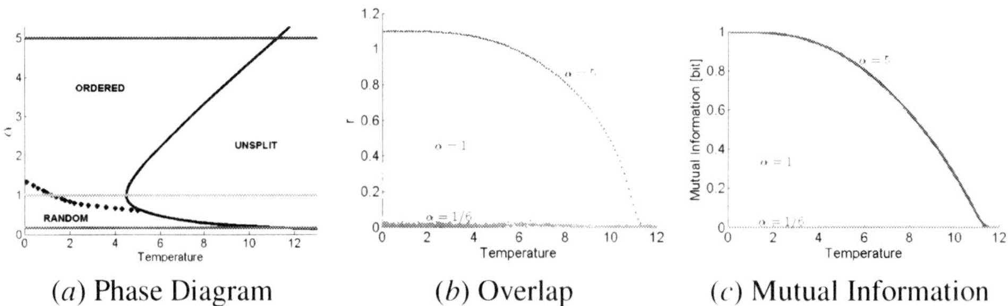

<table>
<tr><td>(a) Phase Diagram</td><td>(b) Overlap</td><td>(c) Mutual Information</td></tr>
</table>

Figure 4: Experimental study of the overlap $r$ and the mutual information $\mathcal{I}_\beta$ in different learnability limits. The problem complexity is kept constant while varying the number of objects per dimension $\alpha$.

centroids onto the distance vector between the true centroids). Computing these order parameters guides to construct the phase diagram. Thereby, we sample $n = 500$ data items from two Gaussian sources with orthogonal centroids $\mu_1, \mu_2$ and equal prior probabilities $\pi_1 = \pi_2 = 1/2$, and fix the variance $\sigma_0$ at $1/2$. We vary $\alpha$ by changing the dimensionality $D$. To keep the Bayes error fixed, we simultaneously adapt the normalized distance. For different values of $\alpha$ we perform Gibbs sampling and infer the estimated centroids $\hat{\mu}_1$ and $\hat{\mu}_2$ at varying temperature. Then we compute the order parameters and thereby obtain the phase diagram shown in Fig. 4($a$) which is consistent with the theoretical and numerical study in Barkai and Sompolinsky (1994):

**Unsplit phase**: $s = r = 0$. For high temperature and large $\alpha$ the estimated cluster centroids coincide, i.e. $\hat{\mu}_1 = \hat{\mu}_2$.

**Ordered split phase**: $s, r \neq 0$. For values of $\alpha > \alpha_c = 4u^{-4}$, the single cluster obtained in the unsplit phase splits into two clusters such that the projection of the distance vector between the two estimated and the two true sources is nonzero.

**Random split phase**: $s \neq 0, r = 0$. For $\alpha < \alpha_c$, the direction of the split between the two estimated centers is random. Therefore, $r$ vanishes in the asymptotic limits. The experiments also find such a meta-stability at low temperatures which correspond to the disordered spin-glass phase in statistical physics.

Therefore, as temperature decreases, different types of phase transitions can be observed:

1. $\alpha \gg \alpha_c$: Unsplit $\rightarrow$ Ordered. We investigate this scenario by choosing $D = 100$ and then $\alpha = 5$. The order parameter $r$ in Fig. 4($b$) shows the occurrence of such a phase transition.

2. $\alpha \gtrsim \alpha_c$: Unsplit $\rightarrow$ Ordered $\rightarrow$ Random. With $n = D = 500$, we then have $\alpha = 1$. The behavior of the parameter $r$ is consistent with the phase sequence "Unsplit $\rightarrow$ Ordered $\rightarrow$ Random" as the temperature decreases. This result is consistent with the previous study in Barkai and Sompolinsky (1994).

3. $\alpha \ll \alpha_c$: Random phase. With the choice of $D = 3000, \alpha = 1/6$ then $r$ is always zero. This means there is almost no overlap between the true and the estimated centroids.

As mentioned before, changing the dimensionality affects the complexity of the problem. Therefore, we adapt the distance between the true centroids to keep the Bayes error fixed. In the following, we study the approximation capacity for each of these phase transitions and compare them with the results we obtain in simulations.


## 6.2. Approximation Capacity of Phase Transition in Learnability Limits

Given the two datasets $\mathbf{X}^{(1)}$ and $\mathbf{X}^{(2)}$ drawn from the same source, we calculate the mutual information between the first and the second datasets according to Eq. 13. We again numerically compute the mutual information $\mathcal{I}_\beta$ for the entire interval of $\beta$ to obtain the approximation capacity (Eq. 9). Figure 4($c$) shows this numerical analysis for the three different learnability limits. The approximation capacity reflects the difference between the three scenarios described above:

1. **Unsplit $\rightarrow$ Ordered**: The centroids are perfectly estimated. The approximation capacity attains the theoretical maximum of 1 bit at low temperature.

2. **Unsplit $\rightarrow$ Ordered $\rightarrow$ Random**: The strong meta-stability for low temperatures prevents communication. The mutual information is maximized at the lowest temperature above this random phase.

3. **Random**: The centroids are randomly split. Therefore, there is no information between the true and the estimated centroids over the entire temperature range. In this regime the mutual information is always 0 over all values of $\beta$.

We extend the study to a hypothesis class of 4 centroids, thus enabling the sampler to overfit. Using Gibbs sampling on $\mathbf{X}^{(1)} \in \mathbb{R}^{500 \times 100}$ (scenario (1)) under an annealing schedule, we compute the clustering $c(\mathbf{X}^{(1)})$. At each temperature, we then compute the mutual information and the transfer costs. In this way, we study the relationship between approximation capacity and generalization error in the asymptotic limits. Figure 5 illustrates the consistency of the costs of the transferred clustering solution with the approximation capacity. Furthermore, in the annealing procedure, the correct model order, i.e. $\hat{K} = 2$, is attained at the temperature that corresponds to the maximal approximation capacity.

## 7. Conclusion

Model selection and model order selection pose critical design issues in all unsupervised learning tasks. The principle of maximum approximation capacity (wASC) offers a theoretically well-founded approach to answer these questions. We have motivated this principle and derived the general form of the capacity. As an example, we have studied the approximation capacity of Gaussian mixture models (GMM). Thereby, we have demonstrated that the choice of the optimal number of Gaussians based on the approximation capacity coincides with the configurations yielding optimal generalization ability. *Weighted approximation set coding* finds the true number of Gaussians used to generate the data.

Weighted approximation set coding is a very general model selection principle which is applicable to a broad class of pattern recognition problems (for SVD see Frank and Buhmann (2011)). We have shown how to use wASC for model selection and model order selection in clustering. Future work will address the generalization of wASC to discrete continuous optimization problems, such as sparse regression, and to algorithms without cost functions.

## Acknowledgments

This work has been partially supported by the DFG-SNF research cluster FOR916, by the FP7 EU project SIMBAD, and by the Zurich Information Security Center.

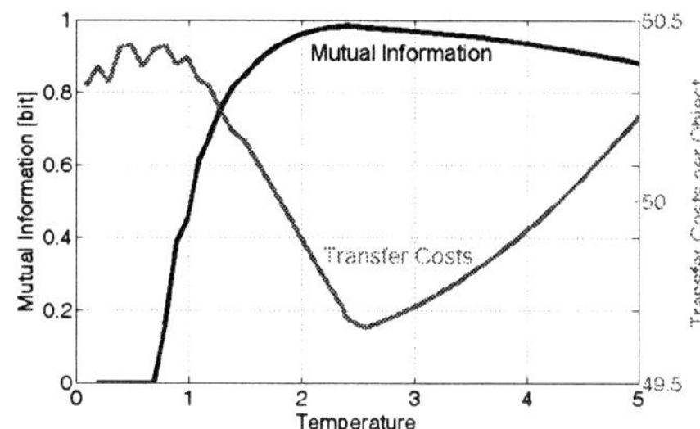


Figure 5: Expected transfer costs and approximation capacity when the number of observations per dimensions $\alpha = 5$. The Gibbs sampler is initialized with four centroids.

## References


N. Barkai and H. Sompolinsky. Statistical mechanics of the maximum-likelihood density estimation. *Phys. Rev. E*, 50(3):1766–1769, 1994.

N. Barkai, H. S. Seung, and H. Sompolinsky. Scaling laws in learning of classification tasks. *Phys. Rev. Lett.*, 70(20):3167–3170, 1993.

Shai Ben-David, Ulrike von Luxburg, and Dávid Pál. A sober look at clustering stability. In G. Lugosi and H.U. Simon, editors, *COLT'06, Pittsburgh, PA, USA*, pages 5–19, 2006.

Joachim M. Buhmann. Information theoretic model validation for clustering. In *International Symposium on Information Theory*, pages 1398 – 1402. IEEE, 2010.

Joachim M. Buhmann. Context sensitive information: Model validation by information theory. In *MCPR 2011*, volume 6718 of *LNCS*, pages 21–21. Springer, 2011.

Kenneth P. Burnham and David R. Anderson. *Model selection and inference: a practical information-theoretic approach, 2nd ed.* Springer, New York, 2002.

Thomas M. Cover and Joy A. Thomas. *Elements of Information Theory.* Wiley, 2006.

Sandrine Dudoit and Jane Fridlyand. A prediction-based resampling method for estimating the number of clusters in a dataset. *Genome Biology*, 3(7), 2002.

Mario Frank and Joachim M. Buhmann. Selecting the rank of SVD by maximum approximation capacity. In *ISIT 2011*. IEEE, 2011.

Mario Frank, Morteza Chehreghani, and Joachim M. Buhmann. The minimum transfer cost principle for model-order selection. In *ECML PKDD '11: Machine Learning and Knowledge Discovery in Databases*, volume 6911 of *Lecture Notes in Computer Science*, pages 423–438. Springer Berlin / Heidelberg, 2011.

Stuart Geman and Donald Geman. Stochastic relaxation, gibbs distributions, and the bayesian restoration of images. *IEEE PAMI*, 6(6):721–741, 1984.

Trevor Hastie, Robert Tibshirani, and Jerome Friedman. *The Elements of Statistical Learning: Data Mining, Inference, and Prediction.* Springer Verlag, New York, 2008.

Tilman Lange, Mikio Braun, Volker Roth, and Joachim M. Buhmann. Stability-based validation of clustering solutions. *Neural Computation*, 16(6):1299–1323, June 2004.

Kenneth Rose. Deterministic annealing for clustering, compression, classification, regression, and related optimization problems. *IEEE PAMI*, 86(11):2210–2239, 1998.

G. Schwarz. Estimating the dimension of a model. *Annals of Statistics*, 6:461–464, 1978.

Susanne Still and William Bialek. How many clusters? an information-theoretic perspective. *Neural Computation*, 16:2483–2506, 2004.

Andreas P. Streich, Mario Frank, David Basin, and Joachim M. Buhmann. Multi-assignment clustering for Boolean data. In *ICML'09*, pages 969–976, 2009.

Naftali Tishby, Fernando C. Pereira, and William Bialek. The information bottleneck method. In *Proc. of the 37-th Annual Allerton Conference on Communication, Control and Computing*, pages 368–377, 1999.

Vladimir N. Vapnik. *Statistical learning theory.* Wiley, New York, 1998.

Aree Witoelar and Michael Biehl. Phase transitions in vector quantization and neural gas. *Neurocomputing*, 72(7-9):1390–1397, 2009.

# Clustering: Science or Art?

**Ulrike von Luxburg**      ULRIKE.LUXBURG@TUEBINGEN.MPG.DE
*Max Planck Institute for Intelligent Systems, Tübingen, Germany*

**Robert C. Williamson**      BOB.WILLIAMSON@ANU.EDU.AU
*Australian National University and NICTA, Canberra ACT 0200, Australia*

**Isabelle Guyon**      ISABELLE@CLOPINET.COM
*ClopiNet, 955 Creston Road, Berkeley, CA 94708, USA*

**Editor:** I. Guyon, G. Dror, V. Lemaire, G. Taylor, and D. Silver

## Abstract

We examine whether the quality of different clustering algorithms can be compared by a general, scientifically sound procedure which is independent of particular clustering algorithms. We argue that the major obstacle is the difficulty in evaluating a clustering algorithm without taking into account the context: why does the user cluster his data in the first place, and what does he want to do with the clustering afterwards? We argue that clustering should not be treated as an application-independent mathematical problem, but should always be studied in the context of its end-use. Different techniques to evaluate clustering algorithms have to be developed for different uses of clustering. To simplify this procedure we argue that it will be useful to build a "taxonomy of clustering problems" to identify clustering applications which can be treated in a unified way and that such an effort will be more fruitful than attempting the impossible — developing "optimal" domain-independent clustering algorithms or even classifying clustering algorithms in terms of how they work.

## 1. Introduction

Knuth (1974) said of computer programming that *"It is clearly an art, but many feel that a science is possible and desirable."* Whether clustering is art or science is an old question: *"Is taxonomy art, or science, or both?"* asked Anderson (1974) whilst reviewing the state of systematic biological taxonomy. He justifiably went on to claim

> *Discussions of taxonomic theory or practice that refer to the concepts "science" and "art" without finer delineation will be less clear and less productive than discussions that first attempt definitions of specific concepts in the panoply of science such as precision or objectivity or repeatability or confidence and then apply these explicitly in evaluating alternatives and specific steps within the taxonomic process* (Anderson, 1974, p. 59).

The purpose of this paper is to provide such a finer delineation taking as our starting point the means by which clustering algorithms are evaluated. Our original motivation was to develop better benchmark challenge problems for clustering. Our reflections on this lead us to grapple with the broader issue of the point and purpose of clustering (and abandon the idea of such benchmarks).

Clustering is "unsupervised classification" or "unsupervised segmentation". The aim is to assign instances to classes that are not defined *a priori* and that are (usually) supposed to somehow reflect the "underlying structure" of the entities that the data represents. In the broader, non machine learning literature it is common to use the word "classification" when talking of clustering (Bowker and Star, 1999; Farris, 1981). "Taxonomy" refers to what machine learners would call hierarchical clustering.

Clustering relates data to knowledge and is a basic human activity. Bowker and Star (1999) have argued how fundamental it is in understanding the world[1]. It affects knowledge representation and discovery (Kwasnik, 1999). It defines infrastructures that have real political significance (Bowker and Star, 1999). It forms the basis for systematic biology, and the need for classification remains ever-present (Ruepp et al., 2004). It is pervasive.

Clustering holds a fascination for many mathematicians and engineers and as a consequence there is a large literature on domain independent clustering techniques. However, this literature rarely finds its way to practitioners and has long been criticized for its lack of relevance: Farris (1981) wrote of a 1976 symposium on "Classification and Clustering" that

> *The technical skill shown by many of the contributors to this symposium might well produce valuable new methods, if it could be directed to problems of systematic importance.*

The lack of appreciation is puzzling. *Supervised* classification techniques *are* widely appreciated (and used) for solving real problems. And clustering seems to simply be "unsupervised classification." However, there is a fundamental difference: supervised clustering can be easily made into a well defined problem with a loss function, which precisely formalizes what one is trying to do (and furthermore can be grounded in a rational way in the real underlying problem). The loss function can be viewed as an *abstraction* of the ultimate end-use problem. The difficulty with unsupervised clustering is that there are a huge number of possibilities regarding what will be done with it and (as yet) no abstraction akin to a loss function which distills the end-user intent.

---

1. The reason why Borges' famous strange classification quoted below strikes us as so bizarre is precisely because it makes us wonder what on earth would it be like to understand the world in that extraordinary manner — "the impossibility of thinking that" (Foucault, 1970).

> These ambiguities, redundancies, and deficiencies recall those attributed by Dr. Franz Kuhn to a certain Chinese encyclopedia called the Heavenly Emporium of Benevolent Knowledge. In its distant pages it is written that animals are divided into (a) those that belong to the emperor; (b) embalmed ones; (c) those that are trained; (d) suckling pigs; (e) mermaids; (f) fabulous ones; (g) stray dogs; (h) those that are included in this classification; (i) those that tremble as if they were mad; (j) innumerable ones; (k) those drawn with a very fine camel's-hair brush; (l) etcetera; (m) those that have just broken the flower vase; (n) those that at a distance resemble flies. (Borges, 1999)

Depending on the use to which a clustering is to be put, the same clustering can either be helpful or useless.

It is often presumed that for any situation where clustering may be used there is a single "right" clustering. ("The goal of data clustering ... is to discover the *natural* grouping(s) of a set of patterns points, or objects" (Jain, 2010, p. 3)[2]) Others take this further and maintain that the right answer is determinable by the data (alone, without reference to intended use): "the data should vote for their preferred model type and model complexity" (Buhmann, 2010). The presumption seems to be based on the notion that categories exist independently of human experience and intent (a sort of Platonic "carving nature at its joints"). Such a doctrine of "natural kinds" has been convincingly discredited (see e.g. Gilmour and Walters (1964)). Philosophical analysis of "natural kinds" reveals substantial difficulties (Bird and Tobin, 2010), see also Lakoff (1987) for a cognitive science perspective. The problems of such "absolute" approaches to clustering are also demonstrated in Kleinberg (2003).

*Users* of classification methods often reject the notion that clustering is a domain-independent subject:

> *I suspect that one of the reasons for the persistence of the view that classification is subject-independent is that classificatory theorists have been largely insulated from sources that would inform them otherwise* (Farris, 1981, p. 213).

The same problem occurs in supervised classification: if one is not prepared to commit to a particular loss function (as a formal codification of the use to which your classifier will be put) one can just estimate the underlying probability distribution. But then one is stuck with the question of *how to judge the quality of such a distribution estimate*. There can be no loss-independent way of doing this that is universally superior to other methods; pace the use of the area under the receiver operating characteristic curve (Hand, 2008).

Many of the arguments about the right way to cluster or how to compare clustering methods are side-effects of the fact that there is a very wide diversity of clustering problems. Even within a particular domain of application (say the classification of biological organisms) there can be very diverse and opposing views as to what constitutes a valuable classification: see the account of Hull (1988) of the battles between *phenetics*, which attempts to classify on the basis of observable characteristics ignoring phylogeny, and *cladistics*, which is avowedly phylogenetic. The arguments between adherents of either of these camps are not resolvable (even in principle) by domain-independent means — their conflict stems from a disagreement regarding what is the real problem that needs solving.

---

2. Ironically, in the same paper Jain recognizes the point, which contradicts any notion of "natural" that "the representation of the data is closely tied with the purpose of the grouping. The representation must go hand in hand with the end goal of the user" (Jain, 2010, p. 11). We do not believe there can be a "true" clustering definable solely in terms of the data — truth is relative to the problem being solved.

The focus of the present paper is on *problem solving* with clustering and how clustering methods are *used* rather than on the algorithmic details of the *techniques*; there are already many comprehensive reviews of techniques available (e.g., Jain et al., 1999; Xu and Wunsch, 2005; Berkhin, 2006). In this paper we do not really care how a clustering algorithm works, as long as it achieves the goal we have set. From this perspective it is pointless to argue whether clustering is *essentially* density-level set estimation, information-compression[3] or an instance of a problem in graph theory. We argue that theoretical or methodological motivations of clustering algorithms alone are insufficient to qualify clustering as a scientific method. Some practitioners think that in the past research focused too much on this methodological side. Johnson (1968, p. 224) expressed his frustration caustically:

> *The theoreticians of numerical taxonomy have enjoyed themselves immensely over the past decade (though not without developing several schools with scant respect for each other!). The mushrooming literature is quite fascinating and new developments tumble after each other. Anyone who is prepared to learn quite a deal of matrix algebra, some classical mathematical statistics, some advanced geometry, a little set theory, perhaps a little information theory and graph theory, and some computer technique, and who has access to a good computer and enjoys mathematics (as he must if he gets this far!) will probably find the development of new taximetric methods much more rewarding, more up-to-date, more 'general', and hence more prestigious than merely classifying plants or animals or working out their phylogenies.*

Arguably, it is the most important part of the scientific process to evaluate whether methods serve the end goals they have been designed for. We believe that there is an urgent need for such evaluation procedures for clustering.

## 2. Deficiencies in current clustering evaluation

In this section we discuss why most of the methods used in the clustering literature to evaluate clustering algorithms are very problematic and do not serve their purpose. The point we want to make is that clustering algorithms cannot be evaluated in a problem-independent way: whether a clustering of a particular data set is good or bad cannot be evaluated without taking into account what we want to do with the clustering once we have it. This insight is remarkably old. Gilmour and Walters (1964, p. 5) quote Mercier (1912):

> *The nature of the classification that we make ... must have direct regard to the purpose for which the classification is required. In as far as it serves the purpose, the classification is a good classification, however*

---

3. Shannon's prophetic words are still true: "the use of a few exciting words like *information, entropy, redundancy*, do not solve all our problems" (Shannon, 1956).

> *'artificial' it may be. In as far as it does not serve this purpose, it is a bad
> classification, however 'natural' it may be.*

In this section we argue that this insight is completely ignored by most of the current literature on clustering. Scanning the current literature on clustering algorithms one will find that one or several of the following methods are typically used to argue for the success of a clustering algorithm. We think that all these methods are insufficient and can be completely misleading.

**Evaluation on artificial data sets.** Clustering algorithms are applied to artificial data sets, for example points drawn from a mixture of Gaussians. Then the clustering results are compared against the "ground truth". Such a procedure can make sense to evaluate the statistical performance of a clustering algorithm under particular assumptions on the data generating process. It cannot be used to evaluate the *usefulness* of the clustering — usefulness cannot be evaluated without a particular purpose in mind.

**Evaluation on classification benchmark data sets.** Clustering algorithms are applied to classification data sets, that is data sets where samples come with class labels. Then the class labels are treated as the ground truth against which the clustering results of different algorithms are compared (using one out of various scores as the (adjusted) Rand index, misclassification error, F-measure, normalized mutual information, variation of information, and so on). High agreement with the ground truth is interpreted as good clustering performance.

We believe this approach is dangerous and misleading: It is an *assumption* that class labels coincide with cluster structure and that the "best" clustering of the data set coincides with the labels. This assumption might be true for some data sets but not for others. There might even exist a more "natural" clustering of the data points that is not reflected in the current class labels. Or, as it is often the case in high-dimensional data, different subspaces of features support completely different clusters. Consider a set of images that is labeled according to whether it contains a car, but a clustering algorithm decides to cluster the images according to whether they are greyscale or color. In such a case, the clustering algorithm discovers a very reasonable clustering, but achieves a very bad classification error. The classification error by itself cannot be used as a valid score to compare clusterings.

**Evaluation on real world data sets.** Sometimes people run their algorithm on a real data set, and then try to convince the reader that the clusters "make sense" in the application; this is claimed by some to be "the best way to evaluate clustering algorithms" (Kogan, 2007, p. 156)[4]. For example, proteins are grouped according to some known structure. This is more or less a qualitative version of the approach using benchmark data sets. It can make sense if the clustering algorithm is intended for use

---

4. Compare Farris (1981, p. 208): "A clustering method is selected in each application for its ability to manufacture a grouping most in accord with the subjective feelings of a 'professional taxonomist.' (That taxonomist, of course, will then claim vindication of his views; they have been verified by an 'objective' method!) One must wonder what value might be attributed to a method chosen primarily for its failure to contradict preconceptions."

in exploratory data analysis in this particular application, but does not carry any further meaning otherwise.

**Internal clustering quality scores.** There exist many internal scores to measure how "good" a clustering is (sum of square distances to cluster centers, ratio of between-cluster to within-cluster similarities, graph cut measures like the normalized cut or the Cheeger cut, likelihood scores, and so on). We argue that all these scores are unsuitable to evaluate the quality of clustering algorithms in an objective way. Such *scores are useful on the level of algorithms* where they can be used as an objective function in an optimization problem, and it is a valid research question how different scores can be optimized efficiently. However, across different algorithms these scores tell only little about the usefulness of the clustering. For every score preferring one clustering over the other one can invent another score which does the opposite. A unique, global, objective score for all clustering problems does not exist.

In supervised classification we are faced with a similar problem. Depending on whether we compare algorithms based on the zero-one loss or the area under the ROC curve, say, we may get different answers (that is different algorithms will be superior depending on the measure used). However, the advantage we have in supervised learning is that we can abstract from the real problem we have to solve by introducing a loss function which can guide the choice of solution. In clustering, this cannot be achieved in a domain-independent function, which makes the situation much worse.

**Universal "Benchmark data sets".** The UCI approach to supervised classification (whereby there is a small fixed collection of data sets, divorced from their real end use, that are used as a simple one-dimensional means of evaluating machine learning solutions) is widely used to compare supervised learning algorithms. However its value can be questioned even in the supervised case. For example, although lip-service is often paid to the idea that a loss function can be derived from utilities (arising in the particular end-use problem one is solving) these benchmark problems are typically only evaluated on a single loss function. Given the diversity of end uses to which clustering is put, any such approach seems hopeless for clustering. One need only look at how problematic it is to study taxonomic repeatability within a particular domain to be daunted (Moss, 1971).

We believe it makes sense to study particular applications of clustering and find out what procedures are good or bad per application. But this is a process of interaction between algorithm developers and practitioners. We need input from practitioners to help judge whether results are good or bad. We do not believe that such a process can be automated — the data sets would need to have a "true classification", and then we are in the situation described above.

## 3. Proposed method of evaluation: measure the *usefulness* for the *particular task* under consideration

As we have discussed above, clusterings or clustering algorithms cannot be evaluated without taking into account the use the clustering will be put to. To sketch how evaluation procedures might look like if they do take into account the use the clustering is put to, let us consider the two following, very distinct scenarios.

### 3.1. Evaluating the usefulness of clustering: Two example scenarios

**Clustering for data pre-processing.**    Often "one does not learn structure for its own sake, but rather to facilitate solving some higher level task" (Seldin and Tishby, 2010). Clustering is often used as an automated pre-processing step in a whole data processing chain. For example, we cluster customers and products to compress the contents of a huge sales data base before building a recommender system. Or we cluster the search results of a search engine query to discover whether the search term was ambiguous, and then use the clustering results to improve the ranking of the answers. In such situations, the whole purpose of clustering is to improve the overall performance of the system. This overall performance can usually be quantified by some *problem-dependent* score.

From a methodological point of view it is relatively straightforward and uncontroversial how clustering can be evaluated in this scenario. One can interpret the clustering as just one element in a whole chain of processing steps. Put more extremely, the clustering (algorithm) is just one more "parameter" which has to be tuned, and this tuning can be achieved similarly as for all other parameters, for example by cross validation over the final outcome of our system as a whole. We do not directly evaluate the "quality" of the clustering, and we are not interested in whether the clustering algorithm discovers "meaningful groups". All we care about is the usefulness of the clustering for achieving our final goal. For example, to build a music recommender system it might be a useful preprocessing step to cluster songs or users into groups to decrease the size of the underlying data set. In this application we do not care whether the clustering algorithm yields a meaningful clustering of songs or users, as long as the final recommender system works well. If it performs better when a particular clustering algorithm is used, this is all we need to know about the clustering step.

If the final application in mind is indeed supervised classification and clustering is performed as a pre-processing step, then one can evaluate the quality of the clustering in terms of the degradation of classification performance, when quantized observations are used in place of the original observations (Bock, 1992), or the improvement in performance arising from the additional information in the class labels, along with the original observations (Candillier et al., 2006).

**Clustering for exploratory data analysis.**    Here, clustering is used to discover aspects of the data which are either completely new, or which are already suspected to exist, or which are hoped not to exist. For example, one can use clustering to define certain sub-categories of diseases in medicine, or as a means for quality control to detect undesirable groupings that suggest experimental artifacts or confounding factors in the data. Exploratory data analysis should "present the data to the analyst such that he can see patterns in the data and formulate interesting hypotheses about the data" (Good, 1983).

As far as we know, no systematic attempt has been made to assess whether clustering in general (or a particular clustering algorithm) is useful for exploratory data analysis. In addition to the technical question concerning how to perform the clustering, this question has a psychological aspect. Ultimately it is a human user who will explore the data and hopefully detect a pattern. One approach to evaluate the performance of a certain clustering algorithm might be to ask humans to use a particular clustering algorithm to generate hypotheses, and later evaluate the quality of the hypotheses on independent data. Major obstacles to this endeavor are how to evaluate whether a hypothesis is "interesting," and how to perform a "placebo clustering" as a null model to compare with.

Data exploration is often performed visually. If clustering and visualization are treated as two independent components of some data exploration software, we believe it likely that the particular choice of the clustering algorithm is not very relevant compared to the design of the human computer interaction interface — the visualization and data manipulation capabilities of the system will likely be responsible for success or failure of the attempt to discover structure in the data. As opposed to treating clustering and visualization independently from each other, it is a promising approach to consider them jointly. For example, it might make sense to sacrifice a bit of accuracy in the clustering algorithm if this leads to a performance gain in the visualization part (consider the t-SNE algorithm of van der Maaten and Hinton (2008) as an example). The evaluation of such a system has to take into account the human user as well.

### 3.2. Can we optimize the "usefulness" directly?

The information bottleneck approach (Tishby et al., 1999) attempts to directly optimize the usefulness of a clustering. It tries to find a cluster assignment of all input points that is as "informative as possible" (in terms of mutual information) about a particular "property of interest". At first glance, this framework seems to be exactly what we are looking for. At second glance, one realizes that it is not so obvious how to implement it in practice. Often it is very hard to quantify a "feature of interest". In the exploratory data analysis setting this seems close to impossible. In the data pre-processing setting, we are interested in a high classification accuracy in the end, which is too abstract a target for the information bottleneck approach (one may as well directly optimize classification performance (Bock, 1992)). Furthermore, the method unjustifiably assumes that Shannon information is "intrinsic" and captures the essence of meaning in the data. There are in fact many different notions of information and even just "gathering information" from the data implicitly presumes a loss function (DeGroot, 1962). Formally the choice of a notion of information is tantamount to the choice of a loss function in a supervised learning problem (Reid and Williamson, 2011). Nevertheless we believe that of all the literature on clustering, the information bottleneck is closest in intent to what we are interested in as it at least tries to take into account "what we are interested in."

### 3.3. How are meta-criteria like clustering stability related to the usefulness?

In a statistical setting, it is assumed that the given data points are samples from some underlying probability distribution. There are many data sets where such an assumption makes sense (customers are samples from the "set of humans"; a particular set of handwritten digits just contains a few instances out of a much larger set of "all possible hand written digits"). In such a setting, it has often been advocated that it is important to ascertain whether a particular clustering just "fits noise" or uncovers "true structure" of the data. There are several different tools that attempt to distinguish between these two cases. Below are four such notions (sorted by increasing stringency).

**Stability.** The same clustering algorithm is applied repeatedly to perturbed versions of the original data. Then a stability score is computed that evaluates whether the results of the algorithm are "stable" or "unstable". If the results are unstable, they are considered unreliable and unsuitable for further use. See von Luxburg (2010) for a large list of references.

**Convergence of clustering algorithms.** The question is whether, for increasing amounts of data, the results of a clustering algorithm converge to a particular solution and whether this solution is reasonable. See Pollard (1981) or von Luxburg et al. (2008) for examples.

**Generalization bounds.** One computes generalization bounds that tell how much the clustering results obtained on a finite sample are different from the ones one would obtain on the full underlying distribution. These bounds are in the tradition of statistical learning theory and depend on the size of the class of models from which the clustering is chosen. See Buhmann (2010) for an approach where the richness of the hypothesis is balanced against the stability of the clustering results and Seldin and Tishby (2010) for PAC-Bayesian generalization bounds for the expected out-of-sample performance of clustering.

**Statistical significance.** Here the goal is to assign confidences scores to clustering results. They should tell how confident we are that the clustering results significantly deviate from some null model of "unclustered" data. In many branches of science it is a strong requirement to report such confidence scores. There exist numerous ways to compute confidence scores in the literature, see Efron et al. (1996) for an example.

None of the criteria listed above directly evaluates the quality of a particular clustering of a particular data set; they always take into account the clustering *algorithm* (respectively, the model class from which the clustering solution was chosen). The purpose of all the criteria mentioned above is to handle the statistical uncertainty in the data.

How are these criteria related the usefulness of a clustering? Their importance also depends on the particular use of the clustering. In the pre-processing setting. statistical considerations do not play any role, as long as the system works. If the clustering algorithm gives different results on different samples, but the system works on either of these results, then we are fine. For example, many people use $k$-means to cluster a huge set of texts, say, into a set of manageable size. If we want to cluster a data set of $10^6$ items into $10^3$ clusters, it does not really matter whether the clusters correspond to true underlying structure — they only have to serve for data compression. We can have a completely different clustering each time we get a new data set, but the overall system might still work fine on any of these data sets.

Interestingly, it is in the setting of "discovering structure" that statistical significance is important. In the exploratory setting, a user does not have infinite time to inspect all sorts of meaningless clusterings, and we cannot hope to generate a meaningful hypothesis from nothing. In this sense, statistical significance is a necessary (but not sufficient) criterion for an algorithm to be useful. Similarly, significance is important in taxonomic applications, for example when defining species in biology. Here we want to define categories based on underlying structure and not on noise. This insight is quite striking: it is the "soft" exploratory data analysis and discovering structure setting where we have the "hardest" statistical requirements for our clustering algorithms.

It is conceivable that one could define the problem of "structure discovery" precisely enough to allow one to then analyse the statistical significance of structures so discovered, but we do not have any concrete ideas concerning this. We imagine that at best it would be "discovery" from a predetermined set of structures. Ideally such a test would condition on the data (Reid, 1995).

## 4. A suggestion for future research

We have seen that clustering is used in a variety of contexts with very different goals and that clustering results cannot be evaluated without taking into account this context. We are left with the question as to what can be done. We take our lead from Tukey (1954):

> *Difficulties in identifying problems have delayed statistics far more than difficulties in solving problems. This seems likely to be the case in the future too. Thus it is appropriate to be as systematic as we can about unsolved problems. ...Different ends require different means and different logical structures. ...While techniques are important in experimental statistics, knowing when to use them and why to use them are more important.*

## 4.1. A systematic catalog of clustering problems

We believe that it would be valuable (and relatively straight-forward) to compile a table of different clustering problems and corresponding evaluation procedures. A more effective approach might be to come up with a way to treat several clustering problems with similar methods, so one does not have to start from scratch for every new application. We believe that the most effective approach would be to systematically build a taxonomy or catalog of clustering problems (Hartigan, 1977). We believe that this should be done in a solution agnostic way: define the problem in a purely declarative manner without trying to say how it should be solved. We conjecture that such a taxonomy will be of considerable help. This proposal is tantamount to the research program "left to the reader as an exercise" in the self-referentially titled *The Botryology of Botryology* (Good, 1977). This proposal is quite different to the several clusterings of clustering *algorithms* that have appeared Jain et al. (2004); Jain (2010); Andreopoulos et al. (2009) and which do not really address an end-users' concerns. We emphasize that our focus is on distinguishing properties of clustering *problems*, not *algorithms*. Hence, for our catalogue of clustering problems it is irrelevant to take into account distinctions like parametric vs. non-parametric, frequentist vs. Bayesian, model-based vs. model free, information-theoretic vs. probabilistic, etc. Of course different algorithms "solve" different problems — but we are suggesting to attempt a classification of problems in a manner independent of how to solve it. Such a declarative approach has proved valuable in computer programming and other branches of engineering.

We do not yet attempt to suggest how this catalog of clustering problems will look in any detail, but do suggest several "dimensions" of clustering applications which may be important in such an endeavor. These dimensions are largely independent of application domain or clustering algorithm.

**Exploratory — confirmatory.** We expect that exploratory/confirmatory distinction (Tukey, 1977) mentioned earlier will be central. While exploratory data analysis is used to discover patterns in data and to formulate concrete hypotheses about the data, confirmatory data analysis deals with the question of how to validate a given hypothesis based on empirical data. Clustering is employed in both contexts. Examples for exploratory uses of clustering are: Detecting latent structure, defining categories in data for later use (e.g. different subcategories of a disease), verifying that there is structure in the data, verifying that no unexpected clusters show up (quality control), verifying that expected clusters are there. For confirmative clustering consider the following example from medicine. Assume that one hypothesizes a particular categorization among patients (e.g., according to a syndrome). Then data are collected from a different feature space (say, gene expressions). The confirmation comes from looking at the clusters in the new space and comparing them to the hypothesized categorization.

**Qualitative — quantitative.** A fundamental distinction is whether the clustering results are used in a quantitative or qualitative context, that is whether we can compute a score to evaluate the overall performance of our system. The exploratory context is often "qualitative". We are interested in certain properties of the clusterings, but not in any final scores. Examples for quantitative uses of clustering are data preprocessing, data compression, or clustering for semi-supervised learning. Here, a final score can be used to evaluate the clustering performance. Note that quantitative clustering is not necessarily the same as confirmatory data analysis. In this sense, the distinction "exploratory — confirmative" is not necessarily the same as "quantitative — qualitative".

**Unsupervised — supervised.** There are different degrees of supervision in clustering problems. Situations where one does not have a clue what one is looking for are rare, and in many cases additional information can be used. For example, in exploratory data analysis the analyst usually has a good idea what he is looking for and he might bias the clustering with a particular choice of features or a particular choice of similarity measure between patterns. Further, the assessment of the usefulness of clustering as a preprocessing is often an iterative process: if clustering is found useful in a supervised task, it may be more likely to be useful in similar tasks. For instance, clustering has become a mainstream technique to build codebooks of phonemes or subwords in speech recognition. In some transfer learning methods, a set of clusters selected as a good preprocessing for one supervised task might used for another similar supervised task. In some transduction learning approaches, clustering might be carried out on both labeled and unlabeled data and the classification of labeled data performed according to the majority of labels found in a given cluster.

**Bias towards particular solutions.** In most uses of clustering, one has a "bias" concerning what one is looking for. This bias affects the type of clusters one tries to construct. In some applications the focus might be to join similar data points; for example, when detecting a chemical compound with similar properties to a given one. In other applications, it might be more important to separate different points, for example to identify emerging topics in a stream of news. Other examples are compact clusters vs. chain-like clusters, peak-based vs. gap-based clusters, flat clustering vs. hierarchical clustering.

**Modeling the data-generating process.** Do we need to find clusters that represent some understanding of how the data were generated? Not necessarily. For instance, in speech recognition it is common to use vector quantization as preprocessing, a method making no assumption about how data were generated. Similarly, in image processing, connected component methods do not attempt to uncover the mechanism by which data were generated. However, in some applications, clustering can be understood as a method for uncovering latent data structure, and prior knowledge may guide users to select the most suitable approach. Below are three examples illustrating how such prior knowledge could be exploited.

1. Phylogenies are usually modeled as a diffusion process. Clustering approaches to such problems usually attempt to uncover the underlying hierarchy, hence are tackled using hierarchical clustering methods.

2. If instead the data were generated by a shallow process, there is no reason to use a hierarchical model. For instance, handwriting used to be taught with a few

methods in the US, hence, handwriting styles could be clustered by assuming just a two level random process: drawing the method, then drawing the writer; Gaussian mixtures or $k$-means type of algorithms are appropriate for such problems.

3. A third type of model is constraint-based and assumes interactions between samples (like in a Markov random field of an Ising model of magnetism). For instance, dress-code can be assumed to emerge from peer-pressure and result in clusters of people dressing in a similar way. Graph partitioning methods may lend themselves to such cases.

Evidently, in each of these three cases using clustering methods that do not attempt to model the data generative process might "work well", if all we are interested in is representing groups efficiently for compression or prediction. But using a data generative model may make a lot of difference from the point of view of gaining insight. For instance, we may want to *predict the consequences of actions* or *devise policies to attain a desired goal*. Consider an example from epidemiology. Patients with the same symptoms may be clustered assuming one of the three models described above. The first model (hierarchical model) can be appropriate if a genetic mutation is responsible for a disease. Then, based on the hierarchical model one can try to trace a population to a common ancestor and use this knowledge to diagnose and treat patients. If patients may have a disease because of one categorical factor of variability (environment, diet, etc.) like in the case of scurvy and lack of vitamin C, then a mixture model might be more appropriate. If the disease is transmitted by contagion, then modeling the data with an interaction model (third case above) is appropriate, and a corresponding disease prevention policy can be devised.

We stress that a catalog of clustering problems is likely to be quite complex, which is no different to, say, a catalog of structural engineering problems. Many real problems require multiple factors dealt with at once (a wall keeps out the rain, wind and noise, but also holds up the roof and provides somewhere to hang a painting). Nevertheless these aspects can be described declaratively (in terms of water-resistance, wind load, sound attenuation, static and dynamic load bearing and visual aesthetics).

## 5. Conclusions

We asked whether clustering was art or science, but concluded that it is meaningless to view clustering as a domain-independent method. We deliberately duck our original question by claiming it unhelpful and irrelevant. If one wants an abstract label, a better one is engineering, which embraces different ways of knowing ("art" and "science"), recognizes the intrinsic psychological component of many problems, has standardized language and problem descriptions to avoid undue technique-focus and most certainly is focussed on solving the end-user problem (Vincenti, 1990).

If clustering researchers want real impact in applications, then it is time to step back from a purely mathematical and algorithmic point of view. What is missing is not "better" clustering algorithms but a problem-centric perspective in order to devise meaningful evaluation procedures.

## Acknowledgements

We are most grateful for inspiring discussions and comments on earlier drafts by the following people (without implying that they agree with our opinions or conclusions): Shai Ben-David, Kristin Bennett, Léon Bottou, Joachim Buhmann, Lawrence Cayton, Christian Hennig, Stefanie Jegelka, Vincent Lemaire, Mark Reid, Volker Roth, Naftali Tishby, and one anonymous reviewer from *Machine Learning Journal*.
RW is supported by the Australian Research Council and NICTA through Backing Australia's Ability.

## References

S. Anderson. Some Suggested Concepts for Improving Taxonomic Dialogue. *Systematic Zoology*, 23(1):58–70, 1974.

B. Andreopoulos, A. An, X. Wang, and M. Schroeder. A roadmap of clustering algorithms: finding a match for a biomedical application. *Briefings in Bioinformatics*, 10:297–314, 2009.

P. Berkhin. A survey of clustering data mining techniques. In Jacob Kogan, Charles Nicholas, and Marc Teboulle, editors, *Grouping Multidimensional Data*, pages 25–71. Springer, Berlin, 2006.

A. Bird and E. Tobin. Natural kinds. In Edward N. Zalta, editor, *The Stanford Encyclopedia of Philosophy (Summer 2010 Edition)*. Stanford University, 2010.

H.H. Bock. A clustering technique for maximizing $\phi$-divergence, noncentrality and discriminating power. In M. Schader, editor, *Analyzing and Modeling Data and Knowledge*, pages 19–36. Springer Verlag, 1992.

J. L. Borges. El idioma analytico de John Wilkins. In *La Nación*. Penguin, London, 8 February 1999. Translated and Republished as "John Wilkins' Analytical Language," pages 229–232 in *The Total Library: Non-fiction, 1922–1986*.

G. C. Bowker and S. Star. *Sorting Things Out: Classification and its Consequences.* MIT Press, Cambridge, Massachusetts, 1999.

J.M. Buhmann. Information theoretic model validation for clustering. In *Proceedings of the IEEE International Symposium on Information Theory (ISIT)*, pages 1398–1402. IEEE, 2010.

L. Candillier, I. Tellier, F. Torre, and O. Bousquet. Cascade evaluation of clustering algorithms. In *Machine Learning: ECML 2006*, pages 574–581. Springer, 2006.

M.H. DeGroot. Uncertainty, Information, and Sequential Experiments. *The Annals of Mathematical Statistics*, 33(2):404–419, 1962.

B. Efron, E. Halloran, and S. Holmes. Bootstrap confidence levels for phylogenetic trees. *Proceedings of the National Academy of Sciences*, 93(23):7085 – 7090, 1996.

J. S. Farris. Classification Among the Mathematicians (Review of "Classification and Clustering," by J. Van Ryzin). *Systematic Zoology*, 30(2):208–214, 1981.

M. Foucault. *The Order of Things: An Archaeology of the Human Sciences.* Random House, 1970.

J.S.L. Gilmour and S.M. Walters. Philosophy and classification. In W.B. Turrill, editor, *Vistas in Botany, Volume IV: Recent Researches in Plant Taxonomy*, pages 1–22. Pergamon Press, Oxford, 1964.

I.J. Good. The botryology of botryology. In J. van Ryzin, editor, *Classification and Clustering: Proceedings of an Advanced Seminar conducted by the Mathematics Research Center, The University of Wisconsin-Madison*, pages 73–94. Academic Press, 1977.

I.J. Good. The philosophy of exploratory data analysis. *Philosophy of Science*, 50(2), 1983.

D. J. Hand. Comment on "The skill-plot: A graphical technique for evaluating continuous diagnostic tests". *Biometrics*, 63:259, 2008.

J. A. Hartigan. Distribution problems in clustering. In J. Van Ryzin, editor, *Classification and Clustering: Proceedings of an Advanced Seminar conducted by the Mathematics Research Center, The University of Wisconsin-Madison.* Academic Press, 1977.

D. L. Hull. *Science as a Process.* University of Chicago Press, 1988.

A. K. Jain, M. N. Murty, and P. J. Flynn. Data clustering: a review. *ACM Comput. Surv.*, 31(3):264 – 323, 1999.

A.K. Jain. Data clustering: 50 years beyond K-means. *Pattern Recognition Letters*, 31 (8):651–666, 2010.

A.K. Jain, A. Topchy, M.H.C. Law, and J.M. Buhmann. Landscape of clustering algorithms. In *Proceedings of the 17th International Conference on Pattern Recognition (ICPR04)*, volume 1, pages 260–263, 2004.

L.A.S. Johnson. Rainbow's End: The Quest for an Optimal Taxonomy. *Proceedings of the Linnean Society of New South Wales*, 93(1):1–45, 1968. Reprinted in *Systematic Zoology*, 19(3), 203–239 (September 1970).

J. Kleinberg. An impossibility theorem for clustering. In S. Thrun S. Becker and K. Obermayer, editors, *Advances in Neural Information Processing Systems 15*, pages 446 – 453. MIT Press, Cambridge, MA, 2003.

D. E. Knuth. Computer Programming as an Art. *Communications of the ACM*, 17(12): 667–673, December 1974.

J. Kogan. *Introduction to Clustering Large and High-Dimensional Data*. Cambridge University Press, 2007.

B.H. Kwasnik. The role of classification in knowledge representation and discovery. *Library Trends*, 48(1):22–47, 1999.

G. Lakoff. *Women, Fire, and Dangerous Things: What Categories Reveal About the Mind*. The University of Chicago Press, 1987.

C. Mercier. *A New Logic*. William Heineman, London, 1912. URL http://www.archive.org/details/newlogic00merciala.

W. W. Moss. Taxonomic repeatability: An experimental approach. *Systematic Zoology*, 20(3):309–330, 1971.

D. Pollard. Strong consistency of k-means clustering. *Annals of Statistics*, 9(1):135 – 140, 1981.

M. D. Reid and R. C. Williamson. Information, divergence and risk for binary experiments. *Journal of Machine Learning Research*, 12:731 – 817, 2011.

N. Reid. The roles of conditioning in inference. *Statisitical Science*, 10(2):138–157, May 1995.

A. Ruepp, A. Zollner, D. Maier, K. Albermann, J. Hani, M. Mokrejs, I. Tetko, U. Guldener, G. Mannhaupt, M. Münsterkötter, and H. Mewes. The FunCat, a functional annotation scheme for systematic classification of proteins from whole genomes. *Nucleic Acids Research*, 32(18):5539, 2004.

Y. Seldin and N. Tishby. PAC-Bayesian Analysis of Co-clustering and Beyond. *Journal of Machine Learning Research*, 11:3595–3646, 2010.

C. E. Shannon. The bandwagon. *IRE Transactions on Information Theory*, 2(3):3, 1956.

N. Tishby, F. Pereira, and W. Bialek. The information bottleneck method. In *Proceedings of the 37th Annual Allerton Conference on Communication, Control and Computing*, pages 368–377, 1999.

J. Tukey. We need both exploratory and confirmatory. *The American Statistician*, 34 (1), 1977.

J. W. Tukey. Unsolved problems of experimental statistics. *Journal of the American Statistical Association*, 49(268):706–731, December 1954.

L. van der Maaten and G. Hinton. Visualizing data using t-SNE. *Journal of Machine Learning Research*, 9:2579–2605, 2008.

W. G. Vincenti. *What Engineers Know and How They Know It: Analytical Studies from Aeronautical History*. The Johns Hopkins University Press, Baltimore, 1990.

U. von Luxburg. Clustering stability: An overview. *Foundations and Trends in Machine Learning*, 2(3):235–274, 2010.

U. von Luxburg, M. Belkin, and O. Bousquet. Consistency of spectral clustering. *Annals of Statistics*, 36(2):555 – 586, 2008.

R. Xu and D. Wunsch. Survey of clustering algorithms. *IEEE Transactions on Neural Networks*, 16(3):645–678, May 2005.

# Transfer Learning by Kernel Meta-Learning

**Fabio Aiolli**             AIOLLI@MATH.UNIPD.IT
*Dept. of Mathematics, University of Padova, Via Trieste 63, 35121 Padova, Italy*

**Editor:** I. Guyon, G. Dror, V. Lemaire, G. Taylor, and D. Silver

## Abstract

A crucial issue in machine learning is how to learn appropriate representations for data. Recently, much work has been devoted to *kernel learning*, that is, the problem of finding a good kernel matrix for a given task. This can be done in a semi-supervised learning setting by using a large set of unlabeled data and a (typically small) set of *i.i.d.* labeled data. Another, even more challenging problem, is how one can exploit partially labeled data of a source task to learn good representations for a different, but related, target task. This is the main subject of *transfer learning*.

In this paper, we present a novel approach to transfer learning based on kernel learning. Specifically, we propose a *kernel meta-learning* algorithm which, starting from a basic kernel, tries to learn chains of kernel transforms that are able to produce good kernel matrices for the source tasks. The same sequence of transformations can be then applied to compute the kernel matrix for new related target tasks. We report on the application of this method to the five datasets of the Unsupervised and Transfer Learning (UTL) challenge benchmark[1], where we won the first phase of the competition.

**Keywords:** transfer learning, kernel meta-learning, unsupervised learning, UTL challenge

## 1. Introduction

Transfer learning (Pan and Yang, 2010; Caruana, 1997) shares some properties with semi-supervised learning: in both cases a large set of unlabeled data and a (generally far smaller) set of labeled data are available. However, in transfer learning, labeled data are only provided for a set of *source* tasks that are related, but different than the *target* task. In this paper, we assume all tasks are defined within a single domain, e.g. face recognition data, handwritten character recognition data, or textual data, just to name a few.

Kernel learning is a state-of-the-art paradigm for semi-supervised learning (Chapelle et al., 2006; Zhu and Goldberg, 2009). The goal of kernel learning is to learn a kernel matrix using available data (labeled and unlabeled) that optimizes an objective function that enforces the agreement between the kernel and the set of i.i.d. labeled data, e.g., by maximizing their alignment (Lanckriet et al., 2004). On the other hand, unlabeled data are used to regularize the generated models by constraining the discrim-

---

1. http://clopinet.com/ul

inant function to be smooth (that is, it should not vary too much on similar examples). However, in transfer learning, the distribution over the training/validation datasets (on examples and their labels) are generally different from the distribution over the target dataset, thus standard kernel learning methods do not directly apply.

In this paper, we explore kernel-based transfer learning and show we can indeed learn something for a task by exploiting other related tasks. In particular, rather than direct learning a kernel for a particular target task, we use source tasks to learn *how* a good kernel can be (algorithmically) generated for *any* task defined over the same domain. In a sense, we propose a *kernel meta-learner* (a learner which learns how to learn kernels from data). To the best of our knowledge this is a novel approach to kernel learning that seems promising for learning kernels and transferring knowledge in multi-task settings. Related work can be found in Pan and Yang (2010) and in the papers cited therein.

Using the technique presented in this paper, we won the first phase of the Unsupervised and Transfer Learning (UTL) challenge. Our algorithm was the best performing on three of five final competition datasets. Although we did not participate in the second phase of the challenge, experiments are presented in this paper demonstrating that the same approach can naturally be adapted to a pure transfer learning setting with competitive results.

**Notation**  We first introduce notation used throughout the paper. Unless otherwise stated, we assume that a dataset is given as an $m \times n$ matrix $X \in \mathbb{R}^{m \times n}$ formed by $m$ rows $X_i$, $i = 1, \ldots, m$, representing $n$-dimensional examples. We use the symbol $\circ$ as the Hadamard product (entry-wise) matrix multiplication. We also denote by $\mathbf{1}$ the column vector where each entry is set to 1, and $\mathbf{0}$ the null column vector, the dimensionality of which should be clear from the context they appear in.

**Background**  In this paper, we mainly focus on positive semi-definite (PSD) matrices, that is the class of real matrices $K \in \mathbb{R}^{m \times m}$ such that $v^\top K v \geq 0$ for any real vector $v \in \mathbb{R}^m$. Given any representation $\{\mathbf{x}_i\}_{i=1,\ldots,m}$ for the examples, it is well known that the kernel matrix $K \in \mathbb{R}^{m \times m}$ formed by the dot products between examples, that is $K(i, j) = \mathbf{x}_i^\top \mathbf{x}_j$, is a PSD matrix. Additionally, for PSD matrices, it is always possible to perform the spectrum decomposition, $K = UDU^\top$, where $U$ is an orthogonal matrix formed by the eigenvectors of $K$, and $D$ is the diagonal matrix with the diagonal formed by the associated (non negative and decreasing) eigenvalues. So, we can always write $K = HH^\top$ where $H = UD^{\frac{1}{2}}$. As a result, any PSD matrix of order $m$ can be seen as a kernel matrix for a dataset of $m$ examples with the examples represented according to the matrix $H$. This highlights an important aspect of kernels, that any kernel matrix induces a representation of examples (e.g. by the matrix $H$) which only considers similarity relations between pairs of examples in the dataset. Furthermore, given a kernel matrix $K$, there can be infinitely many representations for the set of examples that have $K$ as their kernel matrix.

**Synopsis**  We briefly describe the UTL competition setting in Section 2. An algorithm for kernel meta-learning is described in Section 3 and the set of kernel transforms we have used for the challenge is described in Section 4. In Section 5, we discuss an adaptation of the basic algorithm for its use on the UTL challenge as well as some tricks that reduce the risk of overfitting. Additional post-challenge results are also reported. In Section 6, we propose a general strategy to learn the optimal sequence of transforms and present the results obtained on the UTL benchmark. Finally, in Section 7, we conclude the paper with some final considerations and subjective ideas concerning future work.

## 2. The UTL Challenge

The UTL challenge benchmark[2] contains data related to five different real-world, multiclass problems. For each of these domains, three datasets (development, valid, final) have been prepared using different subsets of the original problem classes. Thus, each dataset contains a sample of a subset of classes of the original domain. Multiple binary tasks were also defined on each of these datasets by splitting the classes in two parts in a number of different ways (obtaining positive and negative labels for the tasks). Here, we briefly describe the UTL challenge setting. Please, refer to Guyon et al. (2011) for more details.

### 2.1. The Competition

For each dataset, a data matrix represented as feature vectors ($m$ examples in rows and $n$ features in columns) was provided to the participants. The goal of the challenge was to produce a new kernel matrix $K \in \mathbb{R}^{m \times m}$ between examples such that the transformed representation would lead to good performance on supervised learning tasks defined over the valid and final datasets. The actual labels of the supervised tasks used by the organizers were unknown to the participants. The evaluation was made using cross validation by partitioning data several times into training and test sets. On each run, a simple (Hebbian) linear classifier defined on the training data (and the associated kernel matrix) is used to build the scoring function. The ranking produced is then evaluated in terms of the Area Under the ROC curve (AUC) and averaged over the random splits. The size of training data varies between 1 and 64 examples and the AUC is plotted on a log scale against the number of examples. Finally, the area under the learning curve (ALC) is used as the overall evaluation metric.

Note that the labels of the supervised tasks used for the evaluation were not available in the first or second phase. Additional labels (from the development set) were made available for transfer learning in the second phase only.

### 2.2. The Hebbian classifier

In this section, we give a brief description of the classifier used in the challenge. Let $X$ be the matrix containing the vectorial representation of the examples (i.e. dataset as rows). For a given task defined over the set of examples $X$, the (Hebbian) linear classifier, or linear scoring function, $\mathbf{f} = X\mathbf{w} \in \mathbb{R}^m$ is constructed by setting $\mathbf{w} = X^\top \mathbf{y}$,

---

2. http://www.causality.inf.ethz.ch/ul_data/DatasetsUTLChallenge.pdf

$\mathbf{y} \in \mathbb{R}^m$, $\mathbf{y}_i = +\frac{1}{m_+}$ (resp. $\mathbf{y}_i = -\frac{1}{m_-}$) if the point $X_i$ is positive (resp. negative) for the task, and $\mathbf{y}_i = 0$ if the point does not belong to the training set of the given task. The values $m_-$ and $m_+$ represent the number of negative and positive training points, respectively.

Note that, the scoring function on the set of points $X$ can be written as

$$\mathbf{f} = X\mathbf{w} = XX^\top\mathbf{y} = K\mathbf{y} = K_{tr}\mathbf{y}_{tr},$$

where $K = XX^\top$ is the kernel matrix, $K_{tr}$ is the subset of kernel rows/columns corresponding to the training data of the task, and $\mathbf{y}_{tr}$ are the corresponding entries in $\mathbf{y}$. In other words, for any example $X_i$, $\mathbf{f}_i$ represents the difference between the average of kernels $K(i,+)$ across positive examples minus the average of kernels $K(i,-)$ across negative examples of the training set for the task. Thus, given a task, the score simply represents the algebraic difference of the similarities of an example with the centroids of positive and negative examples of the task.

This type of classifier has a number of nice properties. Firstly, the ranking of examples induced by the classifier does not depend on the scaling of the kernel matrix, that is, defining $K' = \alpha K$, with $\alpha > 0$ a scalar, does not change the ranking produced by the corresponding scoring function. Secondly, if we add a constant value to every entry of a kernel matrix, $K' = K + \beta 11^\top$, the values of the scoring function do not change. In fact, $\mathbf{f} = K'\mathbf{y} = K\mathbf{y} + \beta 11^\top\mathbf{y} = K\mathbf{y} = K_{tr}\mathbf{y}_{tr}$, since $1^\top\mathbf{y} = 0$ by construction. These two properties make it possible to standardize the kernel without changing the ranking produced. Thus, an equivalent kernel taking values in $[0, 1]$ can be obtained by using the linear transformation $K' = (max(K) - min(K))^{-1}(K - min(K)11^\top)$. We found this type of standardization useful as a preliminary step prior to discretization of the kernel matrix, a step that was required before the submission to the challenge server.

Finally, this linear classifier can also be seen as a strongly regularized version of an SVM variant (see for example Aiolli et al. (2008) for details) and it makes this kind of classifiers suitable when very few training data are used.

## 3. The Kernel Meta Learning (KML) Algorithm

The basic idea of this paper is to learn a chain of kernel transformations that, starting from an initial kernel matrix, leads to a better kernel for a target dataset. For this, a set of available validation (or source) supervised tasks are used to train the learner. The expected output of this procedure is an unsupervised "algorithm", or a sequence of operations, to perform on a given dataset and related kernel matrix. It is important to stress that we are not interested in the kernel matrices computed by this algorithm (in fact those kernel matrices cannot be used over different tasks and data) but we mainly focus on the sequence of operations used to obtain them from data.

We propose a greedy algorithm starting with a seed kernel on the available source data, and iteratively transforming it so that each transform results in a kernel with improved performance on the available source tasks. Labeled data is used to find the optimal transformation parameters during each step. Our intuition is that, by keeping the number of parameters involved in this optimization small, the method should generalize well on new tasks. It should produce chains of kernel transformations that are suitable for other related tasks as well.

Many strategies can be used to implement the idea above and optimize the parameters of these kernel transformations, and in Section 6 we propose a general strategy that finds an optimal sequence of such transformations. In this section, we describe the initial strategy we used in the first phase of the UTL challenge.

Assume an initial set of $m$ examples (a dataset) as rows in a matrix $X \in \mathbb{R}^{m \times n}$. Starting from a linear kernel matrix $K(1) = XX^\top$ computed on this dataset, each step $t = 1, \ldots, \bar{t}$ of the algorithm computes a new kernel matrix $K'(t)$ by transforming the kernel $K(t)$ using one of a set of given operators (possible sets of operators will be described in detail in the following section). The next (perturbed) kernel $K(t + 1)$ will be produced as a convex combination of $K(t)$ and $K'(t)$, i.e. $K(t + 1) = (1 - a) * K(t) + a * K'(t)$. The combination coefficient $0 \leq a \leq 1$ is determined by validating over the set of labeled examples. Once a good kernel has been obtained for the validation tasks, or a predefined number of steps is reached, the algorithm stops and outputs the optimal sequence of transformations along with their combination parameters. Then, when the computation of a kernel matrix for another (related) task is needed, the same (unsupervised) sequence of transformations can be applied on the new set of unlabeled data.

Note that, since in the UTL challenge the labeled examples for the validation set were not directly available, a raw (in fact manual) validation was performed on each step (i.e. by submitting kernels and looking at the validation set results). In addition to the general algorithm we have given here, its real application to the challenge and methods to reduce the risk of overfitting on validation tasks will be explained in detail in Section 5.

## 4. Kernel Transforms

For the algorithm in Section 3 to work we must use kernel transforms that do not require direct access to feature vectors. Fortunately, we can exploit the *kernel trick* (Schölkopf et al., 1999), where pattern-pattern similarities can indirectly represent examples. An additional contribution of the present work is to characterize some types of data transformations that have this characteristic. Below, we describe in detail the four classes of kernel transformations we used in the UTL challenge. In particular, we consider: a subset of affine transformations, transformations of the kernel spectrum, polynomial transformations, and an algorithmic transformation based on Hierarchical Agglomerative Clustering (HAC). All the proposed transforms will produce PSD matrices when they are applied to PSD matrices.

### 4.1. Affine Transformations: Centering and Normalization in Feature Space

Consider a kernel matrix $K \in \mathbb{R}^{m \times m}$ and any matrix $X \in \mathbb{R}^{m \times n}$, such that $K = XX^\top$. Here, we show how a set of affine transformations on the rows $\{X_i\}_{i=1,\ldots,m}$ of $X$ can directly be performed by transforming the kernel matrix and hence they are suitable for use by our algorithm. Specifically, if we take any transformation of the form:

$$X_i' = \beta_i X_i + \gamma^\top X \tag{1}$$

with $\beta \in \mathbb{R}^m$, $\gamma \in \mathbb{R}^m$, then it can be shown that the following holds:

$$K' = X'X'^\top = (\beta 1^\top) \circ K \circ (1\beta^\top) + 1\gamma^\top K + K\gamma 1^\top + (\gamma^\top K\gamma)11^\top.$$

Hence, every linear transformation like this can be given as a kernel transformation. Well-known instances of this class of transformations are briefly described next.

A first simple transformation is the centering of examples in feature space. In this case, we have $X_i' = X_i - \frac{1}{m}\sum_{j=1}^m X_j$, which corresponds to the setting in Eq. 1 when $\beta = 1$ and $\gamma = -\frac{1}{m}1$. By applying the formula above, we get the well known centering kernel transformation (see Shawe-Taylor and Cristianini, 2004),

$$K_c = K - \frac{1}{m}11^\top K - \frac{1}{m}K11^\top + \frac{1}{m^2}(1^\top K1)11^\top. \tag{$T_c$}$$

Similar transforms exist for other kinds of data processing. Pattern normalization, for example, corresponds to the same class of transformations given above with $\beta_i = 1/\sqrt{K(i,i)}$ and $\gamma = 0$. In this case, $K'(i,i) = X_i'X_i'^\top = \|X_i'\|^2 = 1$ and trace$(K') = m$, the number of examples in the set.

We notice some interesting facts about kernel centering. If $K = XX^\top$ is centered, then for every $\alpha \in \mathbb{R}^+$ we have $\alpha K = \alpha XX^\top = (\sqrt{\alpha}X)(\sqrt{\alpha}X)^\top$, which is also centered. Furthermore, the sum of centered kernels $K' = \sum_{j=1}^s K_j$ is also a centered kernel since it can be seen as the kernel obtained by concatenating the individual feature space representations, $X_i' = [X_{i,1}, X_{i,2}, \ldots, X_{i,s}]$ and is clearly centered. From these two facts, it follows that any linear combination of centered kernels is also a centered kernel.

Similar results can be given for the trace of kernel matrices. For instance, any matrix obtained as a convex combination of matrices with the same trace $t$, will have trace $t$. Also, any convex combination of normalized kernels is a normalized kernel.

## 4.2. Kernel Spectrum Transformation

Another variety of valid kernels can be computed by modifying the spectrum of a given kernel matrix through a function with codomain in $\mathbb{R}^+$. Considering a positive semi-definite matrix (a kernel) $K$, this can always be written as $K = \sum_{i=1}^m \lambda_i \mathbf{u}_i \mathbf{u}_i^\top$ where $\{\lambda_1, .., \lambda_m | \lambda_i \geq 0, \lambda_i \geq \lambda_{i+1}\}$ are the eigenvalues (in decreasing order) and $\{\mathbf{u}_1, .., \mathbf{u}_m\}$ the corresponding eigenvectors of $K$. Our proposal here is to transform the eigenvalues via a function $\sigma(\lambda)$ taking values in $\mathbb{R}^+$, i.e.

$$K_\sigma = \sum_{i=1}^m \sigma(\lambda_i)\mathbf{u}_i \mathbf{u}_i^\top. \tag{$T_\sigma$}$$

Different types of *spectrum transformation* functions can be defined on the spectrum of the eigen-decomposition of a kernel matrix. The two we have used for the UTL challenge[3] are:

**Step:** $\sigma(\lambda) = 1$ when $\lambda \geq \epsilon\lambda_1$, $0 \leq \epsilon \leq 1$, and $\sigma(\lambda) = 0$ otherwise. This transformation corresponds to the principal directions.

**Power:** $\sigma(\lambda) = \lambda^q$ where $q \geq 0$. This transformation corresponds to exponentiating the kernel matrix, i.e. $K' = K^q$.

Interestingly, the effect obtained when using *Power* as the spectrum transformation function is related to a softer version of the (kernel) PCA on the features of the kernel $K$. In fact, after performing an orthogonalization of the features, the weight of components having small eigenvalues are relatively reduced (when $q > 1$) or increased (when $q < 1$). Note that the complexity of the obtained kernel decreases with $q$, as the higher we set $q$, the smaller will be the number of significant directions we are using. Viceversa, when $q$ tends to 0, the transformed matrix tends to the identity matrix and the data has orthogonal representations. This transform also has interesting connections with diffusion maps (Coifman and Lafon, 2006).

In order to better analyze the effect of this transform, we consider the spectral decomposition of the matrix $K = UDU^\top$, where $U$ contains the eigenvectors $\{\mathbf{u}_j\}_{j=1,\ldots,m}$ as columns and $D$ is a diagonal matrix with the eigenvalues of $K$ in decreasing order. This decomposition exposes the new representations of the examples. Specifically, we have

$$\mathbf{x}'_i = [\ \sqrt{\sigma(\lambda_1)}\mathbf{u}_{1i};\ldots;\ \sqrt{\sigma(\lambda_m)}\mathbf{u}_{mi}].$$

It is apparent that the above corresponds to a reweighting of the components. When $q$ is large, more emphasis is given to the most important components and when $q$ is small, all the components have more equal importance.

Note that when this transformation is used on a centered kernel matrix, the transformed matrix is also centered since all features are scaled by the same amount, while the trace changes according to trace(K') = $\sum_{i=1}^{m} \sigma(\lambda_i)$.

### 4.3. Polynomial-Cosine Kernel Transformation

So far, the proposed transformations do not modify the original feature space, rather they perform only linear transformations of the feature vectors. An alternate, non linear, way to transform the feature vectors is to apply a polynomial transformation. For this, we propose the following kernel transformation:

$$K_\pi(i,j) = \frac{\big(\cos(\mathbf{x}_i,\mathbf{x}_j)+u\big)^p}{(1+u)^p} = \frac{1}{(1+u)^p}\left(\frac{K(i,j)}{\sqrt{K(i,i)}\ \sqrt{K(j,j)}}+u\right)^p \qquad (T_\pi)$$

---

3. Note that, since the kernel obtained after this transformation is not normalized, consistently with other transformations presented in this paper, a subsequent normalization of the obtained kernel can be performed. In fact, this is what we did in the challenge.

where $p \in \mathbb{N}$ and $u \geq 0$. Specifically, with this transformation, the original kernel is used to compute cosine similarity in feature space, and a polynomial transformation is computed on the result. Finally, the normalization term makes $K_\pi(i,i) = 1$ for every $i$. It is easy to show that this is a valid kernel using the closure properties of kernels (see e.g. Shawe-Taylor and Cristianini (2004)). One effect of the polynomial kernel, and this transformation in particular, is to further deemphasize similarities of dissimilar examples.

## 4.4. Algorithmic Transformations: a Kernel based on HAC

A clustering resulting from an unsupervised algorithm can be used to define a kernel. This is another way to exploit a similarity or kernel function to generate a new kernel. Hierarchical Agglomerative Clustering (HAC) is a popular method for clustering data. It starts by treating each pattern as a singleton cluster, subsequently, merging pairs of clusters until a single cluster contains all patterns. To do this, it requires (*i*) a pattern-pattern similarity function and (*ii*) a merge strategy that decides which pair of clusters to merge based on a cluster-cluster similarity function. In our case, the kernel to be transformed is used as the pattern-pattern similarity matrix. Popular cluster-cluster similarity measures are often based on *single-linkage, complete-linkage,* or *average-linkage* strategies. In the single-linkage (resp. complete-linkage) strategy, the similarity between two clusters is defined as the similarity of the most (resp. least) similar members. In the average-linkage strategy, the similarity between two clusters is defined as the average of similarities between all members of the two clusters.

A generalization of these measures can be defined as:

$$S(c_1, c_2) = \frac{\sum_{\mathbf{x}_i \in c_1, \mathbf{x}_j \in c_2} K(i,j) \cdot e^{\eta K(i,j)}}{\sum_{\mathbf{x}_i \in c_1, \mathbf{x}_j \in c_2} e^{\eta K(i,j)}}, \tag{2}$$

where $\eta \in \mathbb{R}$. The single-linkage strategy corresponds to $\eta = +\infty$, the complete-linkage strategy corresponds to $\eta = -\infty$, and the average-linkage strategy is obtained by setting $\eta = 0$.

We now propose a kernel defined on the basis of agglomerative clustering. Let $C_t \in \mathbb{R}^{m \times m}, t \in \{1, \ldots, m\}$, be the matrix with binary entries such that $C_t(i,j) = 1$ whenever the examples $i$ and $j$ are in the same cluster at the $t$-th agglomerative step, and $0$ otherwise. Clearly, $C_1 = I$, as this refers to the initial clustering where each example represents a different cluster, and $C_m = \mathbf{1}\mathbf{1}^\top$ is the matrix with all entries set to 1, as in this case there is a single cluster. Finally, the HAC kernel can be defined by:

$$K_h = \frac{1}{m} \sum_{t=1}^{m} C_t. \tag{$T_h$}$$

In this way $K_h(i,j)$ represents the fraction of times the examples $i$ and $j$ are assigned to the same cluster in the HAC agglomerative process. It is a valid kernel since $K_h(i,j) = \frac{1}{m}\mathbf{x}_i^\top \mathbf{x}_j$, where $\mathbf{x}_i$ is one possible representation of the $i$-th example consisting of a binary vector having a component for each node of the dendrogram generated through the agglomerative process. Specifically, we have $\mathbf{x}_{is} = 1$ whenever the node $s$ belongs to the dendrogram path starting from the root and ending on the leaf corresponding to the example $i$. It is also possible to show this kernel is proportional to the depth in the HAC dendrogram of the *Lowest Common Ancestor* (LCA) of the two examples.

## 4.5. Other kernel transforms

We conclude this section with additional examples of classes of transformations that were not used in the challenge but could be studied and added in the future.

A first interesting transform that can be used is $K' = KAK$ with $A$ an opportune PSD matrix. It is possible to show that KPCA (Schölkopf et al., 1998) is an instance of this transformation obtained by setting $A = U_k D_k^{-1} U_k^\top$, where $U_k$ are the first $k$ columns of the spectral decomposition of $K$, and $D_k \in \mathbb{R}^{k \times k}$ is the associated submatrix of $D$. Moreover, for any $n > 0$ and $C \in \mathbb{R}^{m \times n}$, the matrix $C^\top KC$ is positive definite and thus, this transformation can be used by our algorithm. Another easy transform can be obtained by using the RBF kernel $K(i, j) = \exp(-\beta(K(i,i) + K(j,j) - 2K(i,j)))$ with $\beta > 0$.

## 5. Adaptation of the KML algorithm to the UTL challenge and Results

In the first phase of the UTL challenge, labeled examples were not directly available and the algorithm described in Section 3 was not applicable as it is. So, we adapted the algorithm by simplifying the validation procedure to make it practical to be performed manually. As a side effect, this simplification reduced the effective hypothesis space and also the danger of overfitting. More specifically, we fixed the order of application of the transforms (based on their "complexity", from low to high complexity transforms) and limited the set of possible parameter choices. Moreover, a transformation was only accepted (i.e. $a > 0$) if it improved the ALC on validation significantly. In this way we avoided overtuning parameters. In this section we give additional details about the above-mentioned criteria.

The kernel centering transform, $T_c$, was only performed at the very beginning (i.e. data preprocessing) and only when its application improved the ALC on validation over a raw linear kernel. After that, the other transforms were validated one by one, starting from $T_\sigma$, then $T_\pi$, and finishing with $T_h$. This procedure was then iterated until there was no significant ALC improvement.

The $T_\sigma$ transformation was validated with parameter $q = 0.2 \cdot q_0, q_0 \in \{1, \ldots, 10\}$. In order to decrease the actual number of values to try, we assumed a convex behavior of the $ALC(q)$ curve and limited ourselves by performing a binary search for the best $q$. The combination coefficient assigned to this type of transform was always a binary value, i.e. $a \in \{0, 1\}$, meaning the transform was simply *accepted* or not. This choice was made depending on whether the resulting improvement was considered significant.

The $T_\pi$ transformation was validated with parameters $p \in \{1, \ldots, 6\}$, and $u \in \{0, 1\}$. We started with $p = 1$ and increased the value until the ALC on validation could not be further improved. Even in this case, the combination coefficient was chosen to be binary, depending on the significance of the real ALC improvement on the validation set.

Finally, the $T_h$ transformation was validated with parameter $\eta \in \{-10, 0, +10\}$. However, we noticed that $\eta = 0$ was almost always the best choice. Moreover, since in this case the transformed kernel is expected to be fairly different than the original, the combination coefficient has been more accurately validated by doing a binary search on the set of values $a = 0.05 \cdot a_0, a_0 \in \{0, \ldots, 20\}$.

In Table 1, detailed results obtained at post-challenge time are reported. In particular, we used the same validation performed during the challenge in order to investigate possible overfitting. Notice that in some (very few) cases, the obtained results can slightly differ from the official results reported for the challenge due to minor differences[4]. The results confirm that the method is quite robust to overfitting. Although this behavior was absolutely expected, it could not be given as granted during the challenge.

## 6. A general strategy for Transfer Learning

As described in previous sections, kernel validation was performed manually for the challenge. In this section, we propose a general (and automatic) strategy for use when binary labels for the source tasks are available. Specifically, validation performs a greedy search over the space of transform sequences. We assume access to a predefined and finite set of transforms.

### 6.1. The TKML strategy: The Algorithm

The pseudo-code in Algorithm 1 describes a general strategy to find the optimal set of transformations according to a given notion of accuracy on source tasks. This procedure will be provided with a set of source datasets, $\mathcal{X}$, and a set of binary tasks, $\mathcal{L}$, defined over the source data. Solely to simplify notation, we assume each source dataset has the same number of binary tasks defined over it. However, it is trivial to adapt it to the case where the number of tasks per dataset varies. Finally, we assume the existence and access to a finite set of predefined transformations in $\mathcal{T}$.

The algorithm maintains a priority queue of transformed kernels and additional information about them: the sequence of transformations that have already been applied, and the evaluation of the kernel produced (e.g. the ALC observed on the associated source tasks). The priority of an element of the queue is defined as a function of the ALCs obtained on the associated tasks (e.g. their average). On each iteration, an element of the queue (i.e. a kernel set, the list of applied transforms, and the evaluation of the transformed kernels) is extracted and all transformations available in the set $\mathcal{T}$ are applied, in turn, to the current kernels. After each transform has been applied, the obtained kernels are evaluated and inserted into the queue if it increases the relative performance.

---

4. In particular, during the challenge some transforms were (erroneously) applied to the integer matrix used for submission instead of applying them to the correct real valued kernel matrix to transform.

---

**Algorithm 1** General Search Strategy for Transfer kernel Meta-learning (TKML).

---

**Input**   : $\mathcal{X} = \{X_1, \ldots, X_s\}$: A set of source data matrices $X_i \in \mathbb{R}^{m \times n}$
         $\mathcal{L} = \{L_1, \ldots, L_s\}$: A set of source binary tasks $L_i \in \mathbb{R}^{m \times q}$
         $\mathcal{T} = \{T_1, \ldots, T_k\}$: A set of $k$ transforms

$\mathcal{K} = \{K_i = X_i X_i^{\top}\}_{i=1,\ldots,s}$ (current set of kernel matrices)
$BestEval = \texttt{Evaluate}(\mathcal{K}, \mathcal{L})$ (best evaluation so far)
$Q = [(\mathcal{K}, [\,], BestEval)]$ (priority queue)
$W^* = [\,]$ (optimal list of transforms so far)

**while not empty** $Q$ **do**
     $(\mathcal{K}, W, E) = Q.\texttt{Dequeue}()$
     **if** $E > BestEval$ **then**
         $BestEval = E$
         $W^* = W$
     **end**
     **foreach** $T_i \in \mathcal{T}$ **do**
         $\mathcal{K}' = \texttt{Transform}(\mathcal{K}, T_i)$
         $Eval = \texttt{Evaluate}(\mathcal{K}', \mathcal{L})$
         **if** $\texttt{AcceptTransform}(Eval, E)$ **then**
            $Q.\texttt{Enqueue}((\mathcal{K}', [W|T_i], Eval))$
         **end**
     **end**
**end**

---

**Output**: $W^*$ : Optimal list of transforms

---

Table 1: Details of the validation procedure performed by the KML algorithm on the five datasets of the UTL challenge (Phase 1). *RawVal* is the ALC result obtained by the linear kernel on the validation set. *BestFin* is the ALC result obtained by the best scoring competitor algorithm (its final rank in parenthesis). For each dataset, the ALC on validation and the ALC on the final datasets are presented. Note that only those transformations accepted by the algorithm ($a > 0$) are reported with their *optimal* parameters.

| AVICENNA | RawVal: 0.1034 BestFin: 0.217428(2) | Val ALC | Fin ALC |
|---|---|---|---|
| $T_c$ | k.s1.t0c1n0 | 0.124559 | 0.172279 |
| $T_\sigma(q = 0.4), a = 1$ | k.s1.t0c1n0.q04n | 0.155804 | 0.214540 |
| $T_\pi(p = 6, u = 1), a = 1$ | k.s1.t0c1n0.q04n.p6u1 | 0.165728 | 0.216307 |
| $T_h(\eta = 0), a = 0.2$ | k.s1.t0c1n0.q04n.p6u1.d0.a02 | 0.167324 | 0.216075 |
| $T_\sigma(q = 1.4), a = 1$ | k.s1.t0c1n0.q04n.p6u1.d0.a02.q1_4n | 0.173641 | **0.223646**(1) |
| HARRY | RawVal: 0.6264 BestFin: **0.806196**(1) | Val ALC | Fin ALC |
| $T_c$ | k.s2.t0c1n0 | 0.627298 | 0.609275 |
| $T_\pi(p = 1, u = 0), a = 1$ | k.s2.t0c1n0.p1u0 | 0.604191 | 0.678578 |
| $T_h(\eta = 10), a = 1$ | k.s2.t0c1n0.p1u0.d10 | 0.861293 | 0.716070 |
| $T_\sigma(q = 2), a = 1$ | k.s2.t0c1n0.p1u0.d10.q2n | 0.863983 | 0.704331(6) |
| RITA | RawVal: 0.2504 BestFin: 0.489439(2) | Val ALC | Fin ALC |
| $T_c$ | k.s3.t0c1n0 | 0.281439 | 0.462083 |
| $T_\pi(p = 5, u = 1), a = 1$ | k.s3.t0c1n0.p5u1 | 0.293303 | 0.478940 |
| $T_h(\eta = 0), a = 0.4$ | k.s3.t0c1n0.p5u1.d0.a04 | 0.309428 | **0.495082**(1) |
| SYLVESTER | RawVal: 0.2167 BestFin: **0.582790**(1) | Val ALC | Fin ALC |
| $T_\sigma(\epsilon = 0.00003), a = 1$ | k.s4.t0c0n0.e000003n | 0.643296 | 0.456948(6) |
| TERRY | RawVal: 0.6969 BestFin: 0.843724(2) | Val ALC | Fin ALC |
| $T_c$ | k.s5.t0c1n0 | 0.712477 | 0.769590 |
| $T_\sigma(q = 2), a = 1$ | k.s5.t0c1n0.q2n | 0.795218 | 0.826365 |
| $T_h(\eta = 0), a = 0.95$ | k.s5.t0c1n0.q2n.d0.a095 | 0.821622 | **0.846407**(1) |

At this point, it is important to note that the effectiveness of this algorithm depends very much on the type, and number, of transforms. In general, we can have up to $|\mathcal{T}|^v$ different sequences of length $v$. This dimension represents the size of the hypothesis space. As hypothesis space size increases, we can expect higher accuracies of the optimal sequence generated by the algorithm. However, there is also the danger of overfitting the source data, producing sequences that will not generalize well to other data and tasks.

One way to keep the size of the hypothesis space small, while maintaining good performance, is to utilize the granularity of the available transformations. In particular, we can use a coarse-grained set $\mathcal{T}$ containing more complex transformations (i.e.

transformations which are composition of other simpler transformations). We will give an example of application of this criterium in the following experiments.

## 6.2. TKML strategy: Experimental Setting

At the end of the challenge, we were provided with labels for both the validation datasets and for the transfer datasets for all problems. We now present additional experiments based on these new datasets. Specifically, we were curious to see if adding transfer labels to the validation process could improve results. For each challenge problem, we used only a subsample of the transfer set and corresponding labels, with size equal to the validation and final datasets.

We applied the general strategy described earlier to find the optimal set of transformations. The evaluation of a sequence of transforms (function Evaluate() in Algorithm 1) is performed by averaging the ALCs obtained on the tasks defined on the two source datasets. The acceptance condition (function AcceptTransform() in Algorithm 1) verifies whether adding a new kernel transform improves the performance of the sequence.

As already stated, a crucial factor is the choice of transforms, $\mathcal{T}$. After preliminary experiments performed on validation datasets and considering the criteria presented in Section 6.1, we have chosen the small set of transforms given in Table 2. Note that most of these transforms actually consist of sequences of simpler transforms. We also have performed experiments using a larger set of transforms and a finer selection of the transformations. As expected, the algorithm tended to overfit some source tasks in this case.

## 6.3. TKML strategy: Results

The initial seed kernel for all five challenge datasets is the linear kernel $K = XX^{\top}$, centered and scaled to trace $m$, the number of examples of the set (i.e. *trace standardization*). The validation and transfer datasets have been used as source datasets. The results for the five datasets are as follows (refer to Table 2 for a detailed description):

- AVICENNA: $T_2, T_3, T_1, T_3$ for 3 times;

- HARRY: $T_2$ for 3 times, $T_3, T_2, T_3$ for 3 times;

- RITA: $T_3$ for 4 times, $T_1, T_3$ for 6 times, $T_1, T_3$;

- SYLVESTER: $T_5, T_3$ for 10 times;

- TERRY: $T_1$ for 2 times, $T_3$ for 3 times, $T_4$.

In Table 3, the results on the final datasets are reported. Interestingly, there is a clear improvement when additional source tasks are used. Sometimes this improvement is dramatic, as in HARRY and SYLVESTER, two datasets where the strategy used in the challenge particularly suffered from overfitting. The new strategy is very competitive with other challenge participant entries as it gives the best results on the RITA and TERRY datasets, and the second best results for AVICENNA and HARRY datasets.

Table 2: The five transforms (the set $\mathcal{T}$ of the TKML strategy) used in our experiments.

| 1 | Spectral $\sigma(\lambda) = \lambda^{\sqrt{2}}$<br>(a) Centering ($T_c$) and trace standardization<br>(b) Spectral Transform with $\sigma(\lambda) = \lambda^{\sqrt{2}}$<br>(c) Normalization |
|---|---|
| 2 | Spectral $\sigma(\lambda) = \lambda^{1/\sqrt{2}}$<br>(a) Centering ($T_c$) and trace standardization<br>(b) Spectral Transform with $\sigma(\lambda) = \lambda^{1/\sqrt{2}}$<br>(c) Normalization |
| 3 | Polynomial Transform with $p = 2$ |
| 4 | HAC Transform with $d = 0$ |
| 5 | Step Spectral<br>(a) Centering ($T_c$) and trace standardization<br>(b) Spectral Transform with $\sigma(\lambda_i) = 1$ whenever $i \leq 5$, 0 otherwise<br>(c) Normalization |

Table 3: Results obtained with the TKML strategy of Section 6 and comparison with Phase 1 and Phase 2 challenge results. All the results refer to the final datasets. ALC($\mathcal{T}_0$) corresponds to the ALC obtained by the algorithm we used in the challenge and described in Section 5. ALC($\mathcal{T}_1$) corresponds to the ALC obtained by the TKML strategy with the set of transformations in Table 2. In parenthesis the rank we would have obtained in final challenge results for Phase 1 and 2. Finally, the last two columns report the best ALC obtained by our competitors in Phase 1 and 2.

| DATASET | ALC($\mathcal{T}_0$) | ALC($\mathcal{T}_1$) | BestALC Phase 1 | BestALC Phase 2 |
|---|---|---|---|---|
| AVICENNA | 0.223646 | 0.221256 (1,2) | 0.218265 | **0.227307** |
| HARRY | 0.704331 | 0.815374 (1,2) | 0.806196 | **0.861933** |
| RITA | 0.495082 | **0.507535** (1,1) | 0.489439 | 0.502948 |
| SYLVESTER | 0.456948 | 0.547636 (3,4) | 0.582790 | **0.593283** |
| TERRY | 0.846407 | **0.849543** (1,1) | 0.843724 | 0.843724 |

## 7. Final remarks

We conclude the paper with some final considerations about the proposed paradigm and future work.

*Suitability of the method for Transfer Learning.* The method proposed in this paper seems particularly well suited for transfer learning tasks, as it tries to learn a set of (unsupervised) kernel transformations. On the other hand, standard methods for semi-supervised learning typically optimize an objective function which needs of i.i.d. labeled data. We advocate that learning a sequence of data transformations should make the obtained solution depend more on the domain and less on the particular tasks used for optimization.

*Needs of few labeled data.* The method is expected to require very few labeled examples, since the labels are used only for validation. Although the ALC measure clearly depends on particular tasks for which it is computed, in the UTL challenge, the ALC measure is computed averaging over several binary tasks. This characteristic is important to prevent possible overfitting with respect to each particular task.

*Generality of the method.* The method is able to combine very different kinds of kernels that individually perform well on varied domains. For example, the HAC based kernel seems to be suited for data with some structure. In particular, this transform has been crucial to obtain the best result on TERRY, a dataset related to textual data. Another example, exponentiating the kernel matrix (via the eigenvalues obtained by its spectral decomposition) produces an orthogonalization of the features together with a reweighting of its components. This transform turned out to be very useful for all the challenge datasets. Specifically, its contribution is apparent when data lay on a subspace of reduced dimensions, or when a decorrelation of the features, as *whitening* (Duda et al., 2000) for example, can be beneficial. Finally, as briefly discussed in this paper, additional kernel transforms can be plugged into the algorithm proposed.

*Low computational burden.* The computational requirements mainly depends on SVD computations needed for the $T_\sigma$ transform. Interestingly, note that computing the $T_\sigma$ transform with different parameters requires only a single SVD. In fact, let $K = UDU^\top$ be the SVD decomposition of $K$, then, any different transform with exponent $q$ can be obtained by a matrix multiplication $K' = HH^\top$ where $H = UD^{\frac{q}{2}}$ (since $D$ is diagonal, this last computation does not affect the computational complexity of the transform and it is fast to compute). We used Scilab[5] for the linear algebra routines, such as the SVD computation and matrix manipulation. C++ has been used for the computation of the HAC based kernel and for combining kernels.

*Future works on Transfer Learning.* A study about the connection between ours and other paradigms, such as deep learning, will be the subject of future work. In fact, our method can be considered a sort of 'deep kernel learning' similar in principle to a deep learning architecture where kernel transformations correspond to the different levels of a deep neural network. However, unlike deep learning, we do not use explicit representation of data as examples are represented on the basis of similarities with other examples. In the near future, we plan to investigate whether this method can be extended to transfer knowledge across varied domains. It would be interesting to define a metric between chains of transformations obtained in such a setting. This metric can be used to decide what type of knowledge could be transferred from one domain to another based on domain similarity.

---

5. http://www.scilab.org/

## Acknowledgments

This work was supported by ATENEO 2009/2011 "Methods for the integration of background knowledge in kernel-based learning algorithms for the robust identification of biomarkers in genomics". We warmly thank the challenge organizers, and Isabelle Guyon in particular, for their support, Dav Zimak and the anonymous reviewers for their useful comments and suggestions.

## References

Fabio Aiolli, Giovanni Da San Martino, and Alessandro Sperduti. A kernel method for the optimization of the margin distribution. In *International Conference on Artificial Neural Networks (ICANN)*, pages 305–314, 2008.

Rich Caruana. Multitask learning. In *Machine Learning*, pages 41–75, 1997.

O. Chapelle, B. Schölkopf, and A. Zien, editors. *Semi-Supervised Learning*. MIT Press, Cambridge, MA, 2006. URL `http://www.kyb.tuebingen.mpg.de/ssl-book`.

R.R. Coifman and S. Lafon. Diffusion maps. *Applied and Computational Harmonic Analysis: Special issue on Diffusion Maps and Wavelets*, 21:5–30, 2006.

R.O. Duda, P.E. Hart, and D.G. Stork. *Pattern Classification (2nd Edition)*. Wiley-Interscience, 2000. ISBN 0471056693.

I. Guyon, G. Dror, V. Lemaire, G. Taylor, and D.W. Aha. Unsupervised and transfer learning challenge. In *International Joint Conference on Neural Networks (IJCNN)*, pages 793–800, 2011.

Gert R. G. Lanckriet, Nello Cristianini, Peter L. Bartlett, Laurent El Ghaoui, and Michael I. Jordan. Learning the kernel matrix with semidefinite programming. *Journal of Machine Learning Research*, 5:27–72, 2004.

Sinno Jialin Pan and Qiang Yang. A survey on transfer learning. *IEEE Transactions on Knoweledge and Data Engineering*, 22(10):1345–1359, October 2010.

B. Schölkopf, C. J. C. Burges, and A. J. Smola, editors. *Advances in Kernel Methods— Support Vector Learning*. MIT Press, Cambridge, MA, 1999.

Bernhard Schölkopf, Alexander Smola, and Klaus-Robert Müller. Nonlinear component analysis as a kernel eigenvalue problem. *Neural Computation*, 10:1299–1319, July 1998.

John Shawe-Taylor and Nello Cristianini. *Kernel Methods for Pattern Analysis*. Cambridge University Press, New York, NY, USA, 2004. ISBN 0521813972.

Xiaojin Zhu and Andrew B. Goldberg. *Introduction to Semi-Supervised Learning.* Synthesis Lectures on Artificial Intelligence and Machine Learning. Morgan & Claypool Publishers, 2009.

# Unsupervised and Transfer Learning Challenge: a Deep Learning Approach


**Grégoire Mesnil**[1,2]                                MESNILGR@IRO.UMONTREAL.CA

**Yann Dauphin**[1]                                      DAUPHIYA@IRO.UMONTREAL.CA

**Xavier Glorot**[1]                                     GLOROTXA@IRO.UMONTREAL.CA

**Salah Rifai**[1]                                       RIFAISAL@IRO.UMONTREAL.CA

**Yoshua Bengio**[1]                                     BENGIOY@IRO.UMONTREAL.CA

**Ian Goodfellow**[1]                                    GOODFELI@IRO.UMONTREAL.CA

**Erick Lavoie**[1]                                      LAVOERIC@IRO.UMONTREAL.CA

**Xavier Muller**[1]                                     MULLERX@IRO.UMONTREAL.CA

**Guillaume Desjardins**[1]                              DESJAGUI@IRO.UMONTREAL.CA

**David Warde-Farley**[1]                                WARDEFAR@IRO.UMONTREAL.CA

**Pascal Vincent**[1]                                    VINCENTP@IRO.UMONTREAL.CA

**Aaron Courville**[1]                                   COURVILA@IRO.UMONTREAL.CA

**James Bergstra**[1]                                    BERGSTRJ@IRO.UMONTREAL.CA

[1] *Dept. IRO, Université de Montréal. Montréal (QC), H2C 3J7, Canada*

[2] *LITIS EA 4108, Université de Rouen. 76 800 Saint Etienne du Rouvray, France*

**Editor:** I. Guyon, G. Dror, V. Lemaire, G. Taylor, and D. Silver


## Abstract


Learning good representations from a large set of unlabeled data is a particularly challenging task. Recent work (see Bengio (2009) for a review) shows that training deep architectures is a good way to extract such representations, by extracting and disentangling gradually higher-level factors of variation characterizing the input distribution. In this paper, we describe different kinds of layers we trained for learning representations in the setting of the Unsupervised and Transfer Learning Challenge. The strategy of our team won the final phase of the challenge. It combined and stacked different one-layer unsupervised learning algorithms, adapted to each of the five datasets of the competition. This paper describes that strategy and the particular one-layer learning algorithms feeding a simple linear classifier with a tiny number of labeled training samples (1 to 64 per class).

**Keywords:** Deep Learning, Unsupervised Learning, Transfer Learning, Neural Networks, Restricted Boltzmann Machines, Auto-Encoders, Denoising Auto-Encoders.


## 1. Introduction

The objective of machine learning algorithms is to discover statistical structure in data. In particular, *representation-learning algorithms* attempt to transform the raw data into a form from which it is easier to perform supervised learning tasks, such as classification. This is particularly important when the classifier receiving this representation as input is linear and when the number of available labeled examples is small. This is the case here with the Unsupervised and Transfer Learning (UTL) Challenge [1].

Another challenging characteristic of this competition is that the training (development) distribution is typically very different from the test (evaluation) distribution, because it involves a set of classes different from the test classes, i.e., both inputs and labels have a different nature. What makes the task feasible is that these different classes have things in common. The bet we make is that *more abstract features of the data are more likely to be shared among the different classes*, even with classes which are very rare in the training set. Another bet we make with representation-learning algorithms and with Deep Learning algorithms in particular is that the structure of the input distribution $P(X)$ is strongly connected with the structure of the class predictor $P(Y|X)$ for all of the classes $Y$. It means that representations $h(X)$ of inputs $X$ are useful both to characterize $P(X)$ and to characterize $P(Y|X)$, which we will think of as parametrized through $P(Y|h(X))$. Another interesting feature of this competition is that the input features are anonymous, so that teams are compared based on the strength of their learning algorithms and not based on their ability to engineer hand-crafted features based on task-specific prior knowledge. More material on Deep Learning can be found in a companion paper Bengio (2011).

The paper is organized as follows. The pipeline going from bottom (raw data) to top (final representation fed to the classifier) is described in Section 2. In addition to the score returned by the competition servers, Section 3 presents other criteria that guided the choice of hyperparameters. Section 4 precisely describes the layers we chose to combine for each of the five competition datasets, at the end of the exploration phase that lasted from January 2011 to mid-April 2011.

## 2. Method

We obtain a deep representation by stacking different single-layer blocks, each taken from a small set of possible learning algorithms, but each with its own set of hyperparameters (the most important of which is often just the dimension of the representation). Whereas the number of possible combinations of layer types and hyperparameters is exponential as depth increases, we used a greedy layer-wise approach (Bengio et al., 2007) for building each deep model. Hence, the first layer is trained on the raw input and its hyper-parameters are chosen with respect to the score returned by the competition servers (on the validation set) and different criteria to avoid overfitting to a particular subset of classes (discussed in Section 3). We then fix the $i^{th}$ layer (or keep only a very small number of choices) and search for a good choice of the $i+1^{th}$ layer, pruning and keeping only a few good choices. Depth is thus increased without an explosion in computation until the model does not improve significantly the performance according to our criteria.

The resulting learnt pipeline can be divided in three types of stages: preprocessing, feature extraction and transductive postprocessing.

---

1. http://www.causality.inf.ethz.ch/unsupervised-learning.php

## 2.1. Preprocessing

Before the feature extraction step, we preprocessed the data using various techniques. Let $\mathcal{D} = \{x^{(j)}\}_{j=1,\ldots,n}$ be a training set where $x^{(j)} \in \mathbb{R}^d$.

**Standardization**    One option is to standardize the data. For each feature, we compute its mean $\mu_k = (1/n)\sum_{j=1}^{n} x_k^{(j)}$ and variance $\sigma_k$. Then, each transformed feature $\tilde{x}_k^{(j)} = (x_k^{(j)} - \mu_k)/\sigma_k$ has zero mean and unit variance.

**Uniformization (t-IDF)**    Another way to control the range of the input is to uniformize the feature values by restricting their possible values to $[0,1]$ (and non-parametrically and approximately mapping each feature to a uniform distribution). We rank all the $x_k^{(j)}$ and map them to $[0,1]$ by dividing the rank by the number of observations sorted. In the case of sparse data, we assigned the same range value (0) for zeros features. One option is to aggregate all the features in these statistics and another is to do it separately for each feature.

**Contrast Normalization**    On datasets which are supposed to correspond to images, each input $d$-vector is normalized with respect to the values in the given input vector (global contrast normalization). For each sample vector $x^{(j)}$ subtract its mean $\mu^{(j)} = (1/d)\sum_{k=1}^{d} x_k^{(j)}$ and divide by its standard deviation $\sigma^{(j)}$ (also across the elements of the vector). In the case of images, this would discard the average illumination and contrast (scale).

**Whitened PCA**    The Karhulen-Loève transform constantly improved the quality of the representation for each dataset. Assume the training set $\mathcal{D}$ is stored as a matrix $X \in \mathcal{M}_{\mathbb{R}}(n,d)$. First, we compute the empirical mean $\mu = (1/n)\sum_{i=1}^{n} X_{i.}$ where $X_{i.}$ denotes row $i$ of the matrix $X$, i.e., example $i$. We center the data $\tilde{X} = X - \mu$ and compute the covariance matrix $C = (1/n)\tilde{X}^T \tilde{X}$. Then, we obtain the eigen-decomposition of the covariance matrix $C = V^{-1}UV$ i.e $U \in \mathbb{R}^d$ contains the eigen-values and $V \in \mathcal{M}_{\mathbb{R}}(d,d)$ the corresponding eigen-vectors (each row corresponds to an eigen-vector). We build a diagonal matrix $U'$ where $U'_{ii} = \sqrt{C_{ii}}$. By the end, the output of the whitened PCA is given by $Y = (X - \mu)VU'$. In our experiments, we used the PCA implementation of the scikits [2] toolbox.

**Feature selection**    In the datasets where the input is sparse, a preprocessing that we found very useful is the following: *only the features active on the training (development)* **and** *test (resp. validation) datasets are retained* for the test set (resp. validation) representations. We removed those whose frequency was low on both datasets (this introduces a new hyper-parameter that is the cut-off threshold, but we only tried a couple of values).

---

2. http://scikits.appspot.com/

## 2.2. Feature extraction

Feature extraction is the core of our pipeline and has been crucial for getting the first ranks during the challenge. Here we briefly introduce each method that has been used during the competition. See also Bengio (2011) along with the citations below for more details.

### 2.2.1. $\mu$-ss-RBM

The $\mu$-spike and slab Restricted Boltzmann Machine ($\mu$-ssRBM) (Courville et al., 2011) is a recently introduced undirected graphical model that has demonstrated some promise as a model of natural images. The model is characterized by having both a real-valued *slab* vector and a binary *spike* variable associated with each hidden unit. The model possesses some practical properties such as being amenable to block Gibbs sampling as well as being capable of generating similar latent representations of the data to the mean and covariance Restricted Boltzmann Machine (Ranzato and Hinton, 2010).

The $\mu$-ssRBM describes the interaction between three random vectors: the visible vector $v$ representing the observed data, the binary "spike" variables $h$ and the real-valued "slab" variables $s$. Suppose there are $N$ hidden units and a visible vector of dimension $D$: $v \in \mathbb{R}^D$. The $i$th hidden unit ($1 \leq i \leq N$) is associated with a binary *spike* variable: $h_i \in \{0, 1\}$ and a real valued vector $s_i \in \mathbb{R}^K$, pooling over $K$ linear filters. This kind of pooling structure allows the model to *learn* over which filters the model will pool – a useful property in the context of the UTL challenge where we cannot assume a standard "pixel structure" in the input. The $\mu$-ssRBM model is defined via the energy function

$$E(v, s, h) = -\sum_{i=1}^{N} v^T W_i s_i h_i + \frac{1}{2} v^T \left( \Lambda + \sum_{i=1}^{N} \Phi_i h_i \right) v$$

$$+ \sum_{i=1}^{N} \frac{1}{2} s_i^T \alpha_i s_i - \sum_{i=1}^{N} \mu_i^T \alpha_i s_i h_i - \sum_{i=1}^{N} b_i h_i + \sum_{i=1}^{N} \mu_i^T \alpha_i \mu_i h_i,$$

in which $W_i$ refers to the $i$th weight matrix of size $D \times K$, the $b_i$ are the biases associated with each of the spike variables $h_i$, and $\alpha_i$ and $\Lambda$ are diagonal matrices that penalize large values of $\|s_i\|_2^2$ and $\|v\|_2^2$ respectively.

Efficient learning and inference in the $\mu$-ssRBM is rooted in the ability to iteratively sample from the factorial conditionals $P(h \mid v)$, $p(s \mid v, h)$ and $p(v \mid s, h)$ with a Gibbs sampling procedure. For a detailed derivation of these conditionals, we refer the reader to (Courville et al., 2011). In training the $\mu$-ssRBM, we use stochastic maximum likelihood (Tieleman, 2008) to update the model parameters.

### 2.2.2. DENOISING AUTOENCODER

Traditional autoencoders map an input $x \in \mathbb{R}^{d_x}$ to a hidden representation $h$ (the learnt features) with an affine mapping followed by a non-linearity $s$ (typically a sigmoid): $h = f(x) = s(Wx + b)$. The representation is then mapped back to input space, initially producing a linear reconstruction $r(x) = W' f(x) + b_r$, where $W'$ can be the transpose of

$W$ (tied weights) or a different matrix (untied weights). The autoencoder's parameters $\theta = W, b, b_r$ are optimized so that the reconstruction is close to the original input $x$ in the sense of a given loss function $L(r(x), x)$ (the reconstruction error). Common loss functions include squared error $\|r(x) - x\|^2$, squared error after sigmoid $\|s(r(x)) - x\|^2$, and sigmoid cross-entropy $-\sum_i x_i \log s(r_i(x)) + (1 - x_i) \log(1 - s(r_i(x)))$. To encourage robustness of the representation, and avoid trivial useless solutions, a simple and efficient variant was proposed in the form of the Denoising Autoencoders (Vincent et al., 2008, 2010). Instead of being trained to merely reconstruct its inputs, a Denoising Autoencoder is trained to *denoise* artificially corrupted training samples, a much more difficult task, which was shown to force it to extract more useful and meaningful features and capture the structure of the input distribution (Vincent et al., 2010). In practice, instead of presenting the encoder with a clean training sample $x$, it is given as input a stochastically corrupted version $\tilde{x}$. The objective remains to minimize reconstruction error $L(r(\tilde{x}), x)$ with respect to clean sample $x$, so that the hidden representation has to help denoise. Common choices for the corruption include additive Gaussian noise, and masking a fraction of the input components at random by setting them to 0 (masking noise).

### 2.2.3. CONTRACTIVE AUTOENCODER

To encourage robustness of the representation $f(x)$ obtained for a training input $x$, Rifai et al. (2011) propose to penalize its *sensitivity* to that input, measured as the Frobenius norm of the Jacobian $J_f(x)$ of the non-linear mapping. Formally, if input $x \in \mathbb{R}^{d_x}$ is mapped by an encoding function $f$ to a hidden representation $h \in \mathbb{R}^{d_h}$, this sensitivity penalization term is the sum of squares of all partial derivatives of the extracted features with respect to input dimensions:

$$\|J_f(x)\|_F^2 = \sum_{ij} \left( \frac{\partial h_j(x)}{\partial x_i} \right)^2.$$

Penalizing $\|J_f\|_F^2$ encourages the mapping to the feature space to be contractive in the neighborhood of the training data. The *flatness* induced by having small first derivatives will imply an *invariance* or *robustness* of the representation for small variations of the input.

While such a Jacobian term alone would encourage mapping to a useless constant representation, it is counterbalanced in auto-encoder training by the need for the learnt representation to allow a good reconstruction of the training examples.

### 2.2.4. RECTIFIERS

Recent works investigated linear rectified activation function variants. Nair and Hinton (2010) used Noisy Rectified Linear Units (NReLU) (i.e. $\max(0, x + N(0, \sigma(x)))$ for Restricted Boltzmann Machines. Compared to binary units, they observed significant improvements in term of generalization performance for image classification tasks.

Following this line of work, Glorot et al. (2011) used the rectifier activation function (i.e. $\max(0, x)$) for deep neural networks and Stacked Denoising Auto-Encoders (SDAE) (Vincent et al., 2008, 2010) and obtained similarly good results.

This non-linearity has various mathematical advantages. First, it naturally creates sparse representations with true zeros which are computationally appealing. In addition, the linearity on the active side of the activation function allows gradient to flow well on the active set of neurons, possibly reducing the vanishing gradients problem.

In a semi-supervised setting similar to that of the Unsupervised and Transfer learning Challenge setup, Glorot et al. (2011) obtained state-of-the-art results for a sentiment analysis task (the Amazon 4-task benchmark) for which the bag-of-words input were highly sparse.

But learning such embeddings for huge sparse vectors with the proposed approach is still very expensive. Even though the training cost only scales linearly with the dimension of the input, it can become too expensive when the input becomes very large. Projecting the input vector to its embedding can be made quite efficient by using a sparse matrix-vector product. However, projecting the embedding back to the input space is quite expensive during decoding because one has to compute a reconstruction (and reconstruction error) for all inputs and not just the non-zeros. If the input dimension is 50,000 and the embedding dimension 1,000 then decoding requires 50,000,000 operations. In order to speed-up training for huge sparse input distributions, we use reconstruction sampling (Dauphin et al., 2011). The idea is to reconstruct all the non-zero elements of the input and a small random subset of the zero elements, and to use importance sampling weights to exactly correct the bias thus introduced.

The learning objective is sampled in the following manner:

$$\hat{L}(\mathbf{x}, \mathbf{z}) = \sum_{k}^{d} \frac{\hat{\mathbf{p}}_k}{\mathbf{q}_k} H(\mathbf{x}_k, \mathbf{z}_k)$$

where $\hat{\mathbf{p}} \in \{0, 1\}^{d_x}$ with $\hat{\mathbf{p}} \sim P(\hat{\mathbf{p}}|\mathbf{x})$. The sampling pattern $\hat{\mathbf{p}}$ is resampled for each presentation of the input and it controls which input unit will participate in the training cost for this presentation. The bias introduced by sampling can be corrected by setting the reweighting term $1/\mathbf{q}$ such that $\mathbf{q}_k = E[\hat{\mathbf{p}}_k|k, \mathbf{x}, \tilde{\mathbf{x}}]$.

The optimal sampling probabilities $P(\hat{\mathbf{p}}|\mathbf{x})$ are those that minimize the variance of the estimator $\hat{L}$. Dauphin et al. (2011) show that reconstructing all non-zeros and a small subset of zeros is a good heuristic. The intuition is that the model is more likely to be wrong on the non-zeros than the zeros. Let $C(\mathbf{x}, \tilde{\mathbf{x}}) = \{k : \mathbf{x}_k = 1 \text{ or } \tilde{\mathbf{x}}_k = 1\}$. Then bit $k$ is reconstructed with probability

$$P(\hat{\mathbf{p}}_k = 1|\mathbf{x}_k) = \begin{cases} 1 & \text{if } k \in C(\mathbf{x}, \tilde{\mathbf{x}}) \\ |C(\mathbf{x}, \tilde{\mathbf{x}})|/d_x & \textit{otherwise} \end{cases} \tag{1}$$

Dauphin et al. (2011) show that the computational speed-up is on the order of $d_{SMP}/d_x$ where $d_{SMP}$ is the average number of units that are reconstructed and $d_x$ is the input dimension. Furthermore, reconstruction sampling yields models that converge as fast as the non-sampled networks in terms of gradient steps (but where each step is much faster).

## 2.3. Postprocessing

The competition servers use a Hebbian classifier. Specifically, the discriminant function applied to a test set matrix $Z$ (one row per example) after training the representation on a training set matrix $X$ (one row per example) is given by

$$f(Z) = ZX^T y$$

where $y_i = 1/n_p$ if training example $i$ is positive, or $-1/n_n$ if training example $i$ is negative, where $n_p$ and $n_n$ are the number of positive and negative training examples, respectively. One classifier per class (one against all) is trained.

This classifier does not have any regularization hyperparameters. We were interested in discovering whether some postprocessing of our features could result in Hebbian learning behaving as if it was regularized. It turns out that in a sense, Hebbian learning is already maximally regularized. Fisher discriminant analysis can be solved as a linear regression problem (Bishop, 2006), and the L2 regularized version of this problem yields this discriminant function:

$$g_\lambda(Z) = Z(X^T X + \lambda I)X^T y$$

where $\lambda$ is the regularization coefficient. Note that

$$\lim_{\lambda \to \infty} \frac{g_\lambda(Z)}{\|g_\lambda(Z)\|} = \frac{f(Z)}{\|f(Z)\|}.$$

Since scaling does not affect the final classification decision, Hebbian learning may be seen as maximally regularized Fisher discriminant analysis. It is possible to reduce Hebbian learning's implicit L2 regularization coefficient to some smaller $\lambda$ by multiplying $Z$ by $(X^T X + \lambda I)^{-1/2}$), but it is not possible to increase it.

Despite this implicit regularization, overfitting is still an important obstacle to good performance in this competition due to the small number of training examples used. We therefore explored other means of avoiding overfitting, such as reducing the number of features and exploring sparse codes that would result in most of the features appearing in the training set being 0. However, the best results and regularization where obtained by a transductive PCA.

### 2.3.1. TRANSDUCTIVE PCA

A Transductive PCA is a PCA transform trained not on the training set but on the test (or validation) set. After training the first $k$ layers of the pipeline on the training set, we trained a PCA on top of layer $k$, either on the validation set or on the test set (depending on whether we were submitting to the validation set or the test set). Regarding the notation used in 2.1, we apply the same transformation with $X$ replaced by the representation on top of layer $k$ of the validation set or the test set i.e $h(X_{valid})$.

This transductive PCA thus only retains variations that are the dominant ones in the test or validation set. It makes sure that the final classifier will ignore the variations present in the training set but irrelevant for the test (or validation) set. In a sense, this is a generalization of the strategy introduced in 2.1 of removing features that were not present in the both training and test / validation sets. The lower layers only keep the directions of variation that are dominant in the training set, while the top transductive PCA only keeps those that are significantly present in the validation (or test) set.

We assumed that the validation and test sets contained the same number of classes to validate the number of components on the validation set performance. In general, one needs at least $k-1$ components in order to separate $k$ classes by a set of one-against-all classifiers. Transductive PCA has been decisive for winning the competition as it improved considerably the performance on all the datasets. In some cases, we also used a mixed strategy for the intermediate layers, mixing examples from all three sets.

### 2.3.2. OTHER METHODS

After the feature extraction process, we were able to visualize the data as a three-dimensional scatter plot of the representation learnt. On some datasets, a very clear clustering pattern became visually apparent, though it appeared that several clouds came together in an ambiguous region of the latent space discovered.

In order to attempt to disambiguate this ambiguous region without making hard-threshold decisions, we fit a Gaussian mixture model with the EM algorithm and a small number of Gaussian components chosen by visual inspection of these clouds. We then used the posterior probabilities of the cluster assignments as an alternate encoding.

K-means, by contrast with Gaussian mixture models, makes a hard decision as to cluster assignments. Many researchers were recently impressed when they found out that a certain kind of feature representation (the "triangle code") based on K-means, combined with specialized pre-processing, yielded state of the art performance on the CIFAR-10 image classification benchmark (Coates et al., 2011). Therefore, we tried K-means with a large number of means and the triangle code as a post-processing step.

In the end, though, none of our selected final entries included a Gaussian mixture model or K-means, as the transductive PCA always worked better as a post-processing layer.

## 3. Criterion

The usual kind of overfitting is due to specializing to particular labeled examples. In this transfer learning context, another kind of overfitting arose: overfitting a representation to particular *classes*. Since the validation set and test set have non-intersecting sets of classes, finding representations that work well on the validation set was not a guarantee for good behavior on the test set, as we learned from our experience with the competition first phase. Note also that the competition was focused on a particular criterion, the Area under the Learning Curve (ALC)[3] which gives much weight to the

---

3. http://www.causality.inf.ethz.ch/ul_data/DatasetsUTLChallenge.pdf

cases with very few labeled examples (1, 2, or 4, per class, in particular, get almost half of the weight). So the question we investigated in the second and final phase (where some training set labels were revealed) was the following: does the ALC of a representation computed on a particular subset of classes correlate with the ALC of the same representation computed on a different set of classes?

Overfitting on a specific subset of classes can be observed by training a PCA separately on the training, validation and test sets on **ULE** (this data set corresponds to MNIST digits). The number of components maximizing the ALC will be different, depending on the choice of the subset of classes. Figure 1($a$) illustrates the effect of the number of components retained on the training, validation and test ALCs. While the best number of components on the validation set would be 2, choosing this number of components for the test set significantly degrades the test ALC.

During the first phase, we noticed the absence of correlation between the validation ALC and test ALC computed on the **ULE** dataset. During the second phase, we tried to reproduce the competition setting using the labels available for transfer with the hope of finding a criteria that would guarantee generalization. The ALC was computed on every subset of at least two classes found in the transfer labels and metrics were derived. Those metrics are illustrated in Figure 1($b$). We observed that the standard deviation seems to be inversely proportional to the generalization accuracy, therefore substracting it from the mean ALC ensures that the choice of hyper-parameters is done in a range where the training, validation and test ALCs are correlated. In the case of a PCA, optimizing the $\mu - \sigma$ criteria correctly returns the best number of PCA components, ten, where the training, validation and test ALCs are all correlated.

It appears that this criterion is a simple way to use the small amount of labels given to the competitors for the phase 2. However, this criterion has not been heavily tested during the competition since we always selected our best models with respect to the validation ALC returned by the competition servers. From the phase 1 to the phase 2, we only explored the space of the hyperparameters of our models using a finer grid.

## 4. Results

For each of the five datasets, **AVICENNA, HARRY, TERRY, SYLVESTER** and **RITA**, the strategy retained for the final winning submission on the phase 2 is precisely described. Training such a deep stack of layers from preprocessing to postprocessing takes at most 12 hours for each dataset once you have found the good hyperparameters. All our models are implemented in Theano (Bergstra et al., 2010), a Python library that allows transparent use of GPUs. During the competition, we used a cluster of GPUs, Nvidia GeForce GTX 580.

### 4.1. AVICENNA

**Nature of the data** It corresponds to arabic manuscripts and consists of $150,205$ training samples of dimension 120.

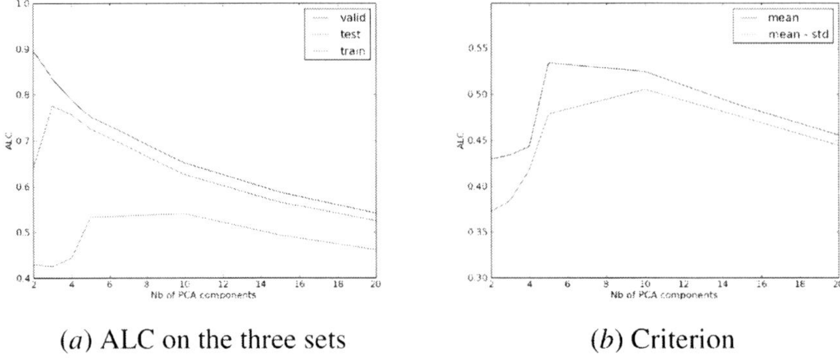

(*a*) ALC on the three sets    (*b*) Criterion

Figure 1: **ULE Dataset** Left: ALC on training, validation and test sets of PCA representations, with respect to the number of principal components retained. Right: Comparison between training, validation and test ALC and the criterion computed from the ALC, obtained on every subset of at least 2 classes present in the transfer labels for different numbers of components of a PCA.

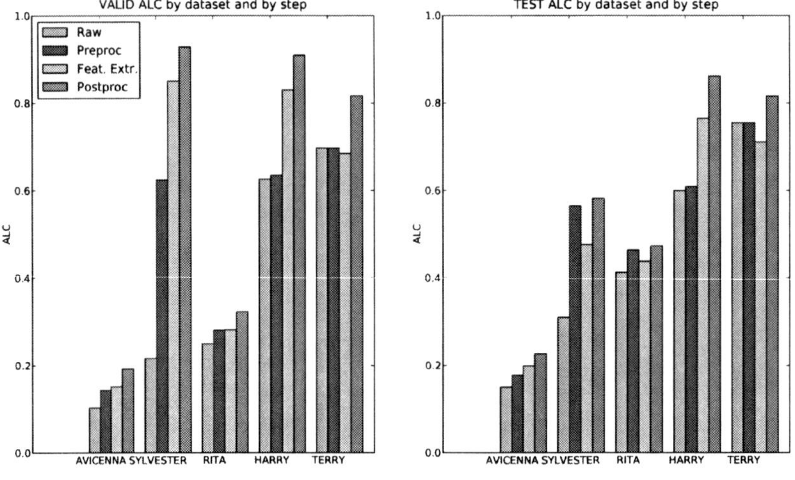

Figure 2: For each data set, we report the Validation and Test ALC after each layer (from raw data to postprocessing). It allows us to see where overfitting arose (**SYLVESTER**) and which of the layers resulted the more important to improve the overall performance.

**Best Architecture**    For preprocessing, we fitted a whitened-PCA on the raw data and kept the first 75 components in order to eliminate the noise from the input distribution. Then, the second layer consisted in a Denoising Autoencoder of 600 hidden units trained with a binomial noise, i.e, each component of the input had a probability $p = 0.3$ of being masked (set to 0). The top layer was a transductive PCA with only the 7 principal components.

**Results**   This strategy ranked first with a validation and final ALC score of 0.1932 and 0.2273 respectively.  Training a contractive auto-encoder gives similar results on the validation set i.e a validation and final ALC score of 0.1930 and 0.1973 respectively.

## 4.2. HARRY

**Nature of the data**   It corresponds to human actions and consists of $69,652$ training samples of dimension $5,000$, which are sparse:  only  2% of the components are typically non-zero.

**Best Architecture**   For the first layer, we uniformized the non-zero feature values (aggregating all features) across the concatenation of the training, validation and test sets. For the second layer, we trained on the union of the 3 sets a Denoising Auto-Encoder with rectifier units and reconstruction sampling. We used the binomial masking noise ($p = 0.5$) as corruption process, the logistic sigmoid as reconstruction activation function and the cross entropy as reconstruction error criterion. The size of the hidden layer was 5000 and we added an $L_1$ penalty on the activation values to encourage sparsity of the representation.  For the third layer, we applied a transductive PCA and kept 3 components.

**Results**   We obtained the best validation ALC score of the competition.  This was also the case for the final evaluation score with an ALC score of 0.861933, whereas the second best obtained 0.754497.  Figure 3 shows the final data representation we obtained for the test (evaluation) set.

## 4.3. TERRY

**Nature of the data**   This is a natural language processing (NLP) task, with $217,034$ training samples of dimension $41,236$, and a high level of sparsity:  only  1% of the components are non-zero in average.

**Best Architecture**   A setup similar to **HARRY** has been used for **TERRY**. For the first layer, we kept only the features that were active on both training and validation sets (and similarly with the test set, for preparing the test set representations). Then, we divided the non-zero feature values by their standard deviation across the concatenation of the training, validation and test set. For the second layer, we trained on the three sets a Denoising Auto-Encoder with rectifier units and reconstruction sampling. We used binomial masking noise ($p = 0.5$) as corruption process, the logistic sigmoid as reconstruction activation function and the squared error as reconstruction error criterion. The size of the hidden layer was 5000 and we added an $L_1$ penalty on the activation values to encourage sparsity of the representation. For the third layer, we applied a transductive PCA and kept the leading 4 components.

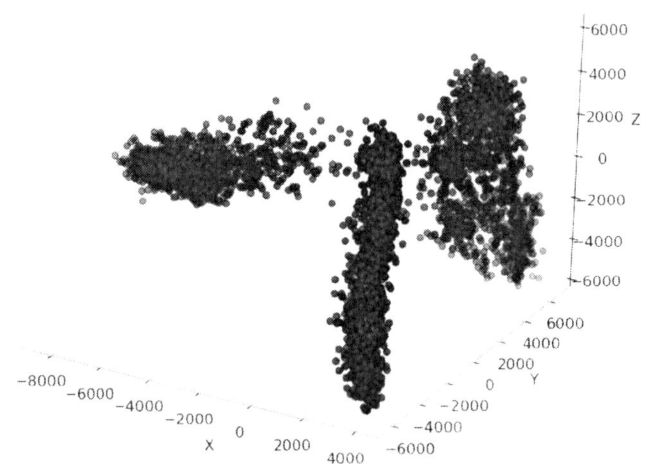

Figure 3: **HARRY** evaluation set after the transductive PCA, the data is nicely clustered, suggesting that the learned preprocessing has discovered the underlying class structure.

**Results**   We ranked second on this dataset with a validation and final score of 0.816752 and 0.816009.

## 4.4. SYLVESTER

**Nature of the data**   It corresponds to ecology data and consists of 572,820 training samples of dimension 100.

**Best Architecture**   For the first layer, we extracted the meaningful features and discarded the apparent noise dimensions using PCA. We used the first 8 principal dimensions as the feature representation produced by the layer because it gave the best performance on the validation set. We also whitened this new feature representation by dividing each dimension by its corresponding singular value (square root of the eigenvalue of the covariance matrix, or corresponding standard deviation of the component). Whitening gives each dimension equal importance both for the classifier and subsequent feature extractors. For the second and third layers, we used a Contractive Auto-Encoder (CAE). We have selected a layer size of 6 based on validation ALC. For the fourth layer, we again apply a transductive PCA.

Figure 4 shows the evolution of the ALC curve for each layer of the hierarchy. Note that at each layer, we only use the top-level features as the representation.

**Results**   This yielded an ALC of 0.85109 for the validation set and 0.476341 for the test set. The difference in ALC may be explained by the fact that Sylvester is the only

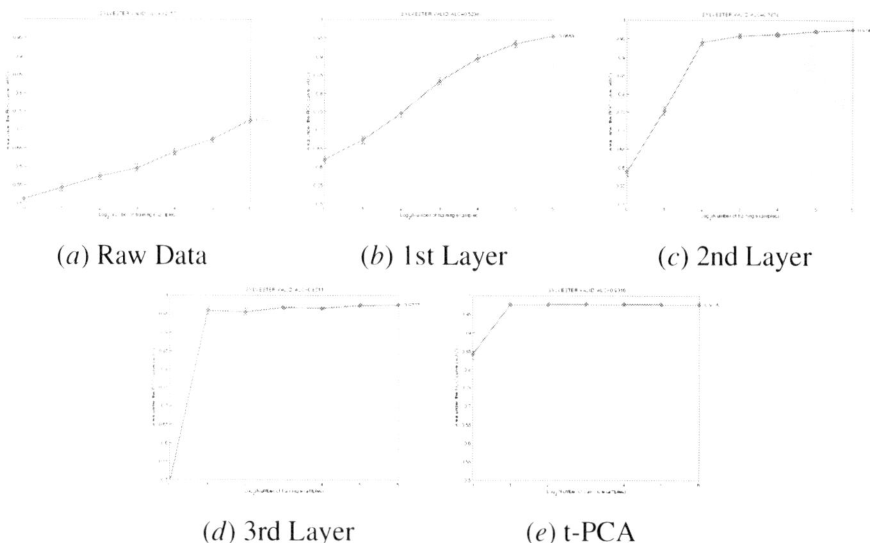

(*a*) Raw Data          (*b*) 1st Layer          (*c*) 2nd Layer

(*d*) 3rd Layer          (*e*) t-PCA

Figure 4:  Validation performance increases with the depth of our feature hierarchy for the **SYLVESTER** dataset.  ALC: Raw Data (0.2167), 1st Layer (0.6238), 2nd Layer (0.7878), 3rd Layer (0.8511), t-PCA(0.9316)

dataset where the test set contains more classes than the validation set and, and thus our assumpptions of equal number of classes might have hurt test performance here.

## 4.5. RITA

**Nature of the data**   It corresponds to the CIFAR RGB image dataset and consists of 111,808 training samples of dimension 7,200.

**Best Architecture**   The $\mu$-ssRBM was initially developed as a model for natural images.  As such, it was a natural fit for the **RITA** dataset.  Their ability to learn the pooling structure was also a clear advantage, since the max-pooling strategy typically used in vision tasks with convolutional networks LeCun et al. (1998) could no longer be employed due to the obfuscated nature of the dataset.

For pre-processing, each image has been contrast-normalized.  Then, we reduced the dimensionality of the training dataset by learning on the first 1,000 principal components. For feature extraction, we chose the number of hidden units to be large enough (1000) while still being computationally efficient on GPUs. The learning rate of $10^{-3}$ and number of training updates (110,000 updates with minibatches of size 16) are chosen such that hidden units have sparse activations through pools of size 9, hovering around 10-25%. The post-processing was consistent with the other datasets: we used the transductive PCA method using only the first 4 principal components.

**Results**   This yielded an ALC score of 0.286 and 0.437 for the validation and final test sets respectively. We also tried to stack 3 layers of contractive auto-encoders directly on the raw data and it achieved a valid ALC of 0.3268. As it appeared actually transductive, we prefered to keep the $\mu$-ssRBM in our competition entries because it was trained on the whole training set.

## 5. Conclusion

The competition setting with different class labels in the validation and the test sets was quite unusual. The similarity between two classes must be sufficient for transfer learning to be possible. More formal assessments of class similarity might be useful in such settings. It is not obvious that the similarity between the subsets of classes chosen for the different datasets in the context of the competition is sufficient for an effective generalization, neither that the similarity between the subsets of classes found in the transfer labels is representative of the similarity between the classes found in the training, validation and test datasets. Finally, for assessing transfer across classes properly would require a larger number of classes. In a future competition, we suggest that both similarity and representativeness (including number of classes) should be ensured in a more formal or empirical way.

On all five tasks, we have found the idea of stacking different layer-wise representation-learning algorithms to work very well. One surprise was the effectiveness of PCA both as a first layer and a last layer, in a transductive setting. As core feature-learning blocks, the contractive auto-encoder, the denoising auto-encoder and spike-and-slab RBM worked best for us on the dense datasets, while the sparse rectifier denoising auto-encoder worked best on the sparse datasets.

## Acknowledgements

The authors acknowledge the support of the following agencies for research funding and computing support: NSERC, RQCHP, CIFAR. This is also funded in part by the French ANR Project ASAP ANR-09-EMER-001.

## References

Yoshua Bengio. Learning deep architectures for AI. *Foundations and Trends in Machine Learning*, 2(1):1–127, 2009. Also published as a book. Now Publishers, 2009.

Yoshua Bengio. Deep learning of representations for unsupervised and transfer learning. In *Workshop on Unsupervised and Transfer Learning (ICML'11)*, June 2011.

Yoshua Bengio, Pascal Lamblin, Dan Popovici, and Hugo Larochelle. Greedy layer-wise training of deep networks. In Bernhard Schölkopf, John Platt, and Thomas Hoffman, editors, *Advances in Neural Information Processing Systems 19 (NIPS'06)*, pages 153–160. MIT Press, 2007.

James Bergstra, Olivier Breuleux, Frédéric Bastien, Pascal Lamblin, Razvan Pascanu, Guillaume Desjardins, Joseph Turian, David Warde-Farley, and Yoshua Bengio.

Theano: a CPU and GPU math expression compiler. In *Proceedings of the Python for Scientific Computing Conference (SciPy)*, June 2010. URL http://www.iro. umontreal.ca/~lisa/pointeurs/theano_scipy2010.pdf. Oral.

Christopher M. Bishop. *Pattern Recognition and Machine Learning*. Springer, 2006.

A. Coates, H. Lee, and A. Ng. An analysis of single-layer networks in unsupervised feature learning. In *Proceedings of the Thirteenth International Conference on Artificial Intelligence and Statistics (AISTATS 2011)*, 2011.

Aaron Courville, James Bergstra, and Yoshua Bengio. Unsupervised models of images by spike-and-slab RBMs. In *Proceedings of the Twenty-eight International Conference on Machine Learning (ICML'11)*, June 2011.

Yann Dauphin, Xavier Glorot, and Yoshua Bengio. Sampled reconstruction for large-scale learning of embeddings. In *Proceedings of the Twenty-eight International Conference on Machine Learning (ICML'11)*, June 2011.

Xavier Glorot, Antoire Bordes, and Yoshua Bengio. Deep sparse rectifier neural networks. In *JMLR W&CP: Proceedings of the Fourteenth International Conference on Artificial Intelligence and Statistics (AISTATS 2011)*, April 2011.

Y. LeCun, L. Bottou, Y. Bengio, and P. Haffner. Gradient based learning applied to document recognition. *IEEE*, 86(11):2278–2324, November 1998.

V. Nair and G. E Hinton. Rectified linear units improve restricted Boltzmann machines. In *Proc. 27th International Conference on Machine Learning*, 2010.

M. Ranzato and G. H. Hinton. Modeling pixel means and covariances using factorized third-order Boltzmann machines. In *Proceedings of the Computer Vision and Pattern Recognition Conference (CVPR'10)*, pages 2551–2558. IEEE Press, 2010.

Salah Rifai, Pascal Vincent, Xavier Muller, Xavier Glorot, and Yoshua Bengio. Contractive auto-encoders: Explicit invariance during feature extraction. In *Proceedings of the Twenty-eight International Conference on Machine Learning (ICML'11)*, June 2011.

Tijmen Tieleman. Training restricted Boltzmann machines using approximations to the likelihood gradient. In William W. Cohen, Andrew McCallum, and Sam T. Roweis, editors, *Proceedings of the Twenty-fifth International Conference on Machine Learning (ICML'08)*, pages 1064–1071. ACM, 2008.

Pascal Vincent, Hugo Larochelle, Yoshua Bengio, and Pierre-Antoine Manzagol. Extracting and composing robust features with denoising autoencoders. In William W. Cohen, Andrew McCallum, and Sam T. Roweis, editors, *Proceedings of the Twenty-fifth International Conference on Machine Learning (ICML'08)*, pages 1096–1103. ACM, 2008.

Pascal Vincent, Hugo Larochelle, Isabelle Lajoie, Yoshua Bengio, and Pierre-Antoine Manzagol. Stacked denoising autoencoders: Learning useful representations in a deep network with a local denoising criterion. *Journal of Machine Learning Research*, 11(3371–3408), December 2010.

# Stochastic Unsupervised Learning on Unlabeled Data

**Chuanren Liu**                                          CHUANREN.LIU@RUTGERS.EDU
*Rutgers, the State University of New Jersey, Newark, NJ 07102, USA*

**Jianjun Xie**                                              JIANJUNXIE@GMAIL.COM
*CoreLogic, 12395 First American Way, Poway, CA 92064, USA*

**Yong Ge**                                                 YONGGE@RUTGERS.EDU
**Hui Xiong**                                               HXIONG@RUTGERS.EDU
*Rutgers, the State University of New Jersey, Newark, NJ 07102, USA*

**Editor:** I. Guyon, G. Dror, V. Lemaire, G. Taylor, and D. Silver

## Abstract

In this paper, we introduce a stochastic unsupervised learning method that was used in the 2011 Unsupervised and Transfer Learning (UTL) challenge. This method is developed to preprocess the data that will be used in the subsequent classification problems. Specifically, it performs $K$-means clustering on principal components instead of raw data to remove the impact of noisy/irrelevant/less-relevant features and improve the robustness of the results. To alleviate the overfitting problem, we also utilize a stochastic process to combine multiple clustering assignments on each data point. Finally, promising results were observed on all the test data sets. Indeed, this proposed method won us the second place in the overall performance of the challenge.
**Keywords:** Stochastic Unsupervised Learning, Clustering, $K$-means, Principal Component Analysis (PCA)

## 1. Introduction

Data preprocessing is usually critical for the success of building classification models. There are many unsupervised learning techniques which can be exploited for data preprocessing in a complementary way. First, clustering techniques target on dividing data objects into different groups such that the objects in the same cluster are more similar to one another than to those from different clusters. Clustering techniques are widely used for summarizing data objects and capturing key data characteristics (Jain and Dubes, 1988). Among various clustering algorithms, $K$-means clustering has been identified as one of the top 10 algorithms in data mining by the IEEE International Conference on Data Mining (ICDM) in December 2006 (Wu et al., 2008).

Also, principal Component Analysis (PCA) (Jolliffe, 2002) is an effective technique for dimension reduction and feature preprocessing. It transforms the data into a new coordinate system such that the greatest variance is achieved by projecting the data into the first coordinate (called the first principal component), the second greatest variance achieved into the second coordinate, and so on. Many researchers combined the $K$-

means and PCA together to achieve more stable results (Ben-Hur and Guyon, 2003). It has been shown that the principal components are the continuous solutions to the discrete cluster membership indicators for $K$-means clustering (Ding and He, 2004).

The 2011 Unsupervised and Transfer Learning Challenge (Guyon et al., 2011) provided a platform for participants to learn good data representations through data preprocessing that can be re-used across tasks by building models that capture regularities of the input space. The representations are evaluated by the organizers on supervised learning target tasks which are unknown to the participants. In the first phase of the challenge, the competitors are given only unlabeled data to learn their data representation. In the second phase of the challenge, the competitors have available, in addition to unlabeled data, a limited amount of labeled data from source tasks distinct from the target tasks.

In this paper, we present the method that we used in the unsupervised learning challenge (first phase). By exploiting the advantages of both PCA and cluster ensemble techniques, we propose a stochastic unsupervised learning method for data processing. This unsupervised learning method is developed to preprocess the data that will be used in the subsequent binary classification problems. There are two challenging issues for the proposed task. First, there is no labeled data in support of this data preprocessing. Without ground truth, it is difficult to identify noisy or irrelevant features. Second, unsupervised learning methods like $K$-means start by randomly choosing initial cluster seeds. The results obtained in this way are not only dependent on the chosen seeds, but can also be locally optimal. For the first issue, we use $K$-means to cluster data represented only with the first $P$ principal components by PCA. In this way, it is expected to remove the negative impact of noisy/irrelevant/less-relevant features. For the second issue, we apply a stochastic strategy to combine clustering results of multiple runs of $K$-means with random initialization. An ensemble of cluster labels is produced for each data point, which is expected to help alleviate the problem of robustness, clustering quality, and overfitting. This stochastic clustering process has been explored in the semi-supervised learning problems (Xie and Xiong, 2011).

The effectiveness of the data representation obtained by unsupervised learning is evaluated by the organizers on supervised learning tasks (i.e. using labeled data not available to the participants) using Hebbian classifier. Specifically, with the training data matrix $\mathbf{X}$ (one row per instance), the classifier computes the wieght $w$ as $\mathbf{X}^T y$, where $y = (y_1, y_2, \cdots, y_n)^T$, $y_i = 1/n_p$ if the $i$th training instance is positive, $y_i = -1/n_n$ otherwise, where $n_p, n_n$ are the number of positive and negative training examples respectively. The test instance $x$ (column vector) will be classified according to the linear discriminant $w^T x$. It is noted that the size of training data $\mathbf{X}$ is very small (no more than 64 per classification problem) in this challenge. The model performance is reported with the metric of Area under the Learning Curve (ALC) which is referred to as the global score. The participants are ranked by ALC for each individual data set. The winner is determined by the best average rank over all data sets for the results of their last complete experiment. We will see the proposed method is effective especially in such a small training set scenario.

Such a linear discriminant classifier assumes the instances lying in the feature space are linearly separable. However, it is not necessarily true in many real-world data sets. For example, Figure 1 illustrates a situation of a mixture model, where the positive instances indicated by the plus marks are surrounded by 3 groups of negative instances indicated by the circles. Noises are indicated by green dots. With clustering algorithms, we can cluster the data set into 4 groups, whose representation becomes linearly separable by the Hebbian classifier. The above can be reflected in the experimental results.

**Overview.** The remainder of this paper is organized as follows. In Section 2, we describe the stochastic unsupervised learning method based on $K$-means and PCA. Section 3 shows the results. In Section 4, we discuss the limitations of the proposed approaches and describe the potential directions for future work.

## 2. The Stochastic Clustering Algorithm

### 2.1. The Algorithm

Algorithm 1 details the common strategy we used for all 5 data sets in the challenge. The output is the final data representation, which is a binary representation of derived cluster labels. If $K$ is 3 for a given data set, the binary representations of label 1, 2 and 3 are (1 0 0), (0 1 0) and (0 0 1), respectively. Therefore, our final data representation will be a bagged $N \times KT$ matrix, where $N$ is the number of examples, $K$ is the number of clusters and $T$ is the number of stochastic iterations. Each data element in the matrix is either 1 or 0. Such a binary representation is chosen to eliminate the numeric meaning of clustering labels which is misleading for the Hebbian classifier.

It is noted that the data set $\mathbf{X}$ is not the raw data set. It is the first $P$ principal components of the raw data set. In the challenge, we used different $P$ and necessary variants of naive PCA for each data set based on online feedback from the validation set. For example, instead of analyzing the principal components of covariance matrix for all features, we also tried to decomposing the correlation matrix, which implies dividing by standard deviation prior to computing the covariances. The transformation of the raw data to retrieved principal components also can be followed by additional processing strategies, such as standardization (to subtract mean and divide deviation for each feature) and weighting (to weight each component by its corresponding eigenvalue).

For the clustering algorithm $K$-means, the number of clusters was also determined based on online feedback during this challenge. For nearly every data set, we found the real number of classes to be predicted. The only exception is SYLVESTER, where the real numbers of classes in the validation set and the final set are 2 and 3 respectively, and we used 3 as the number of clusters. In addition to $K$, we also used different distance/similarity metrics for different data sets. Basically, for low dimensional $\mathbf{X}$, Euclidean distance is used. Otherwise in the high dimensional case, cosine similarity is preferred. Cosine performs an implicit instance-level standardization, i.e., the instance vector is normalized to be of unit length. We found feature-level standardization could also improve the clustering results, such as HARRY. For TERRY, which is in text

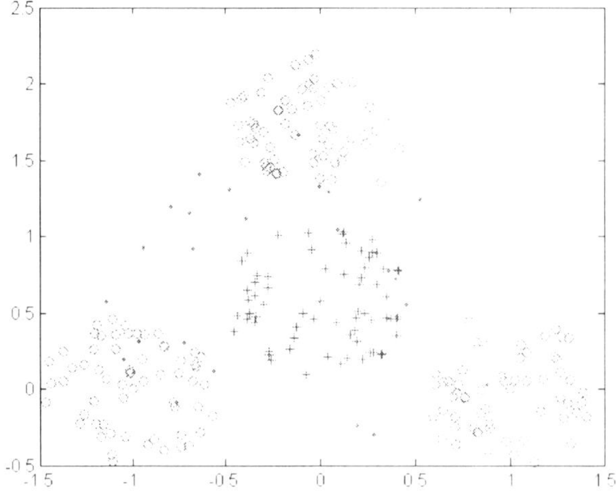

Figure 1: An example of a mixture model.

recognition domain, the well-known TF-IDF transformation is used prior to computing cosine similarities.

Another parameter in Algorithm 1 is the number of stochastic iterations. The motivation of the stochastic process is to settle the overfitting phenomenon. Although the binary matrix generated from only one clustering solution can be directly fed to the classifier, the final classification result will vary a lot with the clustering solution. By combining several different clustering solutions, we found the final classification result could be improved better than that based on any single clustering solution. Ideally, the final result will converge along with the increasing number of stochastic iterations. When the result becomes stable on the validation set, we believe the overfitting problem on the final set has also been circumvented. We analyzed the results of $20, 40, 60, 80$ and $100$ iterations. For most of the data sets, stable results are observed on the validation set after 60 iterations. Thus, in our submission to the challenge, we set the number of stochastic iterations as 100. Details of these variations on each data set will be described in Section 3.

## 2.2. Cluster Assumption

In fact, the proposed algorithm maps the data from the original data space to the space discovered by the underlying clustering algorithm, in the hope that the class does not change in regions of high density within clusters. Such a cluster assumption can be explained using cluster kernels. Specifically, with the achieved clustering solution

$$l : z \mapsto l(z) \in \{1, 2, \cdots, K\}$$

---

**Algorithm 1** The Stochastic Clustering Algorithm

---

**Input** Data set $\mathbf{X}$, the number of clusters $K$, the number of stochastic iterations $T$

**Output** Data set $\mathbf{Y}$

1. For $t = 1, 2, \cdots, T$

    (a) Randomly choose $K$ seeds from $\mathbf{X}$ for $K$-means to generate clusters. Denote the clustering assignments by $\mathbf{I}$.

    (b) Transform $\mathbf{I}$ to binary format, i.e. for each assignment

    $$i \mapsto \boldsymbol{e}_i = (e_{i1}, e_{i2}, \cdots, e_{iK})$$

    where $e_{ij} = \begin{cases} 1 & i = j \\ 0 & i \neq j \end{cases}$. Denote the binary matrix by

    $$\mathbf{B_t} = \begin{pmatrix} \boldsymbol{e}_{I_1} \\ \boldsymbol{e}_{I_2} \\ \vdots \\ \boldsymbol{e}_{I_N} \end{pmatrix}$$

    where $I_n$ is the assignment of the $n$th instance.

2. Combine $\mathbf{B_t}, t = 1, 2, \cdots, T$ together as $\mathbf{Y} = (\mathbf{B_1}|\mathbf{B_2}|\cdots|\mathbf{B_T})$.

---

where $l(z)$ is the clustering assignment for any clustered instance $z$, the mapping function is

$$\phi(z) = ([l(z) = 1], [l(z) = 2], \cdots, [l(z) = K])^T,$$

and the inner product kernel

$$\phi(x)^T \phi(z) = [l(x) = l(z)]$$

will be used by Hebbian classifier. By combining multiple clustering solutions together, the inner product of $x$ and $z$ in the mapped space is $\sum_{t=1}^{T} [l_t(x) = l_t(z)]$, where $l_t$ is the $t$th clustering solution. Such a combination is also used in the study of consensus clustering (Hu and Sung, 2005).

## 2.3. An Illustrative Example

To illustrate the effectiveness and rationale of the proposed algorithm, especially of the clustering component, we analyzed the results of 100 runs of $K$-means for the toy data set ULE whose true labels are available. For each clustering solution, in addition to mean squared error (MSE) as the clustering criterion function, we also computed

ALC and purity (the fraction of correctly classified data when all data in each cluster is classified as the majority class in that cluster). As shown by the representative solutions in Table 1, one can see that better classification results really come along with better clustering solutions. The best ALC value of 0.83764 is achieved with the lowest MSE value of 0.9102239 and the highest purity value of 0.93872. More interestingly, by combining all 100 clustering solutions, we can achieve an ALC value of 0.86642, which is significantly better than that of the best single solution. Such an ensemble effect is the key motivation of the proposed algorithm.

## 3. Results

In this section, we provide an empirical study of the proposed stochastic unsupervised learning method. In most of the data sets studied, the proposed method achieves better performances than that of the raw data and PCA.

### 3.1. AVICENNA: Arabic manuscripts

The results on the AVICENNA data set are shown in Table 2. It seems difficult to get good results on either the validation set or the final set, for the best global score on the leader board turns out less than 0.2 for the validation set. This is the only data set in our experiments where the PCA itself has better global scores on the final data set than $K$-means. We believe this is due to the label overlaps in this data set; that is, one example can belong to multiple classes.

The learning curves of the three scenarios (raw data, PCA and $K$-Means) are shown in Figure 2. The PCs are standardized for this data set such that each feature has zero mean and unit variance. We did notice that the $K$-means underperforms PCA during the first phase of the challenge through the on-line feedback on the validation set. Therefore, we chose the PCA results as our final experiment. However, we did some improvements on $K$-means during the second phase. We found that if we first did a record level normalization on each variables (this is equivalent to a Term-Frequency transformation in document classification), then did PCA on the normalized variables

Table 1: A Comparison of MSE, purity and ALC.

| MSE | Purity | ALC |
|---|---|---|
| 0.9102239 | 0.93872 | 0.83764 |
| 0.9782505 | 0.65332 | 0.51611 |
| 0.9959307 | 0.64429 | 0.51231 |
| 1.0049960 | 0.63550 | 0.48539 |
| 1.0049971 | 0.63550 | 0.48545 |
| 1.0050026 | 0.63501 | 0.48563 |
| 1.0050027 | 0.63452 | 0.48586 |
| 1.0050034 | 0.63599 | 0.48262 |

Table 2: The Results on AVICENNA.

|  | Validation | Final | Algorithm details |
|---|---|---|---|
| Raw Data | 0.1034 | 0.1501 | Original data |
| PCA | 0.1386 | 0.1906 | First 50 standardized PCs from Covariance Matrix and First 50 standardized PCs from Correlation Matrix |
| K-Means | 0.1668 | 0.1511 | Stochastic $K$-means on first 100 standardized PCs. Cluster number = 5. |

and stochastic $K$-means on the first 100 PCs, we could lift the global score on the validation set from 0.1386 to 0.1668. Unfortunately, this improvement on the validation set did not hold on the final set. Our $K$-means results on the final set actually dropped to 0.1511 from 0.1906.

## 3.2. HARRY: Human action recognition

Table 3 lists our experimental results on both PCA and $K$-means on the HARRY data set. In the table, we can observe that PCA works very well on the validation set. The first 5 weighted PCs (weighted by the corresponding eigenvalue of each principal component) can achieve a 0.8056 global score in the validation set. For this data, which is high dimensional and very sparse, the stochastic $K$-means on standardized data works better than that on PCs. We used the Cosine similarity as the distance measure in $K$-means clustering. The number of clusters is set to 3. In Table 3, we can see that, while PCA works pretty well, the stochastic $K$-means without PCA works much better. Such a phenomenon was also observed in TERRY, which is also high dimensional and very sparse.

The learning curves on both the validation and the final sets are illustrated in Figure 3. We can see that the improvements on the validation set do hold well on the final set.

## 3.3. RITA: Object recognition

RITA is another difficult data set in addition to AVICENNA. Our experimental results on both PCA and $K$-means on the RITA data set are shown in Table 4. The first 50 principal components achieve ALC = 0.2834, 0.4622 on the validation set and the final set, respectively. The stochastic $K$-means gives the best results. We find that Euclidian

Table 3: The Results on HARRY.

|  | Validation | Final | Algorithm details |
|---|---|---|---|
| Raw Data | 0.6264 | 0.6017 | Original data |
| PCA | 0.8056 | 0.6243 | First 5 weighted PCs from correlation matrix |
| K-Means | 0.9085 | 0.7357 | Stochastic $K$-means on standardized data. Cluster number = 3. |

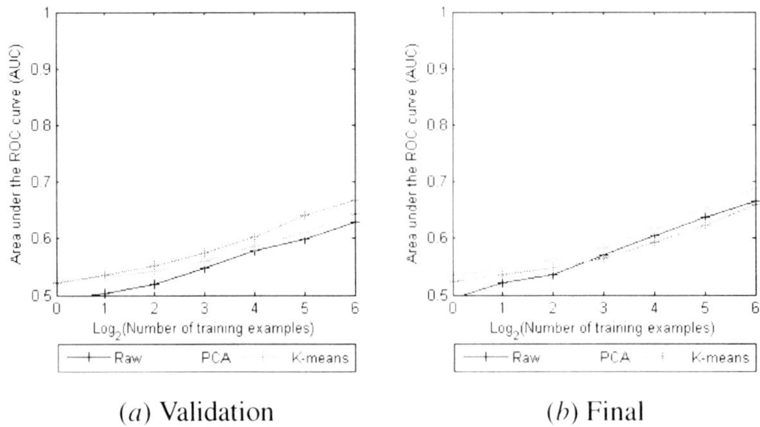

(*a*) Validation    (*b*) Final

Figure 2: The learning curve on AVICENNA

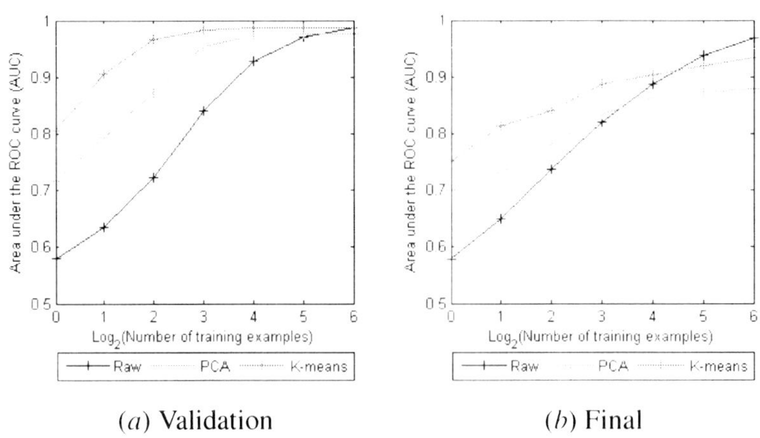

(*a*) Validation    (*b*) Final

Figure 3: The Learning curves on HARRY

Table 4: The results on RITA.

|          | Validation | Final  | Algorithm details                                           |
|----------|-----------|--------|-------------------------------------------------------------|
| Raw Data | 0.2504    | 0.4133 | Original data                                               |
| PCA      | 0.2834    | 0.4622 | First 50 PCs from covariance matrix                         |
| K-Means  | 0.3737    | 0.4782 | Stochastic $K$-means on standardized 50 PCs. Cluster number = 3. |

distance is better than the Cosine similarity in this case. The number of clusters is set to 3. The learning curves on both validation and final sets are illustrated in Figure 4. We can see that the improvements on the global score from PCA and $K$-means over the raw data mainly come from the beginning of the learning curve, which may correspond to small number of training samples.

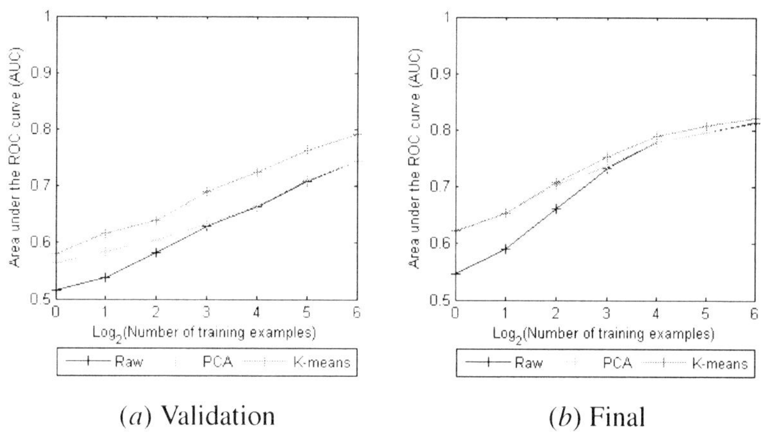

(*a*) Validation          (*b*) Final

Figure 4: The learning curve on RITA

### 3.4. SYLVESTER: Ecology

Table 5 lists our experimental results on both PCA and $K$-means on the SYLVESTER data set. SYLVESTER has only 100 features and is not sparse. In the table, we can see that the first 7 principal components can do much better than the original data.

Table 5: The results on SYLVESTER.

|          | Validation | Final  | Algorithm details                                           |
|----------|-----------|--------|-------------------------------------------------------------|
| Raw Data | 0.2167    | 0.3095 | Original data                                               |
| PCA      | 0.5873    | 0.4436 | First 7 standardized PCs from correlation matrix            |
| K-Means  | 0.7146    | 0.5828 | Stochastic $K$-means on standardized 15 PCs. Cluster number = 3. |

The stochastic $K$-means using $K = 3$ further improves the PCA results from 0.4436 to 0.5828 on the final set. Indeed, our result on the final set was ranked No. 1 in the first phase of the challenge.

Also, Figure 5 shows the learning curves on both the validation set and the final set. In the figure, a similar trend of performances can be observed as in Table 5.

### 3.5. TERRY: Text recognition

Table 6: The results on TERRY.

|          | Validation | Final  | Algorithm details |
|----------|-----------|--------|-------------------|
| Raw Data | 0.6969    | 0.7550 | Original data |
| PCA      | 0.7949    | 0.8317 | First 5 PCs from covariance matrix |
| K-Means  | 0.8176    | 0.8437 | Stochastic $K$-means on TF-IDF data. Cluster number = 5. |

Table 6 lists our experimental results on both PCA and $K$-means on the TERRY data set. This is another high dimensional and very sparse data set similar to HARRY. We find the PCA can generate much better results than the original data like those of HARRY. The first 5 principal components can achieve a global score of 0.8317 on the final set.

However, standardization does not help the clustering anymore. The TERRY data set is from the text recognition domain, where the TF-IDF weight (term frequency-inverse document frequency) is often used for information retrieval and text mining. This weight is a statistical measure used to evaluate how important a word is to a document in a collection or corpus. Thus, instead of standardization, we first did TF-IDF transformation on the raw data, then did stochastic $K$-means using $K = 5$ based on the Cosine similarity.

The learning curves on the validation and final sets are illustrated in Figure 6. We can see that the improvements on the validation set hold well on the final set. The greatest lift on the learning curve over raw data happens in the middle range of the $x$-axis. Our results on the final set is ranked No. 2 in the challenge.

## 4. Discussion

In this section, we first show a comparison of the proposed method with the overall winner. Then, we conclude this study by discussing its limitations and potential extensions.

First, by combining PCA and $K$-means, the proposed stochastic unsupervised learning method achieves the stable results on most of the data sets. Table 7 lists our global scores on all 5 data sets against those of the overall winner. We are ranked 1st on SYLVESTER, 2nd on TERRY and 3rd on HARRY. Although the results on AVI-CENNA and RITA are not impressive in rank compared to others, they are within about

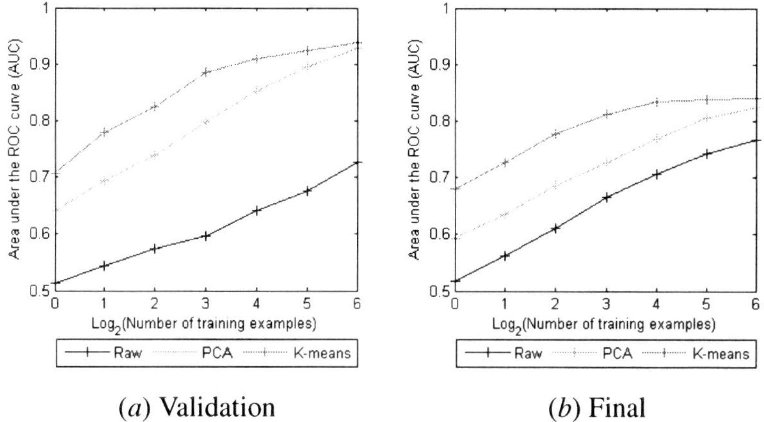

(*a*) Validation        (*b*) Final

Figure 5: The learning curve on SYLVESTER.

Table 7: A Comparison of our results with the overall winner's results. The winner is determined by the average rank on all 5 final data sets. Our results are ranked No. 2.

| Data | Winner-Valid | Winner-Final | Winner-Rank | Our-Valid | Our-Final | Our-Rank |
|---|---|---|---|---|---|---|
| AVICENNA | 0.1744 | 0.2183 | 1 | 0.1386 | 0.1906 | 6 |
| HARRY | 0.8640 | 0.7043 | 6 | 0.9085 | 0.7357 | 3 |
| RITA | 0.3095 | 0.4951 | 1 | 0.3737 | 0.4782 | 5 |
| SYLVESTER | 0.6409 | 0.4569 | 6 | 0.7146 | 0.5828 | 1 |
| TERRY | 0.8195 | 0.8465 | 1 | 0.8176 | 0.8437 | 2 |

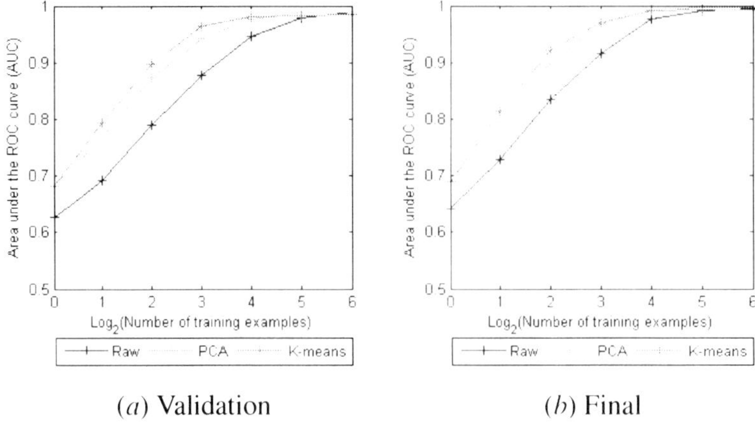

(*a*) Validation        (*b*) Final

Figure 6: The learning curve on TERRY.

0.02 in ALC from the winner's results. Our overall performance in rank was placed 2nd in the challenge.

Indeed, the performance of the proposed method significantly depends on its component: $K$-means clustering. Although we employed a stochastic strategy of cluster ensemble, the inherent characteristics of $K$-means still have impact on the final results. For instance, since $K$-means tends to favor globular clusters with similar sizes (Xiong et al., 2006, 2009), it cannot handle some of the data sets in the challenge that have different shapes or sizes of clusters. Also, $K$-means is very sensitive to data density. In the case that data have various densities, some density based clustering algorithms, such as DBSCAN (Ester et al., 1996), could be used in the proposed method. Moreover, some fuzzy clustering methods could be used to handle the data with overlapping labels (Nock and Nielsen, 2006), such as AVICENNA. Finally, when the labels of the data sets are available, we can explore the relationship between the quality of the clustering results and the accuracy of the final classification results. Such information may help to make informed decision in both generation and combination phases of cluster ensemble.

## Acknowledgments

First, we would like to thank the challenge organizers for setting up an excellent platform and providing real-world data for this challenge. This research was supported in part by National Science Foundation (NSF) via grant number CCF-1018151.

## References

A. Ben-Hur and I. Guyon. Detecting stable clusters using principal component analysis. *Methods in Molecular Biology*, 224:159–182, 2003.

C. Ding and X. He. K-means clustering via principal component analysis. In *Proceedings of the twenty-first international conference on Machine learning*, page 29. ACM, 2004.

M. Ester, H.P. Kriegel, J. Sander, and X. Xu. A density-based algorithm for discovering clusters in large spatial databases with noise. In *Proceedings of the 2nd International Conference on Knowledge Discovery and Data mining*, volume 1996, pages 226–231. Portland: AAAI Press, 1996.

I. Guyon, G. Dror, V. Lemaire, G. Taylor, and D. W. Aha. Unsupervised and transfer learning challenge. In *Proc. IJCNN*, 2011.

T. Hu and S.Y. Sung. Consensus clustering. *Intelligent Data Analysis*, 9(6):551–565, 2005.

A.K. Jain and R.C. Dubes. *Algorithms for clustering data*. Prentice-Hall, Inc., 1988.

I. Jolliffe. Principal component analysis. *Encyclopedia of Statistics in Behavioral Science*, 2002.

S. Lloyd. Least squares quantization in pcm. *IEEE Transactions on Information Theory*, 28(2):129–137, 1982.

R. Nock and F. Nielsen. On weighting clustering. *IEEE transactions on pattern analysis and machine intelligence*, pages 1223–1235, 2006.

X. Wu, V. Kumar, J. Ross Quinlan, J. Ghosh, Q. Yang, H. Motoda, G.J. McLachlan, A. Ng, B. Liu, P.S. Yu, et al. Top 10 algorithms in data mining. *Knowledge and Information Systems*, 14(1):1–37, 2008.

Jianjun Xie and Tao Xiong. Stochastic semi-supervised learning on partially labeled imbalanced data. In *Proc. AISTATS Workshop on Active Learning and Experimental Design*, pages 85–98, 2011.

H. Xiong, J. Wu, and J. Chen. K-means clustering versus validation measures: a data distribution perspective. In *Proceedings of the 12th ACM SIGKDD international conference on Knowledge discovery and data mining*, pages 779–784. ACM, 2006.

H. Xiong, J. Wu, and J. Chen. K-means clustering versus validation measures: a data-distribution perspective. *IEEE Transactions on Systems, Man, and Cybernetics, Part B: Cybernetics*, 39(2):318–331, 2009.

# Transfer Learning with Cluster Ensembles

**Ayan Acharya**[1]    MASTERAYAN@GMAIL.COM
**Eduardo R. Hruschka**[1,2]    ERH@ICMC.USP.BR
**Joydeep Ghosh**[1]    GHOSH@ECE.UTEXAS.EDU
**Sreangsu Acharyya**[1]    SREANGSU@GMAIL.COM

[1] *University of Texas (UT) at Austin, USA*
[2] *University of Sao Paulo (USP) at Sao Carlos, Brazil*


**Editor:** I. Guyon, G. Dror, V. Lemaire, G. Taylor, and D. Silver

## Abstract

Traditional supervised learning algorithms typically assume that the training data and test data come from a common underlying distribution. Therefore, they are challenged by the mismatch between training and test distributions encountered in transfer learning situations. The problem is further exacerbated when the test data actually comes from a different domain and contains no labeled example. This paper describes an optimization framework that takes as input one or more classifiers learned on the source domain as well as the results of a cluster ensemble operating solely on the target domain, and yields a consensus labeling of the data in the target domain. This framework is fairly general in that it admits a wide range of loss functions and classification/clustering methods. Empirical results on both text and hyperspectral data indicate that the proposed method can yield superior classification results compared to applying certain other transductive and transfer learning techniques or naïvely applying the classifier (ensemble) learnt on the source domain to the target domain.

**Keywords:** Transfer Learning, Ensembles, Classification, Clustering.

## 1. Introduction

Transfer learning emphasizes the transfer of knowledge across related domains and tasks (Silver and Bennett, 2008). and distributions that are similar but not the same. This contribution deals with learning scenarios where training and test distributions are different, as they represent (potentially) related but not identical tasks. In addition it is also assumed that the training and test domains involve the same set of class labels, which are *only available from the training domain*. There are certain application domains such as the problem of land-cover classification of spatially separated regions studied in this paper, where the setting of this paper is appropriate.

The literature on transfer learning is fairly rich and varied (*e.g.*, see Pan and Yang (2010); Silver and Bennett (2008) and references therein), with much work done in the past 15 years (Thrun and Pratt, 1997). The tasks may be learnt simultaneously (Caruana, 1997) or sequentially (Bollacker and Ghosh, 2000). Typically these methods assume that if the target problem involves classification, then at least some labeled examples are available for the target task, which is not the case here. To address this added challenge, we leverage the theory of both classifier and cluster ensembles, which is a new aspect, though there is a recent paper that uses a single clustering to modify the weights of base classifiers in an ensemble in order to provide some transfer learning capability (Gao et al., 2008).

Recently we formulated an optimization based approach called $\mathbf{C^3E}$ (Consensus between Classification and Clustering Ensembles) (Acharya et al., 2011) — which can be used to aid weak classifiers via additional clustering results. This work was aimed at situations where the weakness is caused by lack of training data. In this paper, we provide a reformulation of $\mathbf{C^3E}$ for transfer learning settings, and then demonstrate its effectiveness via empirical studies.

**Notation**. Vectors and matrices are denoted by bold faced lowercase and capital letters, respectively. Scalar variables are written in italic font. A set is denoted by a calligraphic uppercase letter. The effective domain of a function $f(y)$, i.e., the set of all $y$ such that $f(y) < +\infty$ is denoted by $dom(f)$, while the interior and the relative interior of a set $\mathcal{Y}$ are denoted by $int(\mathcal{Y})$ and $ri(\mathcal{Y})$, respectively. Also, for $\mathbf{y}_i, \mathbf{y}_j \in \mathbb{R}^k$, $\langle \mathbf{y}_i, \mathbf{y}_j \rangle$ denotes their inner product.

## 2.  Description of $\mathbf{C^3E}$ for Transfer Learning

The overall framework of $\mathbf{C^3E}$, depicted in Fig.1, employs one or more classifiers learnt on the source domain and one or more "clusterers" (clustering algorithms) applied to the target domain. So without lack of generality we can assume the presence of both a classifier ensemble as well as a cluster ensemble. Suppose an ensemble of classifiers has been previously induced from the source domain. The target domain is represented by a separate set $X = \{\mathbf{x}_i\}_{i=1}^n$, that has not been used to build the ensemble of classifiers and *does not contain any labeled information*.

The ensemble of classifiers is first employed to estimate the initial class probabilities for every instance $\mathbf{x}_i \in X$. These probability distributions are stored as a set of vectors $\{\boldsymbol{\pi}_i\}_{i=1}^n$. The objective of our approach is to improve upon these estimated class probability assignments with the help of a cluster ensemble applied to the target domain. From this point of view, the cluster ensemble provides supplementary constraints for classifying the instances of $X$, with the rationale that similar instances are more likely to share the same class label. Each of the $\boldsymbol{\pi}_i$'s is of dimension $k$ so that, in total, there are $k$ classes denoted by $C = \{C_\ell\}_{\ell=1}^k$. In order to capture the similarities between the instances of $X$, $\mathbf{C^3E}$ also takes as input a similarity (co-association) matrix $\mathbf{S}$. Each entry of this matrix corresponds to the relative co-occurrence of two instances in the same cluster (Strehl and Ghosh, 2002) — considering all the data partitions that form the cluster ensemble induced from $X$. Note that $\mathbf{C^3E}$ can also receive as input a *proximity matrix* obtained from computing pair-wise similarities between instances and a *cophenetic matrix* resulting from running a hierarchical clustering algorithm. To summarize, $\mathbf{C^3E}$ receives as inputs a set of vectors $\{\boldsymbol{\pi}_i\}_{i=1}^n$ and the similarity matrix $\mathbf{S}$. After processing these inputs, $\mathbf{C^3E}$ outputs a consolidated classification — represented by a set of vectors $\{\mathbf{y}_i\}_{i=1}^{n}$, where $\mathbf{y}_i = \hat{P}(C \mid \mathbf{x}_i)$ — for every instance in $\mathcal{X}$. This procedure is described in more detail in the sequel.

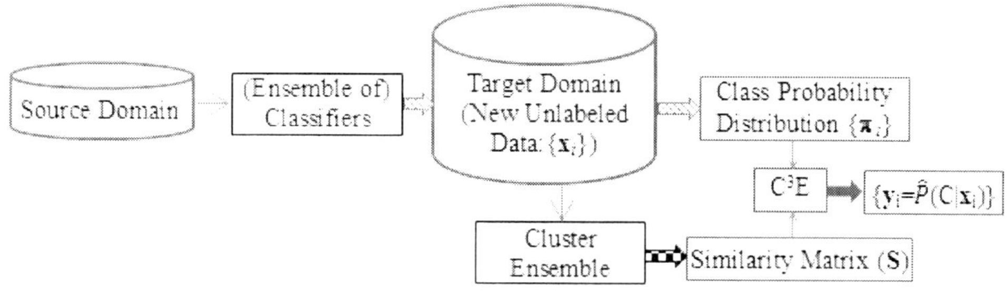


Figure 1: Overview of $\mathbf{C^3E}$ for Transfer Learning.

## 2.1. $\mathbf{C^3E}$ Algorithm

Consider that $r_1$ classifiers, indexed by $q_1$, and $r_2$ clusterers, indexed by $q_2$, are employed to obtain a consolidated classification. The following steps (A-C) outline the proposed approach. Steps A and B can be seen as preliminary steps to get the inputs for $\mathbf{C^3E}$, and Step C is, in fact, the $\mathbf{C^3E}$ Algorithm.

**Step A - Obtain input from classifier ensemble.** The output of classifier $q_1$ for instance $\mathbf{x}_i$ is a $k$-dimensional class probability vector $\pi_i^{(q_1)}$. From the set of such vectors $\{\pi_i^{(q_1)}\}_{q_1=1}^{r_1}$, an average vector can be computed for $\mathbf{x}_i$ as:

$$\pi_i = \frac{1}{r_1} \sum_{q_1=1}^{r_1} \pi_i^{(q_1)}. \tag{1}$$

**Step B - Obtain input from cluster ensemble.** After applying $r_2$ clustering algorithms (clusterers) to $\mathcal{X}$, a similarity (co-association) matrix $\mathbf{S}$ is computed. Assuming each clustering is a hard data partition, the similarity between two instances $\mathbf{x}_i$ and $\mathbf{x}_j$ is simply given by $s_{ij} = r_{ij}/r_2$, where $r_{ij}$ is the number of clustering solutions in which these two instances lie in the same cluster and $r_2$ is the number of clustering solutions[1]. Note that such similarity matrices are byproducts of several cluster ensemble solutions.

**Step C - Obtain consolidated results from $\mathbf{C^3E}$.** Having defined the inputs for $\mathbf{C^3E}$, the problem of combining ensembles of classifiers and clusterers can be posed as an optimization problem whose objective is to minimize $J$ in (2) w.r.t. the set of probability vectors $\{\mathbf{y}_i\}_{i=1}^{n}$, where $\mathbf{y}_i = \hat{P}(C \mid \mathbf{x}_i)$, i.e., $\mathbf{y}_i$ is the new and hopefully improved estimate of the aposteriori class probability distribution for a given instance in $\mathcal{X}$.

$$J = \sum_{i \in \mathcal{X}} \mathcal{L}(\pi_i, \mathbf{y}_i) + \alpha \sum_{(i,j) \in \mathcal{X}} s_{ij} \mathcal{L}(\mathbf{y}_i, \mathbf{y}_j) \tag{2}$$

---

1. A similarity matrix can also be defined for soft clusterings — e.g., see (Punera and Ghosh, 2008).

The quantity $\mathcal{L}(\cdot,\cdot)$ denotes a (non-negative) loss function. Informally, the first term in Eq. (2) captures dissimilarities between the class probabilities provided by the ensemble of classifiers and the output vectors $\{\mathbf{y}_i\}_{i=1}^n$. This term tries to drive the $\mathbf{y}_i$'s towards $\pi_i$'s. The second term encodes the cumulative weighted dissimilarity between all possible pairs $(\mathbf{y}_i, \mathbf{y}_j)$. The weights to these pairs are assigned in proportion to the similarity values $s_{ij} \in [0,1]$ of matrix $\mathbf{S}$. Intuitively, if the objective function $J$, given in Eq. 2, is minimized over $\{\mathbf{y}_i\}_{i=1}^n$ and $s_{ij}$ is high for a pair of instances $(\mathbf{x}_i, \mathbf{x}_j)$, then $\mathcal{L}(\mathbf{y}_i, \mathbf{y}_j)$ tends to go down, implying that $\mathbf{y}_i$ and $\mathbf{y}_j$ are more in agreement with each other. The coefficient $\alpha \in \mathbb{R}_+$ controls the relative importance of classifier and cluster ensembles. Therefore, minimizing the objective function over $\{\mathbf{y}_i\}_{i=1}^n$ involves combining the evidence provided by the ensembles in order to build a more consolidated classification. Note that the final clustering, and consequently the similarity matrix computed from cluster ensemble, is pretty robust compared to individual clustering results as it has been empirically shown in (Strehl and Ghosh, 2002).

The approach taken in this paper is quite general in that any Bregman divergence (Banerjee et al., 2005) can be used as the loss function $\mathcal{L}(\cdot,\cdot)$ in Eq. (2). Bregman divergences include a large number of useful loss functions such as the well-known squared loss, hinge loss, logistic loss, KL divergence and I-divergence. A specific Bregman Divergence (*e.g.* KL-divergence) can be identified by a corresponding convex function $\phi$ (*e.g.* negative entropy for KL-divergence), and hence be written as $d_\phi(\mathbf{y}_i, \mathbf{y}_j)$. Using this notation, the optimization problem can be rewritten as:

$$\min_{\{\mathbf{y}_i\}_{i=1}^n} \left[ \sum_{i \in \mathcal{X}} d_\phi(\pi_i, \mathbf{y}_i) + \alpha \sum_{(i,j) \in \mathcal{X}} s_{ij} d_\phi(\mathbf{y}_i, \mathbf{y}_j) \right]. \tag{3}$$

All Bregman divergences have the remarkable property that the single best (in terms of minimizing the net loss) representative of a set of vectors, is simply the expectation of this set (!) provided the divergence is computed with this representative as the second argument of $d_\phi(\cdot,\cdot)$ — see **Theorem 1** in the sequel for a more formal statement of this result. Unfortunately this simple form of the optimal solution is not valid if the variable to be optimized occurs as the first argument. In that case, however, one can work in the (Legendre) dual space, where the optimal solution has a simple form — see Banerjee et al. (2005) for details. Re-examining Eq. (3), we notice that the $\mathbf{y}_i$'s to be minimized over occur both as first and second arguments of a Bregman divergence. Hence optimization over $\{\mathbf{y}_i\}_{i=1}^n$ is not available in closed form.

We circumvent this problem by creating two copies for each $\mathbf{y}_i$ — the left copy, $\mathbf{y}_i^{(l)}$, and the right copy, $\mathbf{y}_i^{(r)}$. The left(right) copies are used whenever the variables are encountered in the first(second) argument of the Bregman divergences. The right and left copies are updated iteratively, and an additional constraint is used to ensure that the two copies of a variable remain close during the updates. First, keeping $\{\mathbf{y}_i^{(l)}\}_{i=1}^n$ and $\{\mathbf{y}_i^{(r)}\}_{i=1}^n \setminus \{\mathbf{y}_j^{(r)}\}$ fixed, the part of the objective function that only depends on $\mathbf{y}_j^{(r)}$ can be written as:

$$J_{[\mathbf{y}_j^{(r)}]} = d_\phi(\boldsymbol{\pi}_j^{(r)}, \mathbf{y}_j^{(r)}) + \alpha \sum_{i^{(l)} \in \mathcal{X}} s_{i^{(l)} j^{(r)}} d_\phi(\mathbf{y}_i^{(l)}, \mathbf{y}_j^{(r)}). \tag{4}$$

Note that the optimization of $J_{[\mathbf{y}_j^{(r)}]}$ in (4) w.r.t. $\mathbf{y}_j^{(r)}$ is constrained by the fact that the left and right copies of $\mathbf{y}_j$ should be equal. Therefore, a soft constraint is added in (4), and the optimization problem now becomes:

$$\min_{\mathbf{y}_j^{(r)}} \left[ d_\phi(\boldsymbol{\pi}_j^{(r)}, \mathbf{y}_j^{(r)}) + \alpha \sum_{i^{(l)} \in \mathcal{X}} s_{i^{(l)} j^{(r)}} d_\phi(\mathbf{y}_i^{(l)}, \mathbf{y}_j^{(r)}) + \lambda_j^{(r)} d_\phi(\mathbf{y}_j^{(l)}, \mathbf{y}_j^{(r)}) \right], \tag{5}$$

where $\lambda_j^{(r)}$ is the corresponding penalty parameter. For every valid assignment of $\{\mathbf{y}_i^{(l)}\}_{i=1}^n$, it can be shown that there is a unique minimizer $\mathbf{y}_j^{(r)*}$ for the optimization problem in (5). For that purpose, a new Corollary is developed from the results of **Theorem 1** Banerjee et al. (2005) which is stated below.

**Theorem 1 (Banerjee et al. (2005))** *Let $Y$ be a random variable that takes values in $\mathcal{Y} = \{\mathbf{y}_i\}_{i=1}^n \subset S \subseteq \mathbb{R}^k$ following a probability measure $v$ such that $\mathbb{E}_v[Y] \in ri(S)$. Given a Bregman divergence $d_\phi : S \times ri(S) \to [0, \infty)$, the optimization problem*

$$\min_{s \in ri(S)} \mathbb{E}_v[d_\phi(Y, s)]$$

*has a unique minimizer given by $\mathbf{s}^* = \boldsymbol{\mu} = \mathbb{E}_v[Y]$.*

**Corollary 2** *Let $\{Y_i\}_{i=1}^n$ be a set of random variables, each of which takes values in $\mathcal{Y}_i = \{\mathbf{y}_{ij}\}_{j=1}^{n_i} \subset S \subseteq \mathbb{R}^d$ following a probability measure $v_i$ such that $\mathbb{E}_{v_i}[Y_i] \in ri(S)$. For a Bregman divergence $d_\phi : S \times ri(S) \to [0, \infty)$, the function of the form $J_\phi(\mathbf{s}) = \sum_{i=1}^m \alpha_i \mathbb{E}_{v_i}[d_\phi(Y_i, \mathbf{s})]$ with $\alpha_i \in \mathbb{R}_+ \; \forall i$ has a unique minimizer given by*

$$\mathbf{s}^* = \boldsymbol{\mu} = \left[ \sum_{i=1}^m \alpha_i \mathbb{E}_{v_i}[Y_i] \right] \Big/ \left[ \sum_{i=1}^m \alpha_i \right].$$

**Proof** *The proof is similar to that of **Theorem 1** as given in Banerjee et al. (2005) but omitted here for space constraints.* ∎

From these results, the unique minimizer of the optimization problem in (5) is obtained as:

$$\mathbf{y}_j^{(r)*} = \frac{\boldsymbol{\pi}_j^{(r)} + \gamma_j^{(r)} \sum_{i^{(l)} \in \mathcal{X}} \delta_{i^{(l)} j^{(r)}} \mathbf{y}_i^{(l)} + \lambda_j^{(r)} \mathbf{y}_j^{(l)}}{1 + \gamma_j^{(r)} + \lambda_j^{(r)}}, \tag{6}$$

where $\gamma_j^{(r)} = \alpha \sum_{i^{(l)} \in \mathcal{X}} s_{i^{(l)} j^{(r)}}$ and $\delta_{i^{(l)} j^{(r)}} = s_{i^{(l)} j^{(r)}} / \left[ \sum_{i^{(l)} \in \mathcal{X}} s_{i^{(l)} j^{(r)}} \right]$. The same optimization in (5) is repeated over all the $\mathbf{y}_j^{(r)}$'s. After the right copies are updated, the objective function is (sequentially) optimized w.r.t. all the $\mathbf{y}_i^{(l)}$'s. Like in the first step, $\{\mathbf{y}_j^{(l)}\}_{j=1}^n \setminus \{\mathbf{y}_i^{(l)}\}$ and $\{\mathbf{y}_j^{(r)}\}_{j=1}^n$ are kept fixed, and the equality of the left and right copies of $\mathbf{y}_i$ is added as a soft constraint, so that the optimization w.r.t. $\mathbf{y}_i^{(l)}$ can be rewritten as:

$$\min_{\mathbf{y}_i^{(l)}} \left[ \alpha \sum_{j^{(r)} \in \mathcal{X}} s_{i^{(l)} j^{(r)}} d_\phi(\mathbf{y}_i^{(l)}, \mathbf{y}_j^{(r)}) + \lambda_i^{(l)} d_\phi(\mathbf{y}_i^{(l)}, \mathbf{y}_i^{(r)}) \right], \tag{7}$$

where $\lambda_i^{(l)}$ is the corresponding penalty parameter. As mentioned earlier, one needs to work in the dual space now, using the convex function $\psi$ (Legendre dual of $\phi$):

$$\psi(\mathbf{y}_i) = \langle \mathbf{y}_i, \nabla\phi^{-1}(\mathbf{y}_i) \rangle - \phi(\nabla\phi^{-1}(\mathbf{y}_i)). \tag{8}$$

One can show that $\forall \mathbf{y}_i, \mathbf{y}_j \in int(dom(\phi))$, $d_\phi(\mathbf{y}_i, \mathbf{y}_j) = d_\psi(\nabla\phi(\mathbf{y}_j), \nabla\phi(\mathbf{y}_i))$ — see Banerjee et al. (2005) for more details. Thus, the optimization problem in (7) can be rewritten in terms of the Bregman divergence associated with $\psi$ as follows:

$$\min_{\nabla\phi(\mathbf{y}_i^{(l)})} \left[ \alpha \sum_{j^{(r)} \in \mathcal{X}} s_{i^{(l)} j^{(r)}} d_\psi(\nabla\phi(\mathbf{y}_j^{(r)}), \nabla\phi(\mathbf{y}_i^{(l)})) + \lambda_i^{(l)} d_\psi(\nabla\phi(\mathbf{y}_i^{(r)}), \nabla\phi(\mathbf{y}_i^{(l)})) \right]. \tag{9}$$

The unique minimizer of the problem in (9) can be computed using **Corollary 2**. $\nabla\phi$ is monotonic and invertible for $\phi$ being strictly convex and hence the inverse of the unique minimizer for problem (9) is unique and equals to the unique minimizer for problem (7). Therefore, the unique minimizer of problem (7) w.r.t. $\mathbf{y}_i^{(l)}$ is given by:

$$\mathbf{y}_i^{(l)*} = \nabla\phi^{-1} \left[ \frac{\gamma_i^{(l)} \sum_{j^{(r)} \in \mathcal{X}} \delta_{i^{(l)} j^{(r)}} \nabla\phi(\mathbf{y}_j^{(r)}) + \lambda_i^{(l)} \nabla\phi(\mathbf{y}_i^{(r)})}{\gamma_i^{(l)} + \lambda_i^{(l)}} \right], \tag{10}$$

where $\gamma_i^{(l)} = \alpha \sum_{j^{(r)} \in \mathcal{X}} s_{i^{(l)} j^{(r)}}$ and $\delta_{i^{(l)} j^{(r)}} = s_{i^{(l)} j^{(r)}} / \left[ \sum_{j^{(r)} \in \mathcal{X}} s_{i^{(l)} j^{(r)}} \right]$.

For the experiments reported in this paper, the generalized I-divergence defined as:

$$d_\phi(\mathbf{y}_i, \mathbf{y}_j) = \sum_{\ell=1}^{k} y_{i\ell} \log(\frac{y_{i\ell}}{y_{j\ell}}) - \sum_{\ell=1}^{k} (y_{i\ell} - y_{j\ell}), \forall \mathbf{y}_i, \mathbf{y}_j \in \mathbb{R}_+^k, \tag{11}$$

has been used. Thus, Eq. (10) can be rewritten as:

$$\mathbf{y}_i^{(l)*,I} = exp \left( \frac{\gamma_i^{(l)} \sum_{j^{(r)} \in \mathcal{X}} \delta_{i^{(l)} j^{(r)}} \nabla\phi(\mathbf{y}_j^{(r)}) + \lambda_i^{(l)} \nabla\phi(\mathbf{y}_i^{(r)})}{\gamma_i^{(l)} + \lambda_i^{(l)}} \right) - 1. \tag{12}$$

Optimization over the left and right arguments of all the data points constitutes one pass (iteration) of the algorithm. These two steps are repeated till convergence. Since, at each step, the algorithm minimizes the objective in (3) which is lower bounded by zero and the minimizer is unique due to the strict convexity of $\phi$, the algorithm is guaranteed to converge. On convergence, all $\mathbf{y}_i$'s are normalized to unit $L_1$ norm, to yield the individual class probability distributions for every instance $\mathbf{x}_i \in \mathcal{X}$. The main steps of $\mathbf{C^3E}$ are summarized in Algorithm 1.

## 2.2. Pedagogical Example

This example illustrates, via a simple experiment, how the supplementary constraints provided by the clustering algorithms can be useful for improving the generalization capability of classifiers using $\mathbf{C^3E}$. Consider the two-dimensional dataset known as *Half-Moon*, which has two classes, each of which represented by 400 instances. From this dataset, 2% of the instances are used for training (source domain), whereas the remaining instances are used for testing (target domain). A classifier ensemble formed by three well-known classifiers (Decision Tree, Linear Discriminant, and Generalized Logistic Regression) is adopted. In order to get a cluster ensemble, a single linkage (hierarchical) clustering algorithm is chosen. The cluster ensemble is then obtained from five data partitions represented in the dendrogram, which is cut for different number of clusters (from 4 to 8). Fig. 2 shows the target data class labels obtained from the standalone use of the classifier ensemble, whereas Fig. 3 shows the corresponding results achieved by $\mathbf{C^3E}$. Comparing Fig. 2 to Fig. 3, one can see that $\mathbf{C^3E}$ does a better job, since the cluster ensemble is able to indicate the class-continuity at the edges of the two moons showing that the information provided by the similarity matrix can improve the generalization capability of classifiers.

## 3. Experimental Evaluation

The real-world datasets employed in the experiments are:

**a) Text Documents** — (Pan and Yang, 2010): From the well-known text collections *20 newsgroup* and *Reuters-21758*, 9 cross-domain learning tasks are generated. The

---

**Algorithm 1 $\mathbf{C^3E}$**

---

**Inputs**: $\{\pi_i\}, \mathbf{S}$. **Output**: $\{\mathbf{y}_i\}$.
**Step 0:** Initialize $\{\mathbf{y}_i^{(r)}\}, \{\mathbf{y}_i^{(l)}\}$ so that $\mathbf{y}_{i\ell}^{(r)} = \mathbf{y}_{i\ell}^{(l)} = \frac{1}{k}$ $\forall \ell \in \{1, 2, \cdots, k\}$.
Loop until convergence:
**Step 1:** Update $\mathbf{y}_j^{(r)}$ using Eq. (6) $\forall j \in \{1, 2, \cdots, n\}$.

**Step 2:** Update $\mathbf{y}_i^{(l)}$ using Eq. (10) $\forall i \in \{1, 2, \cdots, n\}$.
End Loop
**Step 3:** Compute $\mathbf{y}_i = 0.5[\mathbf{y}_i^{(l)} + \mathbf{y}_i^{(r)}]$ $\forall i \in \{1, 2, \cdots, n\}$.
**Step 4:** Normalize $\mathbf{y}_i$ $\forall i \in \{1, 2, \cdots, n\}$.

---

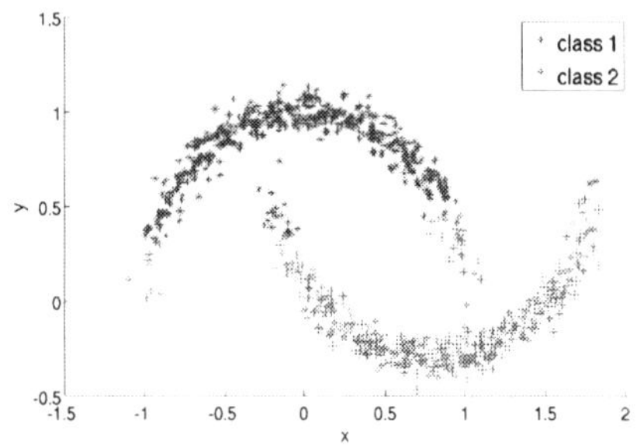


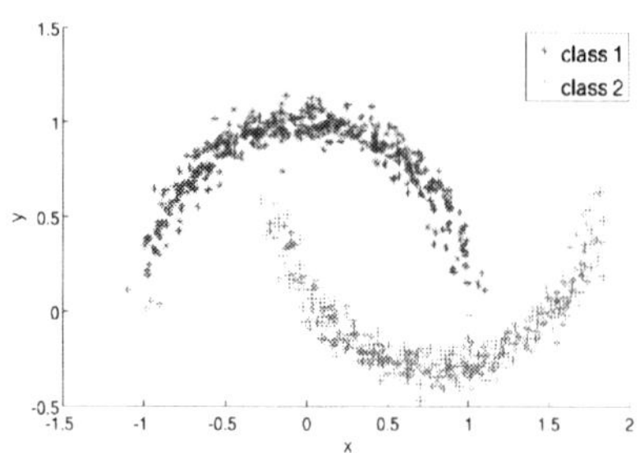


Figure 2: Results from Classifier Ensemble.

Figure 3: Results from $C^3E$.

two-level hierarchy in both of these datasets is exploited to frame a learning task involving a top category classification problem with training and test data drawn from different sub categories — *e.g.*, to distinguish documents from two top newsgroup categories (rec and talk), the training set is built from "`rec.autos`", "`rec.motorcycles`", "`talk.politics`", and "`talk.politics.misc`", and the test set is formed from the sub-categories "`rec.sport.baseball`", "`rec.sport.hockey`", "`talk.politics.mideast`", and "`talk.religions.misc`" (see Dai et al. (2007) for more details).

**b) Botswana** — (Rajan et al., 2006): This is an application of transfer learning to the pixel-level classification of remotely sensed images, which provides a real-life scenario where such learning will be useful (in contrast to the contrived setting of text classification, which is chosen as it has previously been used, e.g. Dai et al. (2007). It is relatively easy to acquire an image, but expensive to label each pixel manually. This is because images typically have about a million pixels and might often represent inaccessible terrain. Thus, typically, only part of an image gets labeled. Moreover, when the satellite again flies over the same area, the new image can be quite different due to change of season, thus a classifier induced on the previous image becomes significantly degraded for the new task. These hyperespectral data sets used are from a $1476 \times 256$ pixel study area located in the Okavango Delta, Botswana. It has nine different land-cover types consisting of seasonal swamps, occasional swamps, and drier woodlands located in the distal portion of the delta. Data from this region for different months (May, June, and July) were obtained by the Hyperion sensor of the NASA EO-1 satellite for the calibration/validation portion of the mission in 2001. Data collected for each month was further segregated into two different areas. While the May scene is characterized by the onset of the annual flooding cycle and some newly burned areas, the progression of the flood and the corresponding vegetation responses are seen in the June and July data. The acquired raw data was further processed to produce 145 features. From each area of Botswana, different transfer learning tasks are generated: the classifiers are trained on either May, June or {May ∪ June} data and tested on either June or July data.

For text data, logistic regression (LR), Support Vector Machines (SVM)[2], and Winnow (WIN) are used as baseline classifiers. For clustering the target data, the well-known CLUTO package (Karypis, 2002) is used (with default settings and two clusters). We also compare $C^3E$ with two transfer learning algorithms from the literature — Transductive Support Vector Machines (TSVM) and the Locally Weighted Ensemble (LWE) (Gao et al., 2008).

For the hyperspectral data, two baseline classifiers are used: the well-known naïve Bayes Wrapper (NBW) and the Maximum Likelihood (ML) classifier (which performs well when used with a best bases feature extractor (Kumar et al., 2001)). The target set instances are clustered by $k$-means, using a varied number of clusters (from 30 to 50). PCA is used for reducing the number of features employed by ML. The parameters of $C^3E$ are manually optimized for better performance.

The results for text data are reported in Table 1. The different learning tasks corresponding to different pairs of categories are listed as "Mode". As it can be seen, $C^3E$ improves the performance of the classifier ensemble (formed by combining WIN, LR and SVM via output averaging) for all learning tasks, except for *O vs Pl*, where apparently the training and test distributions are similar. Also, the $C^3E$ accuracies are much better than those achieved by both TSVM and LWE in most of the datasets. Except for WIN, the performances of the base classifiers and clustereres (and hence of $C^3E$) are quite invariant, thereby resulting in very low standard deviations. For the experiments, Bayesian Logistic Regression `http://www.bayesianregression.org/` is used for running the logistic regression classifier, LIBSVM (`http://www.csie.ntu.edu.tw/~cjlin/libsvm/`) for **SVM**, SNoW Learning Architecture `http://cogcomp.cs.illinois.edu/page/software_view/1` for **Winnow**, and SVM[light] `http://svmlight.joachims.org/` for transductive **SVM**. The posterior class probabilities from **SVM** are also obtained using the LIBSVM package with linear kernel. For SNoW, "-S 3 -r 5" is used and the remaining parameters of all the packages are set to their default values. The values of $(\alpha, \lambda)$ are set as $(0.01, 0.1)$ for the transfer learning tasks corresponding to 20 *Newsgroup* datasets. For *Reuters*-21578, the values of the parameters $(\alpha, \lambda)$ are set as $(0.01, 0.1)$, $(0.00001, 0.1)$, and $(0.1, 0.1)$ for *O vs Pe*, *O vs Pl*, and *Pe vs Pl*, respectively (see Table 1). For hyperspectral data, Table 2 reports the results. Note that $C^3E$ provides consistent accuracy improvements for both NBW and ML.[3] The column "PCs" indicate the number of principal components used for dimension reduction while training/testing with ML classifier.

## 4. Conclusions

We described an optimization framework that takes the outputs of a cluster ensemble applied to the target task to moderate posterior probability estimates provided by classifiers previously induced on a related (source domain) task, so that they are better adapted to the new context. The framework is quite general and has shown very

---

2. The posterior class probabilities are obtained by using the package LIBSVM.
3. Standard deviations of the accuracies from NBW and ML were close to 0 and hence not shown.

Table 1: Classification of Text Data — *20 newsgroup* and *Reuters-21758*.

| Dataset | Mode | WIN | LR | SVM | Ensemble | TSVM | LWE | C$^3$E |
|---|---|---|---|---|---|---|---|---|
| 20 Newsgroup | C vs S | 66.61 | 67.17 | 67.02 | 69.58 | 76.97 | 77.07 | **94.61** |
| | R vs T | 60.43 | 68.79 | 63.87 | 65.98 | 89.95 | 87.46 | **92.78** |
| | R vs S | 80.11 | 76.51 | 71.40 | 77.39 | 89.96 | 87.81 | **94.19** |
| | S vs T | 73.93 | 72.16 | 71.51 | 75.11 | 85.59 | 81.99 | **96.39** |
| | C vs R | 89.00 | 77.36 | 81.50 | 85.18 | 89.64 | 91.09 | **96.75** |
| | C vs T | 93.41 | 91.76 | 93.89 | 93.48 | 88.26 | 98.90 | **98.90** |
| Reuters-21758 | O vs Pe | 70.57 | 66.19 | 69.25 | 73.30 | 76.94 | 76.77 | **81.81** |
| | O vs Pl | 65.10 | 67.87 | 69.88 | 69.21 | **70.08** | 67.59 | 68.92 |
| | Pe vs Pl | 56.75 | 56.48 | 56.20 | 57.59 | 59.72 | 59.90 | **68.61** |

Table 2: Classification of Hyperspectral Data — *Botswana*.

| Data | Source-Target | NBW | NBW+C$^3$E | ML | ML+C$^3$E | $\alpha$ | $\lambda$ | PCs |
|---|---|---|---|---|---|---|---|---|
| Area1 | may-june | 70.68 | **73.58** ($\pm$0.42) | 74.47 | **82.52** ($\pm$0.52) | 0.0070 | 0.1 | 9 |
| | may-july | 61.85 | **62.22** ($\pm$0.29) | 58.58 | **66.47** ($\pm$0.53) | 0.0001 | 0.2 | 12 |
| | june-july | 70.55 | **73.50** ($\pm$0.17) | 79.71 | **82.44** ($\pm$0.26) | 0.0070 | 0.1 | 127 |
| | may/june-july | 75.53 | **81.42** ($\pm$0.31) | 85.78 | **86.25** ($\pm$0.23) | 0.0010 | 0.1 | 123 |
| Area2 | may-june | 66.10 | **70.08** ($\pm$0.28) | 70.16 | **81.48** ($\pm$0.43) | 0.0040 | 0.1 | 9 |
| | may-july | 61.55 | **63.74** ($\pm$0.14) | 52.78 | **65.05** ($\pm$0.22) | 0.0001 | 0.2 | 12 |
| | june-july | 54.89 | **59.93** ($\pm$0.53) | 75.62 | **77.12** ($\pm$0.37) | 0.0050 | 0.1 | 80 |
| | may/june-july | 63.79 | **63.96** ($\pm$0.16) | 77.33 | **80.97** ($\pm$0.23) | 0.0080 | 0.1 | 122 |

promising results. An extensive study across a wide variety of problem domains will further reveal its capabilities as well as potential limitations, and is worth undertaking in light of what we have observed so far.

A promising venue for future work involves investigating how to perform the automatic selection of the parameter $\alpha$. To that end, strategies based on methods like Covariate Shift (Sugiyama et al., 2007) may be useful. Covariate Shift assumes that the training and test distributions are known or, more realistically, can be estimated from data, which is a difficult problem. Note that C$^3$E does not require the densities to be known. In all of our controlled experiments, we tuned the parameter $\alpha$ using some cross-validation in the training set. This approach may not be a suitable practice for transfer learning applications, where ideally we should have some mechanism to select $\alpha$ based on the density differences between the source and the target domains. This leads to the analogous difficulties faced as when using Covariate Shift. We shall note that, unlike covariate shift, C$^3$E takes similarity information from the target domain and does not require the conditional distribution of classes given instances to be same in both source and target domains. Finally, for high-dimensional data, the use of a cluster ensemble is additionally attractive as it can potentially project the data onto lower subspaces, and this aspect is worth exploring further.

## Acknowledgments

This work has been supported by NSF Grants (IIS-0713142 and IIS-1016614) and by the Brazilian Research Agencies FAPESP and CNPq. We thank the anonymous reviewers for their insightful comments and Dr. Goo Jun for providing the data and code for experiments with hyperspectral data.

## References

A. Acharya, E. R. Hruschka, J. Ghosh, and S. Acharyya. $C^3E$: A Framework for Combining Ensembles of Classifiers and Clusterers. In *10th International Workshop on Multiple Classifier Systems*, pages 86–95. LNCS Vol.6713, Springer, 2011.

A. Banerjee, S. Merugu, I. Dhillon, and J. Ghosh. Clustering with Bregman divergences. *Jl. Machine Learning Research (JMLR)*, 6:1705–1749, October 2005.

K. D. Bollacker and J. Ghosh. Knowledge transfer mechanisms for characterizing image datasets. In *Soft Computing and Image Processing*. Physica-Verlag, Heidelberg, 2000.

R. Caruana. Multitask learning. *Mach. Learn.*, 28:41–75, July 1997.

W. Dai, G. Xue, Q. Yang, and Y. Yu. Co-clustering based classification for out-of-domain documents. In *Proceedings of KDD*, pages 210–219, New York, NY, USA, 2007.

J. Gao, W. Fan, J. Jiang, and J. Han. Knowledge transfer via multiple model local structure mapping. In *Proceedings of KDD*, pages 283–291, 2008.

George Karypis. *CLUTO - A Clustering Toolkit*. Dept. of Computer Science, University of Minnesota, May 2002.

S. Kumar, J. Ghosh, and M. M. Crawford. Best-bases feature extraction algorithms for classification of hyperspectral data. *IEEE TGRS*, 39(7):1368–79, 2001.

S. J. Pan and Q. Yang. A survey on transfer learning. *IEEE TKDE*, 22:1345–1359, 2010.

K. Punera and J. Ghosh. Consensus based ensembles of soft clusterings. In *Applied Artificial Intelligence*, volume 22, pages 780–810, August 2008.

S. Rajan, J. Ghosh, and M. M. Crawford. Exploiting class hierarchies for knowledge transfer in hyperspectral data. *IEEE TGRS*, 44(11):3408–3417, 2006.

D. L. Silver and K. P. Bennett. Guest editor's introduction: special issue on inductive transfer learning. *Mach. Learn.*, 73:215–220, December 2008.

A. Strehl and J. Ghosh. Cluster ensembles – a knowledge reuse framework for combining multiple partitions. *Jl. Machine Learning Research (JMLR)*, 3 (Dec):583–617, 2002.

Masashi Sugiyama, Matthias Krauledat, Klaus-robert Müller, and Yoshua Bengio. Covariate shift adaptation by importance weighted cross validation. *Journal of Machine Learning Research*, 8, 2007.

S. Thrun and L.Y. Pratt, editors. *Learning to Learn*. Kluwer Academic, 1997.

# Divide and Transfer: an Exploration of Segmented Transfer to Detect Wikipedia Vandalism

**Si-Chi Chin**                                                        SI-CHI-CHIN@UIOWA.EDU
*Information Science, The University of Iowa*

**W. Nick Street**                                                    NICK-STREET@UIOWA.EDU
*Management Sciences Department, The University of Iowa*

**Editor:** I. Guyon, G. Dror, V. Lemaire, G. Taylor, and D. Silver

## Abstract

The paper applies knowledge transfer methods to the problem of detecting Wikipedia
vandalism detection, defined as malicious editing intended to compromise the integrity of the content of articles. A major challenge of detecting Wikipedia vandalism
is the lack of a large amount of labeled training data. Knowledge transfer addresses
this challenge by leveraging previously acquired knowledge from a source task. However, the characteristics of Wikipedia vandalism are heterogeneous, ranging from a
small replacement of a letter to a massive deletion of text. Selecting an informative
subset from the source task to avoid potential negative transfer becomes a primary
concern given this heterogeneous nature. The paper explores knowledge transfer
methods to generalize learned models from a heterogeneous dataset to a more uniform dataset while avoiding negative transfer. The two novel *segmented transfer (ST)*
approaches map unlabeled data from the target task to the most related cluster from
the source task, classifying the unlabeled data using the most relevant learned models.
**Keywords:** Transfer learning, classifier reuse, Wikipedia vandalism detection

## 1. Introduction

*Transfer learning* discusses how to transfer knowledge across different data distributions, providing solutions when labeled data are scarce or expensive to obtain. Motivated by the problem of Wikipedia vandalism detection (Potthast and Gerling, 2007;
Chin et al., 2010), this paper investigates the question: *how do we transfer a classifier
trained to detect vandalism in one article to another?* We introduce two novel *segmented transfer (ST)* approaches to learn from a labeled but diverse source task, which
exhibits a wide-ranging distribution of both positive and negative examples over the
feature space, and then selectively transfer the classifier to predict an unlabeled and
more uniform target task. Our methods are also tested when transferring between articles with similar distributions.

Our work is related to the source task selection problem, investigating methods to
enhance transfer learning performance and to minimize negative transfer. We concentrate specifically on transfer at the knowledge level, i.e. the reuse of learned classifiers
from a source task, as opposed to transfer at level of instances, priors, or functions as

exemplified by Pan and Yang (2010). We investigate two methods to exploit a single source task to predict a target task with no available labels. To improve knowledge transfer, it is useful to identify an effective method to transfer knowledge from the source task to the target task. In this paper, we assume that *perhaps not all the source task is useful* and *perhaps not all the target task can learn from the available source task*. Our work aims to address the following questions:

- If not all the source task is related to the target task, how do we select the most relevant subset from the source task?

- If not all the target task can be explained or learned from the source task, how do we identify the subset from the target task that can benefit from most the knowledge transfer?

*Wikipedia*, the online encyclopedia, is a popular and influential collaborative information system. The collaborative nature of authoring has exposed it to malicious editing, or vandalism, defined as "any addition, removal, or change of content made in a deliberate attempt to compromise the integrity of Wikipedia[1]." Wikipedia vandalism detection, an adversarial information retrieval task, is a recently emerging research area. The goal of the task is to determine, for each newly edited revision, whether it could be a vandalism instance and to create a ranked list of probable vandalism edits to alert Wikipedia users (usually the stewards for an article). However, determining if an edit is malicious is challenging and acquiring reliable class labels is non-trivial. To classify a new and unlabeled dataset, it is useful to leverage knowledge from prior tasks.

Wikipedia vandalism instances exhibit heterogeneous characteristics. A vandalism instance can be a large-scale editing or a small change of stated facts. Each type of vandalism may demonstrate different feature characteristics and an article may contain more instances of one type of vandalism than others. Moreover, the distribution of different types of vandalism may vary from article to article. For example, 'Microsoft' article may contain higher ratio of graffiti instances whereas 'Abraham Lincoln' article may be more vulnerable to misinformation instances. The heterogeneous nature of Wikipedia vandalism detection could potentially introduce negative transfer (Rosenstein et al., 2005). It requires a selective mechanism to assure the quality of knowledge transfer, for example, leveraging knowledge about "graffiti" instances from the source task to detect graffiti, as opposed to other types of vandalism instances, in the target task. To resolve the problem of a heterogeneous source task, we introduce two methods to identify the informative segments from the source task in the absence of class labels.

In this paper, instead of learning from multiple source, we focus on the problem setting in which only a single source task is available. Both the source and target task have the same input and output domains, but their samples are drawn from different populations. Each sample in both the source and target task is a revision of a given Wikipedia article, preprocessed into a feature space representing a collection of statistical language model features. The output labels indicate whether the article is a vandalism instance.

---

[1]. http://en.wikipedia.org/wiki/Wikipedia:Vandalism

We organize the rest of paper as follows. Section 2 introduces the two *segmented transfer* approaches. Section 3 describes the experimental setups, including the datasets, the features, and the six experimental settings. Sections 4 and 5 present the experimental results and evaluations. In Section 6, we discuss related work, and we conclude the paper with future directions.

## 2. Segmented Transfer (ST)

In this paper, we propose *segmented transfer (ST)* to enrich the capability of transfer learning and to address the issue of potential negative transfer. The goal of ST is to identify and learn from the most related segment, a subset from the training samples, from the source task. Our motivation comes from two assumptions:

- Not all of the source task is useful, and

- Not all of the target task can benefit from the available source task.

We propose the *source task segmented transfer (STST)* and the *target task segmented transfer (TTST)* approaches to address each assumption and summarize the two approaches in Table 1.

**Source task segmented transfer (STST)**   The STST approach clusters the source task, assigning cluster membership to the target task. In Figure 1, the labeled source task is first segmented into clusters. Each cluster has its own classifier. We then assign cluster membership to the unlabeled target task and transfer the classifier trained from the corresponding cluster of the source task. Because the distribution of the feature space is different between the source and target tasks, it is likely that some source task data will not be used. The approach aims to transfer knowledge acquired only from the related segment to minimize negative transfer.

**Target task segmented transfer (TTST)**   The TTST approach clusters the target task, assigning cluster membership to the source task. The goal of the TTST is to differentiate

Table 1: Tabular comparison of STST and TTST

|  | STST | TTST |
|---|---|---|
| Primary assumption: | Not all the source task is useful | Not all the target task can benefit from the available source task |
| Train cluster models at: | Source task | Target task |
| Assign cluster membership to: | Target task | Source task |
| Max number of classifiers: | Number of clusters found in the source task | Number of clusters found in the target task |
| Transferred object: | Classifiers trained from the source task | |

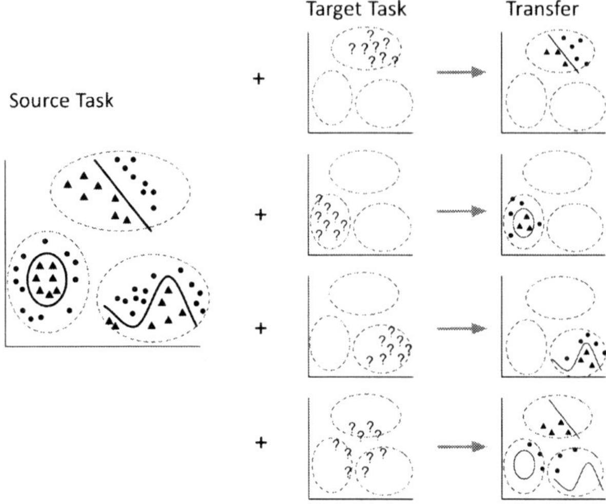



Figure 1: Flowchart of source task segmented transfer (STST).

samples that can be better learned from the provided source task. In Figure 2, the unlabeled target task is first segmented into clusters. We then assign cluster membership to the labeled source task and train a classifier for each cluster. Finally, the classifiers are transferred to the corresponding clusters in the target task. As shown in Figure 2, some data from the target task may not be well learned because of the lack of an appropriate source task.

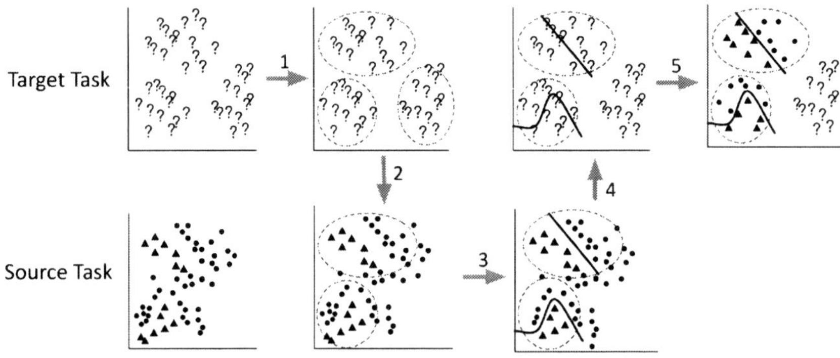



Figure 2: Flowchart of target task segmented transfer (TTST).

## 3. Experiments

This section describes the datasets used for experiments, the input feature space, the six experimental settings, and the cluster membership assignment distributions for each setting.

DATASET DESCRIPTION    In four of the experiments, we clustered and trained on the Webis Wikipedia vandalism (Webis) corpus (Potthast and Gerling, 2007) and tested on the revision history of the "Microsoft" and "Abraham Lincoln" articles on Wikipedia (Chin et al., 2010). The other two experiments use Microsoft as the source task and transfer to the Lincoln article.

The Webis dataset contained randomly sampled revisions of different Wikipedia articles, drawn from different categories. The Microsoft and Lincoln datasets contained the revision history of those articles. Although class labels were available for both datasets, the class information was ignored during the clustering and was used to build classifiers and to demonstrate the performance of the two methods. Table 2 is a tabular description of the three datasets. The AUC and AP scores for the Microsoft and Lincoln dataset were computed by 10-fold cross validation using the provided class labels using an SVM classifier with RBF kernel. The parameters $\gamma$ and $C$ were chosen empirically to achieve the best performance.

Table 2: Dataset description

|  | Positive | Negative | Total |
|---|---|---|---|
| **Webis** | 301 | 639 | 940 |
| **Microsoft** | 268 | 206 | 474 |
| **Lincoln** | 178 | 223 | 401 |

FEATURE DESCRIPTION    All three datasets used features generated by statistical language modeling (SLM) using the CMU SLM toolkit (Clarkson and Rosenfeld, 1997). SLM computes the distribution of tokens in natural language text and assigns a probability to the occurrence of a string $S$ or a sequence of $m$ words. The *evallm* tool evaluates the language model dynamically, providing statistics such as perplexity, number of $n$-gram hits, number of OOV (out of vocabulary), and the percentage of OOV from a given text. The *evallm* tool generates features separately from the diff data for the new revision and the full new revision to build classifiers. In addition to the 18 attributes (9 for the diff and 9 for the full revision) generated from SLM, three features: ratio of insertion, ratio of change, and ratio of deletion, were added to the set of attributes. The 21 attributes were generated for each revision in the dataset. Table 3 summarizes features used for the classification.

EXPERIMENTAL SETUP AND CLUSTERING ALGORITHM    Table 4 describes six experimental settings. STST and TTST each have three experiments with different combinations of the source and target task. We used the Weka (Hall et al., 2009) implementation of clustering, using the Expectation Maximization (EM) algorithm to optimize Gaussian mixture models to cluster the source and target tasks. Using cross validation, the EM algorithm determined the number of clusters to generate. To evaluate the ranked results from the experiments, we used AUC and Average Precision (AP). The ranked list was sorted by the probability of the predictions generated by SVM classifiers.

Table 3: Definition of Features

| Feature | Definition |
| --- | --- |
| word_num($d$) | Number of known words (from *diff*) |
| perplex($d$) | Perplexity value (from *diff*) |
| entropy($d$) | Entropy value (from *diff*) |
| oov_num($d$) | Number of unknown words (from *diff*) |
| oov_per($d$) | Percentage of unknown words (from *diff*) |
| bigram_hit($d$) | Number of known bigrams (from *diff*) |
| bigram_per($d$) | Percentage of known bigrams (from *diff*) |
| unigram_hit($d$) | Number of known unigrams (from *diff*) |
| unigram_per($d$) | Percentage of known unigrams (from *diff*) |
| ratio_a | Ratio of added text from previous revision |
| ratio_c | Ratio of changed text from previous revision |
| ratio_d | Ratio of deleted text from previous revision |

Table 4: Six experimental settings for STST and TTST

| Method | Exp | Source Task | Target Task |
| --- | --- | --- | --- |
| STST | 1 | Webis | Microsoft |
| STST | 2 | Webis | Lincoln |
| STST | 3 | Microsoft | Lincoln |
| TTST | 4 | Webis | Microsoft |
| TTST | 5 | Webis | Lincoln |
| TTST | 6 | Microsoft | Lincoln |

CLUSTER MEMBERSHIP DISTRIBUTION   This paragraph describes the cluster memberships and the distributions of positive and negative instances for the six experimental settings. Tables 5 and 6 present the cluster assignment distribution for STST. In Experiments 1 and 2, the source Webis dataset is segmented into 16 clusters (see Table 5). The target Microsoft and Lincoln datasets are mapped to 9 and 8 of these clusters respectively. The results of cluster assignment confirm the assumption that not all the source task is useful for the target task. However, the source task can still be fully exploited. In Experiment 3, as shown in Table 6, all the source task (Microsoft) instances are useful for the target task (Lincoln), both of which were determined to contain three clusters.

Table 7 shows the cluster assignment distributions for the TTST approach (Experiments 4, 5, and 6). The distribution shows that sometimes part of the target task would not have available source task to learn from. For example, in Experiment 4, the source task is only useful for cluster 2 of the target task; in Experiment 5, it is only useful for cluster 1.

Table 5: Cluster membership distributions for Experiments 1 and 2.

| Source cluster | Source Task | Target Task | |
| | Webis<br>Data Distri. $(+,-)$ | Microsoft (Exp:1)<br>Data Distri. $(+,-)$ | Lincoln (Exp:2)<br>Data Distri. $(+,-)$ |
|---|---|---|---|
| 1 | 75 (9,66) | 43 (22,21) | 48 (27,21) |
| 2 | 24 (1,23) | 192 (116,76) | 85 (41,44) |
| 3 | 16 (10,6) | 153 (80,73) | 215 (86,129) |
| 4 | 25 (8,17) | | 18 (6,12) |
| 5 | 46 (24,22) | 49 (20,29) | |
| 6 | 40 (35,5) | 16 (16,0) | 11 (5,6) |
| 7 | 41 (3,38) | 2 (2,0) | 1 (1,0) |
| 8 | 130 (9,121) | | |
| 9 | 63 (50,13) | | |
| 10 | 43 (9,34) | 1 (0,1) | |
| 11 | 75 (2,73) | | |
| 12 | 43 (6,37) | | |
| 13 | 62 (28,34) | 17 (12,5) | 22 (11,11) |
| 14 | 60 (60,0) | | |
| 15 | 149 (8, 141) | | |
| 16 | 48 (39,9) | 1 (0,1) | 1 (1,0) |
| Total | 940 (301,639) | 474 (268, 206) | 400 (178,223) |

## 4. Experimental results

This section describes the experimental results for STST and TTST. Our results show that the two proposed approaches improved the ranking, moving more actual vandalism

Table 6: Cluster membership distribution for Experiment 3.

| Exp | Source cluster | Source Task<br>Microsoft<br>Data Distri. $(+, -)$ | Target Task<br>Lincoln<br>Data Distri. $(+, -)$ |
|---|---|---|---|
| 3 | 1 | 344 (186, 158) | 357 (146, 211) |
| | 2 | 125 (80, 45) | 42 (30,12) |
| | 3 | 5 (2,3) | 2 (2,0) |
| | Total | 474 (268,206) | 401 (178,223) |

Table 7: Cluster membership distribution for Experiments 4, 5, and 6

| Exp | Target cluster | Target Task<br>Data Distri. $(+, -)$ | Source Task<br>Data Distri. $(+, -)$ |
|---|---|---|---|
| 4 | 1 | 344 (186, 158) | 0 |
| | 2 | 125 (80, 45) | 940 (301,639) |
| | 3 | 5 (2,3) | 0 |
| | Total | 474 (268,206) | 940 (301,639) |
| 5 | 1 | 56 (36,20) | 940 (301,639) |
| | 2 | 115 (45,70) | 0 |
| | 3 | 230 (97,133) | 0 |
| | Total | 401 (178,223) | 940 (301,639) |
| 6 | 1 | 56 (36,20) | 159 (93,66) |
| | 2 | 115 (45,70) | 121 (56,65) |
| | 3 | 230 (97,133) | 194 (119,75) |
| | Total | 401 (178,223) | 474 (268,206) |

instances to the top of the ranked list. Table 8 shows the performance of the baseline, a direct transfer without either STST or TTST, using an SVM classifier with linear and RBF kernels. In this section, results that outperform the baseline are marked with a †.

Table 8: Baseline performance.

| Exp | Classifier | AUC | AP |
|---|---|---|---|
| 1 and 4 | SVM w/ linear kernel (C=1) | 0.5333 | 0.6002 |
| | SVM w/ RBF kernel (C=1, $\gamma = 0.1$) | 0.5466 | 0.5862 |
| 2 and 5 | SVM w/ linear kernel (C=1) | 0.5276 | 0.4528 |
| | SVM w/ RBF kernel (C=0.8, $\gamma = 0.16$) | 0.5396 | 0.4454 |
| 3 and 6 | SVM w/ linear kernel (C=500) | 0.6089 | 0.6134 |
| | SVM w/ RBF kernel (C=500, $\gamma = 0.02$) | 0.6215 | 0.6021 |

## 4.1. STST Evaluation

Table 9 shows the experimental results for the STST approach. We compared the performance of STST with the best performance for direct transfer, i.e. train on the source task and transfer directly to the target task, using the SVM classifier with RBF kernel (see Table 8). The results indicate that the STST approach consistently outperforms the baseline across the three experiments.

Table 9: Experiment results for STST

| Experiment 1 | | Experiment 2 | | Experiment 3 | |
|:---:|:---:|:---:|:---:|:---:|:---:|
| AUC | AP | AUC | AP | AUC | AP |
| 0.5541† | 0.6095† | 0.5519† | 0.5063† | 0.6883† | 0.6514† |

## 4.2. TTST Evaluation

Table 10 shows the experimental results for the TTST approach. As shown in Table 7, only cluster 2 in Experiment 4 and cluster 1 in Experiment 5 have the source task to learn from. Therefore, presumably, the classifier trained for the assigned cluster in the target task will perform better on the assigned cluster than on other clusters.

The results in Experiment 4 support the assumption. The performance of cluster 2 is much higher than cluster 1 when we used the same classifier trained from the source task for both clusters. Although the cluster 3 in Experiment 4 has high AUC and AP results, it is noted that the size of the cluster is quite small and the results might be insignificant.

Experiment 5 presents mixed results on AUC and AP. We observe that the AP, but not the AUC, is higher in cluster 1, to which all the source task was assigned. In general, AP is more sensitive to the order at the top of the ranked list whereas AUC evaluates the overall number of correctly ranked pairs. In the case that AP is higher but not AUC, it indicates that the algorithm performs better at the top of the list; however, it doesn't create more correctly ranked pairs. To support this observation, we evaluated the results using Normalized Discounted Cumulative Gain (NDCG) at the rank position 5 and 10. Figure 3 shows that cluster 1 outperforms the other two clusters. The results suggest the occurrence of negative transfer when the learned classifier was used on less related datasets. The results also demonstrate how negative transfer could be minimized when the target task only learned from more informative segments in the source task.

In Experiment 6, all three clusters from the target task (Lincoln) have assigned instances from the source task (Microsoft). The combined result (the 'Total' row) outperforms the baseline (i.e., direct transfer of a classifier trained from the entire source task).

Table 10: Experiment results for TTST, breakdown by cluster.

| Experiment 4 | | | Experiment 5 | | | Experiment 6 | | |
|---|---|---|---|---|---|---|---|---|
| # | AUC | AP | # | AUC | AP | # | AUC | AP |
| 1 | 0.5082 | 0.5503 | 1 | 0.4472 | **0.6346†** | 1 | **0.6792†** | **0.7959†** |
| 2 | **0.6569†** | **0.7201†** | 2 | 0.4942 | 0.3641 | 2 | **0.6288†** | 0.495 |
| 3 | **0.8333†** | **0.8333†** | 3 | **0.5603†** | 0.4393 | 3 | **0.738†** | **0.6637†** |
| | | | | | | Total | **0.6627†** | **0.6426†** |

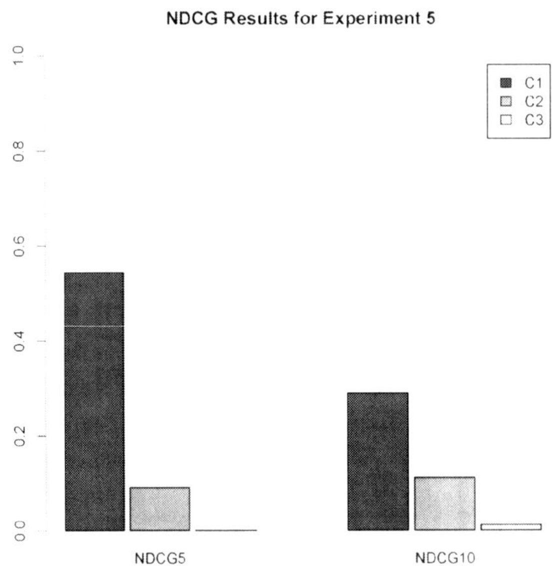


Figure 3: NDCG results for Experiment 5

## 5. Related Work

SOURCE TASK SELECTION   Research on multi-task learning and reinforcement learning has discussed the measurement of task relatedness and the selection of related tasks (Silver and McCracken, 2003; Ben-David and Borbely, 2008; Taylor et al., 2007). In this paper, we focus on the problem setting in which only a single source task is available, and no labels are available for the target task. Research has used semi-supervised learning methods such as EM algorithm combined with NaiveBayes classifier (Nigam et al., 2000) and co-clustering (Dai et al., 2007) to improve text classifiers. In contrast with current research, our approach does not require labeled data in the target task, selecting the source task segments solely based on the feature distribution.

A common approach to select a related source task is to measure the relatedness between the source and target task. Silver and Mercer (1996) employed a learning rate as a function to measure relatedness between a source task and the target task.

Kuhlmann and Stone (2007) constructed a graph for each task based on the elements and rules and then compared the graph isomorphism to select similar tasks. Thrun and O'Sullivan (1998) clustered multiple source tasks into a hierarchy. Their method transferred knowledge from the cluster most related to the target task to emphasize the knowledge among similar and discriminating instances. The authors used class label information to construct clusters, matching the target distribution of a given task with the most similar cluster.

CLASSIFIER REUSE   Knowledge transfer emphasizes the reuse of previously acquired knowledge, i.e. the classifiers, from the source task to the target task. A common approach to reuse classifiers is to select among candidate solutions from the source tasks. Zhang et al. (2005) constructed an ensemble of decision trees trained from related tasks to improve prediction on the problem with limited labeled data. Eaton and Desjardins (2006) developed an ensemble framework where each member classifier focuses on one resolution level. The multi-resolution learning facilitates transfer between related tasks. Yang et al. (2007) described methods to select auxiliary classifiers from a set of existing classifiers. The authors used the EM algorithm to estimate the distribution respective to each between-class score distribution, creating "pseudo" labels to evaluate each classifier and then selecting classifiers of average precision scores.

By comparison, our approach aims to transfer classifiers learned only from the related segment, as opposed to the entire set, of the source task. The experimental results demonstrate the promise of the proposed segmented transfer approach.

## 6. Conclusion and Future Work

In this paper, we investigated two *segmented transfer* approaches to transfer knowledge while avoiding negative transfer. The objective of the proposed approach is to address the heterogeneous characteristics of Wikipedia vandalism. We clustered the source and the target task to map unlabeled data from the target task to the most related cluster from the source task, classifying the unlabeled data using the most relevant learned models. Our results show enhanced performance (e.g. AUC and AP) on ranking the probable vandalism instances. In the future, we will explore the soft clustering method, assign each instance in the target task a probability of cluster membership, and combine predictions. We will also consider enhancing the methods' ability to avoid negative transfer by implementing an overall "relatedness" measure, so that points in the target task are not classified using distant clusters.

## Acknowledgments

Our special thanks go to anonymous reviewers and ICML'11 conference attendees for their constructive feedback to improve this work. This publication was made possible by Grant Number UL1RR024979 from the National Center for Research Resources (NCRR), a part of the National Institutes of Health (NIH). Its contents are solely the responsibility of the authors and do not necessarily represent the official views of the CTSA or NIH.

## References

S. Ben-David and R. S. Borbely. A notion of task relatedness yielding provable multiple-task learning guarantees. *Machine Learning*, 73:273–287, December 2008.

S.-C. Chin, W. N. Street, P. Srinivasan, and D. Eichmann. Detecting Wikipedia vandalism with active learning and statistical language models. In *Proceedings of the 4th Workshop on Information Credibility*, WICOW '10, page 3–10, 2010.

P. Clarkson and R. Rosenfeld. Statistical language modeling using the CMU-Cambridge toolkit. In *Fifth European Conference on Speech Communication and Technology*, pages 2707–2710, 1997.

W. Dai, G.-R. Xue, Q. Yang, and Y. Yu. Co-clustering based classification for out-of-domain documents. In *Proceedings of the 13th ACM SIGKDD International Conference on Knowledge Discovery and Data Mining*, KDD '07, page 210–219, 2007.

E. Eaton and M. Desjardins. Knowledge transfer with a multiresolution ensemble of classifiers. In *ICML Workshop on Structural Knowledge Transfer for Machine Learning*, 2006.

M. Hall, E. Frank, G. Holmes, B. Pfahringer, P. Reutemann, and I. H. Witten. The WEKA data mining software: An update. *SIGKDD Explor. Newsl.*, 11:10–18, November 2009. ISSN 1931-0145.

G. Kuhlmann and P. Stone. Graph-based domain mapping for transfer learning in general games. *Machine Learning: ECML 2007*, page 188–200, 2007.

K. Nigam, A.K. McCallum, S. Thrun, and T. Mitchell. Text classification from labeled and unlabeled documents using EM. *Machine Learning*, 39(2):103–134, 2000.

S. J. Pan and Q. Yang. A survey on transfer learning. *IEEE Transactions on Knowledge and Data Engineering*, 22(10):1345–1359, 2010.

M. Potthast and R. Gerling. Wikipedia vandalism corpus Webis-WVC-07, 2007. URL `http://www.uni-weimar.de/medien/webis/research/corpora`.

M. T. Rosenstein, Z. Marx, L. P. Kaelbling, and T. G. Dietterich. To transfer or not to transfer. In *NIPS'05 Workshop, Inductive Transfer: 10 Years Later*, 2005.

D. L. Silver and P. McCracken. Selective transfer of task knowledge using stochastic noise. In Y. Xiang and B. Chaib-draa, editors, *Advances in Artificial Intelligence*, volume 2671 of *Lecture Notes in Computer Science*, page 994–994. Springer Berlin/Heidelberg, 2003.

D. L. Silver and R. E. Mercer. The parallel transfer of task knowledge using dynamic learning rates based on a measure of relatedness. In *Connection Science Special Issue: Transfer in Inductive Systems*, pages 277–294, 1996.

M. E. Taylor, G. Kuhlmann, and P. Stone. Accelerating search with transferred heuristics. In *ICAPS-07 Workshop on AI Planning and Learning*, September 2007.

S. Thrun and J. O'Sullivan. Clustering learning tasks and the selective cross-task transfer of knowledge. In *Learning to Learn*, page 235–257. Kluwer, 1998.

J. Yang, R. Yan, and A. G. Hauptmann. Cross-domain video concept detection using adaptive SVMs. In *Proceedings of the 15th International Conference on Multimedia*, MULTIMEDIA '07, page 188–197, New York, NY, USA, 2007.

Y. Zhang, W.N. Street, and S. Burer. Sharing classifiers among ensembles from related problem domains. In *Fifth IEEE International Conference on Data Mining*, pages 522–529. IEEE Computer Society, 2005.

# Self-measuring Similarity for Multi-task Gaussian Process

**Kohei Hayashi**                                                    HAYASHI.KOHEI@GMAIL.COM
**Takashi Takenouchi**                                                    TTAKASHI@IS.NAIST.JP
*Graduate School of Information Science,*
*Nara Institute of Science and Technology,*
*8916-5 Takayama, Ikoma, Nara, 630-0192, Japan*

**Ryota Tomioka**                                                    TOMIOKA@MIST.I.U-TOKYO.AC.JP
**Hisashi Kashima**                                                    KASHIMA@MIST.I.U-TOKYO.AC.JP
*Department of Mathematical Informatics,*
*The University of Tokyo,*
*7-3-1 Hongo, Bunkyo-ku, Tokyo, 113-8656, Japan*


**Editor:** I. Guyon, G. Dror, V. Lemaire, G. Taylor, and D. Silver

## Abstract

Multi-task learning aims at transferring knowledge between similar tasks. The multi-task Gaussian process framework of Bonilla *et al.* models (incomplete) responses of $C$ data points for $R$ tasks (e.g., the responses are given by an $R \times C$ matrix) by using a Gaussian process; the covariance function takes its form as the product of a covariance function defined on input-specific features and an inter-task covariance matrix (which is empirically estimated as a model parameter). We extend this framework by incorporating a novel similarity measurement, which allows for the representation of much more complex data structures. The proposed framework also enables us to exploit additional information (e.g., the input-specific features) when constructing the covariance matrices by combining additional information with the covariance function. We also derive an efficient learning algorithm which uses an iterative method to make predictions. Finally, we apply our model to a real data set of recommender systems and show that the proposed method achieves the best prediction accuracy on the data set.

**Keywords:** Multi-task Gaussian process, conjugate gradient, collaborative filtering.

## 1. Introduction

Multi-task learning (Caruana, 1997) is a machine learning framework that aims to improve performance through the learning of multiple tasks at the same time, and sharing the information of each task. An application of multi-task learning is a recommender system. For example, a recommender system of movies makes recommendations of movies which a user may prefer based on the history of the user's preferences. Since the total number of movies is much larger than the number of movies which one user has watched in the past, one user's preferences are insufficient for accurate prediction. Multi-task learning enhances the prediction performance by regarding each user as a relevant task and by sharing the preferences information of the users whose preferences are similar to each other. Such techniques are called collaborative filtering and are widely applied in recommender systems.

Several methods for multi-task learning have been proposed (Pan and Yang, 2010), which includes a method based on a Gaussian process (GP) called the multi-task GP (Bonilla et al., 2008). GP models can make predictions as a distribution, and they provide not only a mean of the prediction but also a variance, which can be used as a reliability of the prediction. Another advantage of the GP approach is that the learning of a model and prediction making are consistently done within the Bayesian framework. As the GP models represent the similarities of each data sample as a covariance matrix, the multi-task GP parametrizes the similarities between tasks and the similarities between data points as two independent covariance matrices, which enables to transfer knowledge among different tasks and data points efficiently. The multi-task GP is a special case of a tensor GP (Yu et al., 2007), and it has various applications such as a robotic manipulation (Chai et al., 2009).

However, if either inputs or task-specific features are not provided, the multi-task GP cannot measure the similarities and thus we cannot construct the corresponding covariance matrix with kernel functions. Since task-specific features are generally difficult to obtain, the method proposed by Bonilla et al. estimates a full covariance matrix over tasks in an empirical Bayesian framework. They also provide a parametric estimation procedure for the covariance with low-rank constraint. The method proposed by Yu et al. (2007) directly estimates the outputs of the GP as parameters, which can be seen as a matrix-factorization-like approximation.

These parametric approaches work well when the model includes the true distribution of the observations, e.g., the observed responses have a low-rank structure. However, such modeling is sometimes too restrictive for real data. In addition, when given responses are sparse and have high dimensionality, i.e., the dimension of the covariance matrix is large compared to the number of the observations, the empirical estimation of the full covariance matrix would be unstable, which may cause a negative effect for prediction.

Another challenge is to improve scalability. The naive computation of the mean of the predictive distribution requires $O(M^3)$ complexity, where $M$ denotes the number of observed responses. When applying the multi-task GP to a large-scale data set, it is inevitable to introduce some approximations such as limitations of kernel functions (e.g., linear kernel) and low-rank approximations of the Gram matrix. The empirical Bayes approach also raises further the computational cost; it is only applicable when the number of tasks or data points is small.

In this paper, we propose a new GP framework for multi-task learning problems. Our main contributions are as follows:

**Self-measuring similarities** We use the responses themselves to measure the similarities between the tasks and the data points. The response-based similarities allow for the flexible representation of more complex data structures. Additional features (e.g., input-specific features) would enhance the prediction accuracy but that is not a requirement in our framework.

**Efficient and exact inference scheme**  We propose an efficient algorithm which computes the predictive mean of the GP with $O(RC(R+C))$ computational cost without any approximations, where $R$ and $C$ denote the number of tasks and data points, respectively.

We evaluate our method in a collaborative filtering problem with a real data set and show that it attains the lowest prediction error.

## 2. Multi-task Gaussian Process

Suppose that we are given $R$ tasks and the $i$-th task has incomplete outputs $x_{i:} \in \mathbb{R}^C$ which may contain unknown values. Our purpose is to predict unobserved elements in a response matrix $X = (x_{1:}^\top, \ldots, x_{R:}^\top)^\top \in \mathbb{R}^{R \times C}$ in which $x_{i:}$ denotes the $i$-th row vector of $X$. In some situations, additional information for data points $S = (s_1, \ldots, s_C)$ and those for tasks $T \equiv (t_1, \ldots, t_R)$ are also given. Figure 1 is a summary of the problem setting, which is particularly called multi-label learning, a special case of multi-task learning.

### 2.1. Notations

We define the "vec" operator which creates a vector "vec $X$" by stacking the column vectors of $X$, i.e.,

$$\text{vec } X = \begin{pmatrix} x_{:1} \\ \vdots \\ x_{:C} \end{pmatrix}$$

where $x_{:k}$ is the $k$-th column vector. We denote $\mathcal{I}$ as an index set of the observed elements, and we have $M = |\mathcal{I}|$ observations $\{x_{ik} | (i,k) \in \mathcal{I}\}$. We also denote $x_{\mathcal{I}}$ as an $M$-dimensional vector which contains the observed elements of $X$ in a certain order

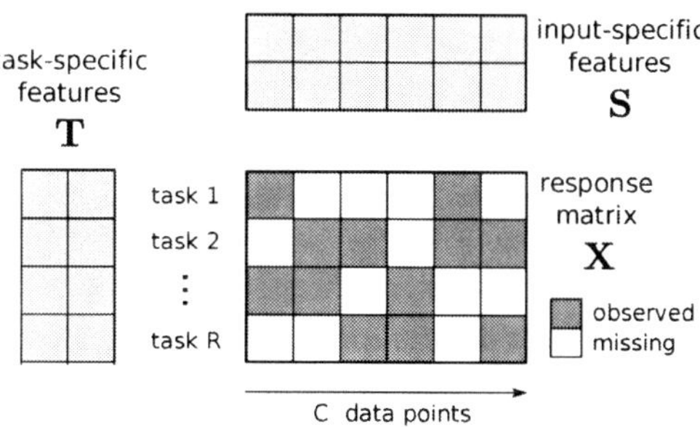


Figure 1:  A problem setting of multi-task learning in this paper. Additional information $S$ and $T$ are not indispensable.

without overlapping. For later convenience, we introduce an observation matrix $P \in \{0,1\}^{M \times RC}$ which removes the unobserved elements, i.e., $x_I = P(\text{vec } X)$.

## 2.2. Modeling, Learning, and Predicting

The multi-task GP models $X$ as

$$x_{ik} = m_{ik} + \mu + \varepsilon_{ik}, \quad \varepsilon_{ik} \sim N(0, \sigma^2) \tag{1}$$

where $\mu$ is a common bias and $\varepsilon_{ik}$ is an i.i.d. spherical Gaussian noise. $m_{ik}$ follows a tensor GP (Yu et al., 2007), which is defined as

$$m \sim \mathcal{GP}(0, \Sigma \otimes \Omega) \tag{2}$$

where $\otimes$ denotes the Kronecker product and both $\Sigma \in \mathbb{R}^{C \times C}$ and $\Omega \in \mathbb{R}^{R \times R}$ are symmetric and PSD. Each $\Sigma$ and $\Omega$ represents a covariance over columns and rows (i.e., data points and tasks), respectively. From the definition, the covariance between $x_{ik}$ and $x_{jl}$ is written as

$$\text{cov}[x_{ik}, x_{jl}] = \Omega_{ij}\Sigma_{kl} + \delta_{ij}\delta_{kl}\sigma^2 \tag{3}$$

where $\delta_{ij}$ is the Kronecker's delta, i.e., $\delta_{ij} = 1$ if $i = j$ otherwise $\delta_{ij} = 0$. Equation (3) shows that the multi-task GP assumes the similarity (i.e., covariance) between the two responses $x_{ik}$ and $x_{jl}$ can be factorized into product of the task similarity $\Omega_{ij}$ and the data point similarity $\Sigma_{kl}$. The joint distribution of $X$ is given by the Gaussian distribution

$$\int p(X \mid M, \mu, \sigma^2) p(M \mid \Sigma, \Omega) dM = N(\text{vec } X \mid \mu, K) \tag{4}$$

where $\mu = (\mu, \ldots, \mu) \in \mathbb{R}^{RC}$ and $K = \Sigma \otimes \Omega + \sigma^2 I$. Note that the multi-task GP is a special case of GP in which the covariance matrix has the structure of the Kronecker product.

By following the GP's framework, we need to determine the covariance matrices $\Sigma$ and $\Omega$ in a nonparametric way. On one hand, Bonilla et al. (2008) construct the covariance $\Sigma$ via a covariance function $k(\cdot, \cdot)$ with the input-specific features $S$ as

$$\Sigma_{kl} = k(s_k, s_l). \tag{5}$$

On the other hand, the task covariance $\Omega$ is estimated as the empirical Bayesian method. We consider obtaining a maximum-likelihood solution $\hat{\Omega}$ by maximizing the log-likelihood function of the observed elements, which is given by marginalizing out the unobserved elements from the joint distribution (4), i.e.,

$$\ln \int N(\text{vec } X \mid \mu, K) \prod_{(i,k) \notin I} dx_{ik}$$
$$= -\frac{1}{2}(x_I - \mu_I)^\top K_I^{-1}(x_I - \mu_I) - \frac{1}{2}\ln\det|K_I| + \text{const.} \tag{6}$$

where $\boldsymbol{\mu}_I = \boldsymbol{P}\boldsymbol{\mu}$ and $\boldsymbol{K}_I = \boldsymbol{P}\boldsymbol{K}\boldsymbol{P}^\top \in \mathbb{R}^{M \times M}$ is a covariance matrix over the observed elements. The common bias $\mu$ is also estimated as a maximum-likelihood solution $\hat{\mu} = \frac{1}{M} \sum_{(i,k) \in I} x_{ik}$.

Given the partially observed response matrix and the input-specific features $\mathcal{D} = \{X, S\}$, the multi-task GP predicts an unobserved response $x_{ab}$ by a corresponding mean of the predictive distribution

$$\mathbb{E}[x_{ab} \mid \mathcal{D}, \hat{\boldsymbol{\Omega}}] = (\boldsymbol{k}_a \otimes \hat{\omega}_b)_I^\top \boldsymbol{K}_I^{-1}(\boldsymbol{x}_I - \hat{\boldsymbol{\mu}}_I) + \hat{\mu} \tag{7}$$

where $\boldsymbol{k}_a = (k(s_a, s_1), \ldots, k(s_a, s_C))^\top$ and $\hat{\omega}_b$ denotes the $b$-th row vector of $\hat{\boldsymbol{\Omega}}$. Since Equation (7) contains the inverse $\boldsymbol{K}_I^{-1}$, the naive computational complexity of the predictive means is $O(M^3)$.

## 3. Multi-task Gaussian Process with Self-measuring Similarity

Now we extend the multi-task GP model. First we introduce a way to construct the covariance matrices from the response matrix itself. Then we derive an efficient learning algorithm using the conjugate gradient method.

### 3.1. Self-measuring Similarities

We construct the covariance matrices by using the responses themselves:

$$\Omega_{ij} = k(\boldsymbol{x}_{i:}, \boldsymbol{x}_{j:}) \quad \text{and} \quad \Sigma_{kl} = g(\boldsymbol{x}_{:k}, \boldsymbol{x}_{:l}), \tag{8}$$

where $k$ and $g$ are arbitrary PSD kernel functions. As previously mentioned that $\boldsymbol{x}_{i:}$ denotes the $i$-th row vector and $\boldsymbol{x}_{:k}$ denotes $k$-th column vector of $X$. We call this idea *Self-measuring Similarity*. The self-measuring similarity allows us to compute the covariance matrices without any additional information. The computation of the self-measuring similarity is simply done with evaluate the values of the covariance functions; it is much faster than the empirical Bayesian approach.

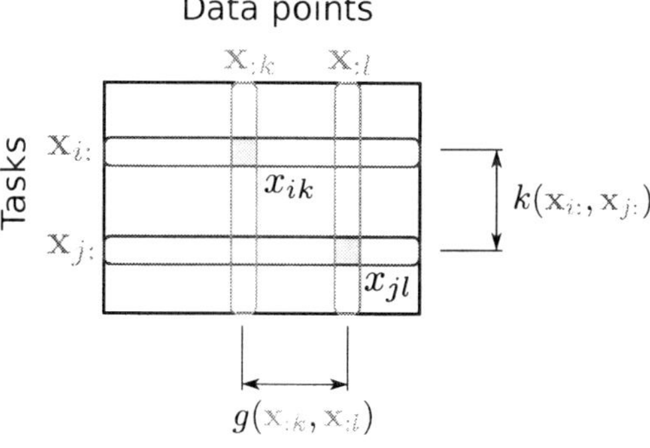


Figure 2: The idea of self-measuring similarities.

We introduce latent variables $\{z_{ik}\}$ for the missing elements $\{x_{ik}|(i,k) \notin I\}$ to compute the kernel function in which the input contains missing values. After learning the latent variables, we use the completed matrix $\tilde{X}$ as kernel inputs, where $\tilde{x}_{ik} = x_{ik}$ if $(i,k) \in I$ otherwise $\tilde{x}_{ik} = z_{ik}$. We use the means of the predictive distribution as estimators of the latent variables, i.e., $\hat{z}_{ik} = \mathbb{E}[x_{ik}|\mathcal{D}]$ in a heuristic way.[1] Note that, however, the EM-like heuristic can be seen as an approximate marginalization (see Appendix A for more details).

Since the values of $\{\hat{z}_{ik}\}$ affect the predictive mean (7) through the kernel functions, the heuristic can be performed iteratively. Note that if the number of the missing elements (i.e. the latent variables) is larger than that of the observed elements, many iterations may cause over-fitting. We avoid this problem by early stopping of the iterations with a randomly picked validation set. For an initial value of $\hat{z}_{ik}$, we use a row-wise mean $\bar{x}_{i\cdot}$ or a column-wise mean $\bar{x}_{\cdot k}$:

$$\bar{x}_{i\cdot} = \frac{1}{M_{i\cdot}} \sum_{k,(i,k)\in I} x_{ik}, \quad \bar{x}_{\cdot k} = \frac{1}{M_{\cdot k}} \sum_{i,(i,k)\in I} x_{ik},$$

where $M_{i\cdot}$ and $M_{\cdot k}$ are the number of observed elements of $x_{i\cdot}$ and $x_{\cdot k}$, respectively.

If we have additional information such as input-specific features $\{s_k|k = 1,\ldots,C\}$ for each data point, we exploit them by combining them with the self-measuring covariance function. For example, we extend the covariance function into a sum form

$$\Sigma_{kl} = g(x_{:k}, x_{:l}) + g'(s_k, s_l) \tag{9}$$

or a product form

$$\Sigma_{kl} = g(x_{:k}, x_{:l})g'(s_k, s_l). \tag{10}$$

Note that if both $g$ and $g'$ are PSD, then the resulting kernel functions are still PSD (Rasmussen and Williams, 2006).

## 3.2. Computation for Prediction

As mentioned, we need to compute the inverse of $K_I$ in Equation (7) for the predictive means, and it requires $O(M^3)$ computational cost. Instead, we solve the linear equation

$$x_I - \hat{\mu}_I = K_I \beta \tag{11}$$

with respect to $\beta$. Note that $K_I$ is positive definite when $\sigma^2 > 0$. After obtaining the solution $\hat{\beta}$, we simply compute the predictive mean of $x_{ab}$ as an inner product

$$\mathbb{E}[x_{ab} \mid \mathcal{D}, \hat{Z}] = (g_a \otimes k_b)_I^{\top} \hat{\beta} + \hat{\mu}. \tag{12}$$

---

1. One of the proper estimation methods for the latent variables is the empirical Bayesian. However, the optimization of the log-likelihood (4) with respect to $\{z_{ik}\}$ is computationally infeasible especially for large-scale data.

We solve Equation (11) using the conjugate gradient method (Shewchuk, 1994), which is an iterative method to solve a linear system in which the matrix is positive definite. Each iteration of the conjugate gradient needs to perform a matrix-vector multiplication; in our case that corresponds to the multiplication of $K_I$ and an $M$-dimensional vector, which requires $O(M^2)$ computation and $O(M)$ memory space.

The computational cost of the multiplication can still be reduced by exploiting a special structure in the matrix. Because of the structure of the Kronecker product in $K_I$, a multiplication of $K_I$ and an $M$-dimensional vector $v$ is rewritten as

$$K_I v = P(\Sigma \otimes \Omega + \sigma^2 I) P^\top v$$
$$= \text{vec } P(\Omega V \Sigma + \sigma^2 V) \tag{13}$$

where $V \in \mathbb{R}^{R \times C}$ is a matrix form of $P^\top v$, i.e., vec $V = P^\top v$. This technique, called *vec-trick* (Vishwanathan et al., 2007; Kashima et al., 2009), reduces the computational complexity from $O(M^2)$ to $O(RC(R+C))$.

Suppose we stop the iteration of the conjugate gradient when the $\ell_2$ error of $\hat{\beta}_l$ (i.e., between the solution at the $l$-th iteration and a true solution $\beta_*$) is less than the error of the initial values $\hat{\beta}_0$ with a tolerance $\epsilon$, i.e., $\|\beta_* - \hat{\beta}_l\|_2 \leq \epsilon \|\beta_* - \hat{\beta}_0\|_2$. Then the maximum number of iterations is bounded

$$l \leq \frac{1}{2} \sqrt{\kappa} \log\left(\frac{2}{\epsilon}\right) \tag{14}$$

where $\kappa$ is the condition number of $K_I$, which is defined as the ratio of the maximum and the minimum eigenvalue of $K_I$. The total cost for obtaining the solution $\hat{\beta}$ is $O(\sqrt{\kappa} RC(R+C))$.

With an observation rate $\alpha$, the number of observations has a relation $M = \alpha RC$. If $X$ is nearly square, i.e., $R \simeq C$, then the computational complexity can be rewritten as $O((\sqrt{\kappa}/\alpha^{\frac{3}{2}})M^{\frac{3}{2}})$, which is much faster than the naive complexity $O(M^3)$. Finally we summarize the entire algorithm as a pseudo code in Algorithm 1.

---

**Algorithm 1** Computation of predictive means with conjugate gradient.

---

1. Initialize $\hat{Z}_0$ with row-means or column-means of $X$

2. For $l = 1$ to maximum number of iterations

    (a) Construct $\Sigma$ and $\Omega$ with $\hat{Z}_{l-1}$ and additional features

    (b) Solve $x_I = K_I \beta$ by the conjugate gradient with a tolerance $\epsilon$

    (c) Compute predictive means $\{\mathbb{E}[x_{ab} \mid \mathcal{D}, \hat{Z}_{l-1}] \mid (a,b) \notin I\}$ of unobserved elements

    (d) Construct $\hat{Z}_l$ from the predictive means

3. Return $\hat{Z}_l$

---

## 4. Experimental Results

We evaluate the proposed method by applying it to a collaborative filtering problem. We use the Movielens 100k data set[2], which contains $100,000$ ratings $x_{ik} \in \{1,2,3,4,5\}$ for $1,682$ movies (data points) labeled by 943 users (tasks). The observation ratio $\alpha$ is $\simeq 0.06$. The data set contains user-specific features $S$ (e.g., age, gender, ...) and movie-specific features $T$ (release date, genre, ...). The data set provides $90,570$ ratings for training and remaining $9,430$ ratings for testing. After learning with the training data set, we evaluate the root-mean-squared-error (RMSE) for the testing data set. All experiments are done with a Xeon 2.93 GHz 8 core machine.

In the experiment, we prepare three forms of covariance functions: a covariance measured by the user-specific and the movie-specific features ("Feature"), a self-measuring covariance ("Self-measuring"), and a combination of them with the product form (10) ("Product"). Note that our model with "Feature" setting is equivalent to the kernel method proposed by Bonilla et al. (2007). We summarize the covariance functions in Table 1.

Table 1: Settings of the covariance functions.

|  | Feature | Self-measuring | Product |
|---|---|---|---|
| $\Sigma_{ij}$ | $k(s_i, s_j)$ | $k(x_{i:}, x_{i:})$ | $k(x_{i:}, x_{i:})k(s_i, s_j)$ |
| $\Omega_{kl}$ | $g(t_k, t_l)$ | $g(x_{:k}, x_{:l})$ | $g(x_{:k}, x_{:l})g(t_k, t_l)$ |

We use the RBF kernel $k(x, x') = g(x, x') = \exp(-\lambda \|x - x'\|^2)$ for the covariance functions. We choose the hyper-parameters as $(\sigma^2, \lambda) = (0.5, 0.001)$ for "Feature" and $(\sigma^2, \lambda) = (0.1, 0.1)$ for {"Self-measuring", "Product"}, selected by three-fold cross validation from candidates $\sigma^2 \in \{1, 0.5, 0.1, 0.05\}$ and $\lambda \in \{10^{-1}, 10^{-2}, 10^{-3}, 10^{-4}, 10^{-5}\}$. We set the tolerance of the conjugate gradient $\epsilon$ as $10^{-3}$. As the initial values of $\{z_{ik}\}$ we use the row-mean for $\Sigma$ and the column mean for $\Omega$.

For fair comparison, the number of the EM-like iteration is determined by early stopping with a validation set randomly drawn 5% of the training set. We compare with standard methods for recommendation system which includes user- and movie-based $K$-nearest neighbour (KNN) with the Pearson correlation and matrix factorization (see Su and Khoshgoftaar (2009) for more details.) We use `MyMediaLite`[3] as these implementations and set the hyper-parameters by following the examples specially recommended for the Movielens 100k dataset.

We summarize the prediction errors in Table 2. The result shows that "Feature" is the worst performance; it suggests that the additional information-based similarity is not enough to capture the observed data structure, and the self-measuring similarity is much more important for the prediction. Nevertheless, the combination with the additional information ("Product") further improves the prediction performance com-

---

2. `http://www.grouplens.org/node/73`
3. `http://www.ismll.uni-hildesheim.de/mymedialite`

Table 2: RMSEs on the Movielens 100k dataset. Lower RMSE indicates higher prediction performance. (Left) The RMSE behaviour of the EM-like heuristic. $l$ indicates the number of iterations. (Right) Existing methods v.s. proposed method with early stopping.

| $l$ | Feature | Self-measuring | Product |
|---|---|---|---|
| 1 | 1.0517 | 0.9431 | 0.9393 |
| 2 | – | 0.9276 | 0.9231 |
| 3 | – | 0.9329 | 0.9292 |
| 4 | – | 0.9439 | 0.9410 |

| Method | RMSE | time |
|---|---|---|
| User-KNN | 0.9507 | 7s |
| Movie-KNN | 0.9354 | 42s |
| Matrix Factorization | 0.9345 | 1m38s |
| Feature | 1.0517 | 7m01s |
| Self-measuring | 0.9308 | 16m22s |
| Product | **0.9256** | 18m25s |

pared to using the self-measuring alone ("Self-measuring"). The left panel of Table 2 shows the EM-like heuristic drastically improves the prediction accuracy at the second iteration, while the third or further iterations produce worse results. The best score ("Product") in the right panel of Table 2 is also the best over other 76 methods listed in `mlcomp.org`[4] as of May, 2011.

## 5. Conclusion

In this paper, we have presented a new framework to solve multi-task problems by using a Gaussian process based on self-measuring similarities. We proposed the efficient algorithm based on the conjugate gradient method with the vec-trick. Our method achieved the best score of the Movielens 100k data set.

## Acknowledgments

We thank the anonymous reviewers for insightful comments and Mauricio Alexandre Parente Burdelis for helpful advices.

## References

Edwin V. Bonilla, Felix V. Agakov, and Christopher K. I. Williams. Kernel multi-task learning using task-specific features. In *Proceedings of the 11th International*

---

4. `http://mlcomp.org/datasets/341`

*Conference on Artificial Intelligence and Statistics*. Omnipress, March 2007. URL `aistats07.pdf`.

Edwin V. Bonilla, Kian M. Chai, and Christopher K. I. Williams. Multi-task Gaussian process prediction. In J. C. Platt, D. Koller, Y. Singer, and S. Roweis, editors, *Advances in Neural Information Processing Systems 20*. MIT Press, Cambridge, MA, 2008. URL `nips08.pdf`.

Rich Caruana. Multitask learning. *Machine Learning*, 28(1):41–75, July 1997. ISSN 08856125. doi: 10.1023/A:1007379606734. URL `http://dx.doi.org/10.1023/A:1007379606734`.

Kian M. Chai, Christopher Williams, Stefan Klanke, and Sethu Vijayakumar. Multi-task gaussian process learning of robot inverse dynamics. In *NIPS 2008*, http://eprints.pascal-network.org/archive/00004640/, 2009.

Hisashi Kashima, Tsuyoshi Kato, Yoshihiro Yamanishi, Masashi Sugiyama, and Koji Tsuda. Link propagation: A fast semi-supervised learning algorithm for link prediction. In *SDM*, pages 1099–1110. SIAM, 2009.

Sinno J. Pan and Qiang Yang. A survey on transfer learning. *Knowledge and Data Engineering, IEEE Transactions on*, 22(10):1345–1359, October 2010. ISSN 1041-4347. doi: 10.1109/TKDE.2009.191. URL `http://dx.doi.org/10.1109/TKDE.2009.191`.

Carl E. Rasmussen and Christopher K. I. Williams. *Gaussian Processes for Machine Learning*. The MIT Press, November 2006. ISBN 026218253X. URL `http://www.amazon.com/exec/obidos/redirect?tag=citeulike07-20&path=ASIN/026218253X`.

J. R. Shewchuk. An introduction to the conjugate gradient method without the agonizing pain. Technical report, Carnegie Mellon University, Pittsburgh, PA, USA, 1994. URL `http://portal.acm.org/citation.cfm?id=865018`.

Xiaoyuan Su and Taghi M. Khoshgoftaar. A survey of collaborative filtering techniques. *Adv. in Artif. Intell.*, 2009:1–19, January 2009. ISSN 1687-7470. URL `http://dx.doi.org/10.1155/2009/421425`.

S. V. N. Vishwanathan, Karsten M. Borgwardt, and Nicol N. Schraudolph. Fast computation of graph kernels. In B. Schölkopf, J. Platt, and T. Hoffman, editors, *Advances in Neural Information Processing Systems 19*, pages 1449–1456. MIT Press, Cambridge, MA, 2007.

Kai Yu, Wei Chu, Shipeng Yu, Volker Tresp, and Zhao Xu. Stochastic relational models for discriminative link prediction. In B. Schölkopf, J. Platt, and T. Hoffman, editors, *Advances in Neural Information Processing Systems 19*, pages 1553–1560. MIT Press, Cambridge, MA, 2007.

## Appendix A.  Interpretation of EM-like Heuristic

The EM-like heuristic can be interpreted as an approximation of the marginalization of the predictive distribution with respect to $Z$. First, given estimated latent variables $\hat{Z}_{l-1}$, we consider to use the predictive distribution $p(X|\mathcal{D}, \hat{Z}_{l-1})$ as a posterior of $Z$, i.e., $p(Z|\mathcal{D}) \approx p(X|\mathcal{D}, \hat{Z}_{l-1})$. If we further approximate the posterior distribution as a delta function $\Delta(Z - \mathbb{E}[X|\mathcal{D}, \hat{Z}_{l-1}])$, then we have

$$\int \mathbb{E}[X \mid \mathcal{D}, Z] p(Z \mid \mathcal{D}) \mathrm{d}Z \approx \int \mathbb{E}[X \mid \mathcal{D}, Z] \Delta(Z - \mathbb{E}[X|\mathcal{D}, \hat{Z}_{l-1}]) \mathrm{d}Z = \hat{Z}_l$$

Note that $(\hat{Z}_l)_{ab}$ is also a predictive mean for an observation $x_{ab}$.

# Transfer Learning for Auto-gating of Flow Cytometry Data


**Gyemin Lee**       GYEMIN@EECS.UMICH.EDU

**Lloyd Stoolman**       STOOLMAN@UMICH.EDU

**Clayton Scott**       CSCOTT@EECS.UMICH.EDU

*University of Michigan, Ann Arbor, MI, USA*



**Editor:** I. Guyon, G. Dror, V. Lemaire, G. Taylor, and D. Silver


## Abstract


Flow cytometry is a technique for rapidly quantifying physical and chemical properties of large numbers of cells. In clinical applications, flow cytometry data must be manually "gated" to identify cell populations of interest. While several researchers have investigated statistical methods for automating this process, most of them falls under the framework of unsupervised learning and mixture model fitting. We view the problem as one of transfer learning, which can leverage existing datasets previously gated by experts to automatically gate a new flow cytometry dataset while accounting for biological variation. We illustrate our proposed method by automatically gating lymphocytes from peripheral blood samples.

**Keywords:** flow cytometry, automatic gating, transfer learning, low-density separation


## 1. Introduction

Flow cytometry is a technique widely used in many clinical and biomedical laboratories for rapid cell analysis (Shapiro, 1994). It plays an important role in the diagnosis of blood-related diseases such as acute or chronic leukemias and malignant lymphomas.

Mathematically, a flow cytometry data can be represented as $\mathcal{D} = \{\mathbf{x}_i\}_{i=1}^{N}$, where $\mathbf{x}_i \in \mathbb{R}^d$ is an attribute vector of the $i$th cell. The attributes include the cell's size (FS), granularity (SS) and expression levels of different antigens (CD45, CD3, CD4, etc.). The number of cells $N$ can range from 10,000 to 1,000,000, and $d$ is usually between 7-12. In clinical settings, each data corresponds to a particular patient, where the cells are typically drawn from a peripheral blood, lymph node, or bone marrow sample.

To make a diagnosis, a pathologist will use a computer to visualize different two-dimensional scatter plots of a flow cytometry data as in Fig. 1. These plots illustrate the presence of several clusters of cells within each dataset. They also illustrate the variation of measured data from one patient to another. This variation arises from both biological (e.g., health condition) and technical (e.g., instrument calibration) sources.

The pathologist will typically visualize a certain type of cell (e.g., lymphocytes in the diagnosis of leukemias) and diagnose based on its shape, range and other distributional characteristics. A necessary preprocessing is to label every cell as belonging to the cell type of interest or not, a process known as "gating." This amounts to assigning binary labels $y_i \in \{-1, 1\}, i = 1, \ldots, N$, to every cell. Fig. 1 indicates lymphocytes with an alternate color. Without gating, cells of other types will overlap with the targeted cell type in the scatter plots used for diagnosis. After gating, only the cells of interest are visualized.

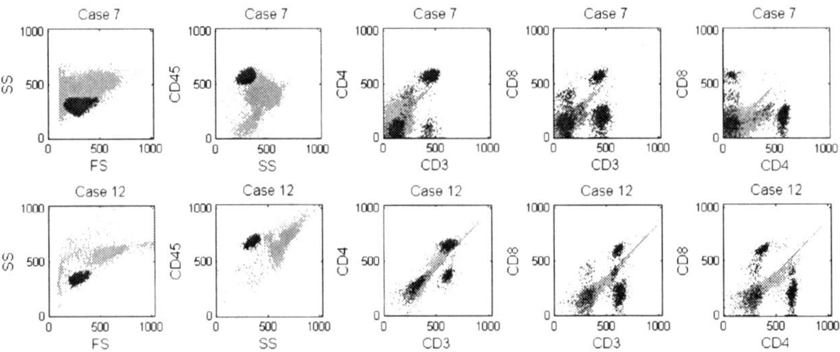

Figure 1: Clinicians analyze flow cytometry data using a series of scatter plots on attributes pairs. The distribution differs from patient to patient, and changes after treatments. Lymphocytes, a type of white blood cell, are marked dark-/blue and others are marked bright/green. These were manually selected by a domain expert.

Unfortunately, in clinical settings gating is still performed manually. It is a labor-intensive task in which a pathologist visualizes the data from different two-dimensional scatter plots, and uses special software to draw a series of boundaries ("gates") to eliminate a portion of the cells outside of the desired type. The person performing gating must utilize specialized domain knowledge together with iterative refinement. Since modern clinical laboratories can see dozens of cases per day, it would be highly desirable to automate this process.

Because of this need, several researchers have tackled the auto-gating problem. A recent survey on flow cytometry analysis revealed that more than 70% of studies focus on auto-gating techniques (Bashashati and Brinkman, 2009). However, the vast majority of approaches rely on a clustering/mixture modeling, using a parametric representation for each cell type (Chan et al., 2008; Lo et al., 2008; Pyne et al., 2009). The mixture modeling approach has a number of difficulties, however. One is that the clusters are typically not elliptical, meaning complex parametric models must be employed, such as skewed Gaussians, leading to challenging inference problems. Another limitation is that human intervention is necessary to interpret the clustering results and to select some of the clusters for the task. Finally, these algorithms are unsupervised, and do not fully leverage expert knowledge.

We propose to view auto-gating as a transfer learning problem. In particular, we assume that a collection of expert-gated datasets are available. Although different datasets have different distributions, there is enough similarity, (e.g., lymphocytes show low levels of SS while expressing high levels of CD45) that this expert knowledge can be transferred to the new data. Our approach is to train classifiers on expert-gated data, and to summarize these classifiers to form a baseline classifier. This baseline is then adapted to the new data by optimizing a "low-density separation" criterion. The transfer learning problem we study is, to our knowledge, a new one, although it has similarities to previously studied transfer learning problems, as well as multi-task learning. These connections are reviewed below.

## 2. Problem Setup

There are $M$ labeled datasets $\mathcal{D}_m = \{(\mathbf{x}_{m,i}, y_{m,i})\}_{i=1}^{N_m}$, $m = 1, \cdots, M$, each a random sample from a distribution $\mathcal{P}_m$. $\mathcal{D}_m$ corresponds to the $m$th flow cytometry dataset and its labels are determined by experts. There is also an unlabeled dataset $\mathcal{T} = \{\mathbf{x}_{t,i}\}_{i=1}^{N_t}$, a random sample from a new distribution $\mathcal{P}_t$ corresponding to a new flow cytometry dataset. The labels $\{y_{t,i}\}_{i=1}^{N_t}$ are not observed. The goal is to assign labels $\{\widehat{y}_{t,i}\}_{i=1}^{N_t}$ to $\mathcal{T}$ so that the misclassification rate is minimized. All the distributions are different, but defined on the same space $\mathbb{R}^d \times \{-1, +1\}$.

## 3. Related Work

As a transfer learning problem, our problem is characterized by having multiple source domains (the expert-gated datasets), and a single target domain (the unlabeled dataset). Using the taxonomy of Pan and Yang (2010), our setting can be described as follows:

(1) the source and target domains are different, because the marginal distributions of $\mathbf{x}$ are different,

(2) the source and target tasks are different, because each dataset requires a different gating,

(3) there are no labeled examples in the target domain.

To the best of our knowledge, previous work has not addressed this combination of characteristics. Many previous works fall under the heading of *inductive transfer learning*, where at least a few labels are given for the target domain (Ando and Zhang, 2005; Rettinger et al., 2006). In *transductive transfer learning* (Arnold et al., 2007), and the related problems of sample selection bias and covariate shift, the source and target tasks are assumed to be the same.

Another closely related area is multi-task learning (Caruana, 1997; Evgeniou and Pontil, 2004). However, our problem contrasts to this line of studies in the sense that our ultimate goal is achieving high performance for the target task only, and not the source tasks.

Toedling et al. (2006) explore using support vector machines (SVMs) for flow cytometry data from multiple patients. They merge all the datasets to form a single large data, and build a classifier on this data. However, due to its size of the combined dataset, the training requires demanding computational and memory resources. Furthermore, this approach ignores the variability among multiple datasets and treats all the datasets as arising from the same distribution. This reduces the problem to standard single-task supervised learning.

## 4. Algorithm

We describe our approach to the problem. In this section, we show how our algorithm summarizes knowledge from the source data and adapts it to the new task.

### 4.1. Baseline Classifier for Summarizing Expert Knowledge

We suppose that the knowledge contained in the source tasks can be represented by a set of decision functions $f_1, \cdots, f_M$. The sign of a decision function $f_m$ provides a class prediction of a data point $\mathbf{x}$: $\widehat{y} = sign(f_m(\mathbf{x}))$. Each $f_m$ is separately learned from each of the $M$ source datasets. Then these decision functions form the pool of knowledge.

In this work, we consider linear decision functions $f(\mathbf{x}) = \langle \mathbf{w}, \mathbf{x} \rangle + b$. Then $f$ defines a hyperplane $\{\mathbf{x} : f(\mathbf{x}) = 0\}$ with a normal vector $\mathbf{w} \in \mathbb{R}^d$ and a bias $b \in \mathbb{R}$. The SVM is among the most widely used methods for learning a linear classifier (Schölkopf and Smola, 2002). It finds a separating hyperplane based on the maximal margin principle. We use the SVM to fit a decision function $f_m$ or a hyperplane $(\mathbf{w}_m, b_m)$ to the $m$th source data $\mathcal{D}_m$.

We devise a baseline classifier $f_0 = \langle \mathbf{w}_0, \mathbf{x} \rangle + b_0$ by letting $(\mathbf{w}_0, b_0)$ be the mean of $(\mathbf{w}_m, b_m)$. Instead of the simple mean, Algorithm 1 uses a robust mean to prevent $f_0$ from being unduly influenced by atypical variations among datasets. Algorithm 2 presents the robust estimator as formulated in Campbell (1980). Here $\psi$ is a weight function corresponding to a robust loss, and we use the Huber loss function. Note that we also robustly estimate the covariance of the $\mathbf{w}_m$, which is used below in Section 4.2.3.

The learning of $f_0$ does not involve $\mathcal{T}$ at all. Thus, it is not expected to provide a good prediction for the target task. Next we describe a way to adapt this baseline classifier to the target data based on the low-density separation principle.

### 4.2. Transferring Knowledge to Target Task

Low-density separation is a concept used extensively in machine learning. This notion forms the basis of many algorithms in clustering analysis, semi-supervised classification, novelty detection and transductive learning. The underlying intuition is that the decision boundaries between clusters or classes should pass through regions where the marginal density of $\mathbf{x}$ is low. Thus, our approach is to adjust the hyperplane parameters so that it passes through a region where the marginal density of $\mathcal{T}$ is low.

#### 4.2.1. Preprocessing

Instrument calibration often introduces different shifting and scaling to each flow cytometry data along coordinate axis. While a typical solution is aligning all datasets via some global $d$-dimensional shift/scale transformation, it is sufficient for our purposes to align datasets in the direction of the baseline normal vector $\mathbf{w}_0$. Specifically, for each

---

**Algorithm 1** Baseline Classifier

**Input:** source task data $\mathcal{D}_m$ for $m = 1, \cdots, M$, regularization parameters $\{C_m\}_{m=1}^{M}$
1: **for** $m = 1$ **to** $M$ **do**
2: $\quad (\mathbf{w}_m, b_m) \leftarrow SVM(\mathcal{D}_m, C_m)$
3: **end for**
4: Robust Mean:
$\quad (\mathbf{w}_0, b_0) \leftarrow Algorithm\ 2(\{(\mathbf{w}_m, b_m)\}_m)$
**Output:** $(\mathbf{w}_0, b_0)$ or $f_0(\mathbf{x}) = \langle \mathbf{w}_0, \mathbf{x} \rangle + b_0$

---

**Algorithm 2** Robust Mean and Covariance

**Input:** $(\mathbf{w}_m, b_m)$ for $m = 1, \cdots, M$
1: Concatenate: $\mathbf{u}_m \leftarrow [\mathbf{w}_m, b_m], \quad \forall m$
2: Initialize: $\mu \leftarrow mean(\mathbf{u}_m), \quad \mathbf{C} \leftarrow cov(\mathbf{u}_m)$

3: **repeat**
4: $\quad d_m \leftarrow \left( (\mathbf{u}_m - \mu)^T \mathbf{C}^{-1} (\mathbf{u}_m - \mu) \right)^{1/2}$
5: $\quad w_m \leftarrow \psi(d_m)/d_m$
6: $\quad$ Update: $\mu^{new} \leftarrow \dfrac{\sum_m w_m \mathbf{u}_m}{\sum_m w_m}$
$$\mathbf{C}^{new} \leftarrow \frac{\sum_m w_m^2 (\mathbf{u}_m - \mu^{new})(\mathbf{u}_m - \mu^{new})^T}{\sum_m w_m^2 - 1}$$
7: **until** Stopping conditions are satisfied
**Output:** $\mu = [\mathbf{w}_0, b_0], \quad \mathbf{C}_0 = \mathbf{C}(1 : d, 1 : d)$

---

dataset, we compute a kernel density estimate (KDE) of the projection onto $\mathbf{w}_0$. Then, we align the target data to each source data using maximum cross-correlation (denoted by $\star$ in Algorithm 3), and modify the baseline bias by the median of these shifts. This new bias $b$ will serve as the initial bias when adapting the baseline to $\mathcal{T}$.

### 4.2.2. VARYING BIAS

We first describe adapting the bias variable to the unlabeled target data $\mathcal{T}$ based on low-density separation. The process is illustrated in Algorithm 4.

To assess whether a linear decision boundary is in a low density region, we count data points near the hyperplane. As the hyperplane moves, this number will be large in a high density region and small in a low density region. In particular, we define a margin, as in SVMs, to be a region of a fixed distance from a hyperplane, say $\Delta$, and count data points within this margin. We use $\Delta = 1$. Given a hyperplane $(\mathbf{w}, b)$, basic linear algebra shows that $\frac{\langle \mathbf{w}, \mathbf{x} \rangle + b}{\|\mathbf{w}\|}$ is the signed distance from $\mathbf{x}$ to the hyperplane. Hence, computing

$$\sum_i \mathbb{I}\{ \frac{|\langle \mathbf{w}, \mathbf{x}_{l,i} \rangle + b|}{\|\mathbf{w}\|} < \Delta \}$$

over a range of $b$ followed by locating a minimizer near the baseline hyperplane gives the desired solution. Algorithm 4 implements this on a grid of biases $\{s_j\}$ and builds $\sum_j c_j \delta(z - s_j)$ where $\delta$ is the Dirac delta. The grid points and the counts at each grid point are denoted by $s_j$ and $c_j$.

Before searching for the minimizing bias, we smooth these counts over the grid by convolving with a Gaussian kernel $k_h(z, z') = \frac{1}{\sqrt{2\pi}h} \exp\left(-\frac{|z-z'|^2}{2h^2}\right)$. The bandwidth $h$ controls the smoothness of the kernel. This operation yields a smooth function $\widehat{p}(z) = \sum_j c_j k_h(z, s_j)$. Running a gradient descent algorithm on this smoothed function $\widehat{p}(z)$ returns a local minimum near 0 if initialized at 0 (the second parameter in Line 8).

To facilitate a streamlined process for practical use, we automatically select the kernel bandwidth $h$ as shown in Algorithm 5. This kernel choice is motivated from the rule of thumb for kernel density estimation suggested in Silverman (1986).

**Algorithm 3** Shift Compensation

**Input:** hyperplane $(\mathbf{w}, b)$, source task data $\{\mathcal{D}_m\}_{m=1}^{M}$, target task data $\mathcal{T}$
1: $z_{t,i} \leftarrow \langle \mathbf{w}, \mathbf{x}_{t,i} \rangle + b, \quad \forall i$
2: **for** $m = 1$ **to** $M$ **do**
3: $\quad z_{m,i} \leftarrow \langle \mathbf{w}, \mathbf{x}_{m,i} \rangle + b, \quad \forall i$
4: $\quad e_m \leftarrow \arg\max_z \mathrm{KDE}(z, z_{t,i}) \; \mathrm{KDE}(z, z_{m,i})$ $\qquad\star$
5: **end for**
6: $b \leftarrow b - median(e_m)$
**Output:** $b$

**Algorithm 4** Bias Update

**Input:** hyperplane $(\mathbf{w}, b)$, target task data $\mathcal{T}$
1: Compute: $\quad z_i \leftarrow \langle \mathbf{w}, \mathbf{x}_{t,i} \rangle + b, \quad \forall i$
2: Build a Grid: $\quad s_i \leftarrow sort\,(z_i)$
3: **for** $j = 1$ **to** $N_t$ **do**
4: $\quad c_j \leftarrow \sum_i \mathbb{I}\{\frac{|z_i - s_j|}{\|\mathbf{w}\|} < 1\}$
5: **end for**
6: $h \leftarrow kernel\ bandwidth\,(\{(s_j, c_j)\}_j)$
7: Smooth: $\quad \widehat{p}(z) \leftarrow \sum_j c_j k_h(z, s_j)$
8: $z^* \leftarrow gradient\ descent\,(\widehat{p}(z), 0)$
9: $b^{new} \leftarrow b - z^*$
**Output:** $b^{new}$ or $f_b(\mathbf{x}) = \langle \mathbf{w}, \mathbf{x} \rangle + b^{new}$

### 4.2.3. VARYING NORMAL VECTOR

We can also adjust the normal vector of a hyperplane. Given a normal vector $\mathbf{w}$, we update the normal vector by $\mathbf{w}^{new} = \mathbf{w} + a_t \mathbf{v}_t$ where $\mathbf{v}_t$ is the direction of change and $a_t$ is the amount of the change. Thus, the new normal vector is from an affine space spanned by $\mathbf{v}_t$.

Now we explain in detail the ways of choosing $\mathbf{v}_t$ and $a_t$. We find a direction of change from the covariance matrix of the normal vectors $\mathbf{w}_1, \cdots, \mathbf{w}_M$ obtained from Algorithm 2. We choose the first principal eigenvector for $\mathbf{v}_t$ after making it orthogonal to $\mathbf{w}_0$, the baseline normal vector, because changes in the direction of $\mathbf{w}_0$ do not affect the decision boundary.

To determine the amount of change $a_t$, we proceed similarly to the method used to update the bias. We count the number of data points inside the margin as the normal vector varies by a regular increment in the direction of $\mathbf{v}_t$. Convolving with a Gaussian kernel smooths these quantities over the range of variation. Then a gradient descent algorithm can spot $a_t$ that leads to a low density solution near the baseline hyperplane. Algorithm 6 summarizes this process.

### 4.2.4. PUTTING IT ALL TOGETHER

Once the normal vector is updated, we build a new hyperplane by combining it with an updated bias so that the hyperplane accords to a low density region of $\mathcal{T}$. Algorithm 7 outlines the overall scheme. In the algorithm, one can repeatedly update the bias and the normal vector. A simple method is running a fixed number of times. In our experience, one round of iteration was sufficient for good solutions.

Although the presented algorithm limits the varying direction of normal vector to a single vector $\mathbf{v}_t$, we can generalize this to multiple directions. To do this, more than one eigenvector can be chosen in Step 5 of Algorithm 7. Then the Gram-Schmidt process (Golub and Van Loan, 1996) generates a set of orthonormal vectors that spans a subspace for a new normal vector. The counting in-margin points in Algorithm 6 can be extended to a multivariate grid with little difficulty.

**Algorithm 5** Kernel Bandwidth

**Input:** grid points and counts $\{(s_k, c_k)\}$

1: $N \leftarrow \sum_k c_k$
2: $\bar{s} \leftarrow \frac{1}{N} \sum_k s_k c_k$
3: $\hat{\sigma} \leftarrow \left( \frac{1}{N-1} \sum_k c_k (s_k - \bar{s})^2 \right)^{1/2}$
4: $h \leftarrow 0.9 \cdot \hat{\sigma} \cdot N^{-1/5}$

**Output:** $h$

---

**Algorithm 6** Normal Vector Update

**Input:** hyperplane $(\mathbf{w}, b)$, direction of change $\mathbf{v}_t$, target task data $\mathcal{T}$

1: **for** $a_k = -0.5$ **to** $0.5$ **step** $0.01$ **do**
2: $\quad \mathbf{w}_k \leftarrow \mathbf{w} + a_k \mathbf{v}_t$
$\quad c_k \leftarrow \sum_i \mathbb{I}\{ \left| \frac{\langle \mathbf{w}_k, \mathbf{x}_{t,i} \rangle + b}{\|\mathbf{w}_k\|} \right| < 1 \}$
3: **end for**
4: $h \leftarrow kernel\ bandwidth(\{(a_k, c_k)\}_k)$
5: Smooth: $\quad g(a) \leftarrow \sum_k c_k k_h(a, a_k)$
6: $a_t \leftarrow gradient\ descent\ (g(a), 0)$
7: $\mathbf{w}^{new} \leftarrow \mathbf{w} + a_t \mathbf{v}_t$

**Output:** $\mathbf{w}^{new}$

---

**Algorithm 7** Set Estimation based on Low-Density Separation

**Input:** source task data $\{\mathcal{D}_m\}_{m=1}^M$, target task data $\mathcal{T}$, regularization parameters $\{C_m\}_{m=1}^M$

1: **for** $m = 1$ **to** $M$ **do**
2: $\quad (\mathbf{w}_m, b_m) \leftarrow SVM(\mathcal{D}_m, C_m)$
3: **end for**
4: Initialize:
$\quad ((\mathbf{w}_0, b_0), \mathbf{C}_0) \leftarrow Alg\ 2(\{(\mathbf{w}_m, b_m)\}_m)$
$\quad \mathbf{v}_0 \leftarrow eig(\mathbf{C}_0)$
5: Normalize:
$\quad \mathbf{w}_t \leftarrow \mathbf{w}_0 / \|\mathbf{w}_0\|, \qquad b_t \leftarrow b_0 / \|\mathbf{w}_0\|$
$\quad \mathbf{v}_t \leftarrow$
$\quad orthonormalize\ \mathbf{v}_0\ with\ respect\ to\ \mathbf{w}_0$

6: Compensate Shift:
$\quad b_t \leftarrow Alg\ 3(\mathbf{w}_t, b_t, \{\mathcal{D}_m\}, \mathcal{T})$
7: Update Bias:
$\quad b_t \leftarrow Alg\ 4(\mathbf{w}_t, b_t, \mathcal{T})$
8: **repeat**
9: $\quad$ Update Normal Vector:
$\quad \mathbf{w}_t \leftarrow Alg\ 6(\mathbf{w}_t, b_t, \mathbf{v}_t, \mathcal{T})$
10: $\quad$ Update Bias:
$\quad b_t \leftarrow Alg\ 4(\mathbf{w}_t, b_t, \mathcal{T})$
11: **until** Stopping conditions are satisfied

**Output:** $(\mathbf{w}_t, b_t)$ or $f_t = \langle \mathbf{w}_t, \mathbf{x} \rangle + b_t$

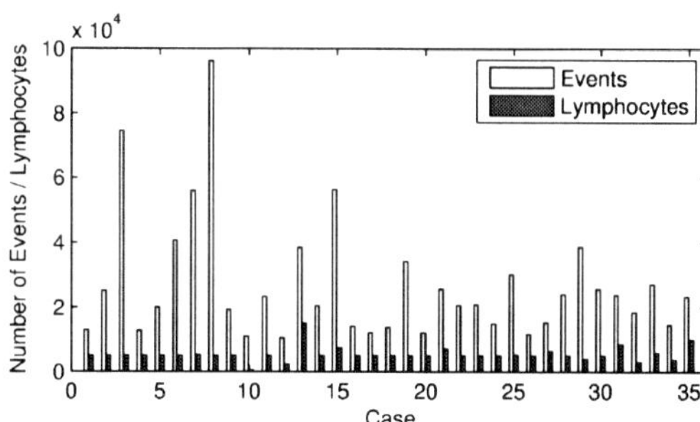


Figure 2: Number of total events and lymphocytes in each flow cytometry dataset.

## 5. Experiments

We demonstrate the proposed methods on clinical flow cytometry data. Specifically, we apply them to detect lymphocytes from peripheral blood samples. In diagnosing diseases such as leukemias, identifying these cells is the first step in most clinical analysis. Since the gating tools for this task are still primitive and unsatisfactory, a streamlined automatic gating procedure will be highly valuable in practice.

For the experiments, peripheral blood sample datasets were obtained from 35 normal patients. These datasets are provided by the Department of Pathology at the University of Michigan. The number of events in a dataset ranges from $10,000$ to $100,000$ with a varying portion of lymphocytes among them (see Fig. 2). An event in a dataset has six attributes (FS, SS, CD45, CD4, CD8 and CD3) and a corresponding binary label ($+1$ for lymphocytes and $-1$ for others) from the manual gates set by experts (see Fig. 1).

For the experiments, we adopt a leave-one-out setting: choose a dataset as a target task $\mathcal{T}$, hide its labels, and treat the other datasets as source tasks $\mathcal{D}_m$. Each $\mathcal{D}_m$ constitutes a binary classification problem with the goal of predicting the correct labels.

On each source data $\mathcal{D}_m$, we trained a SVM classifier $f_m$. We used LIBSVM package (Chang and Lin, 2001) with the default setting ($C_m = 1$). Then we applied the algorithms described in Section 4 and evaluated the prediction accuracy on the unlabeled target data $\mathcal{T}$. The considered transfer learning algorithms are:

- $f_0$ : baseline classifier with no adaptation, referred to as "baseline."
- $f_b$ : classifier adapted to $\mathcal{T}$ by varying the bias-only, referred to as "bias."
- $f_t$ : classifier adapted to $\mathcal{T}$ by varying both the direction and the bias, referred to as "dir. and bias."

In addition to the above classifiers, we compared the error rates from the following classifiers as points of reference:

- `Pooling` : A SVM is trained after merging all source data as in Toedling et al. (2006).
- `Transductive` : A transductive SVM is also trained on the merged source data and the target data using SVM-light package (Joachims, 1999).
- $f_m$ for $m = 1, \cdots, M$ : Each $f_m$ learned from a source data $\mathcal{D}_m$ is applied straight to $\mathcal{T}$. This emulates a supervised learning setup with a train sample $\mathcal{D}_m$ and a test sample $\mathcal{T}$ while implicitly assuming $\mathcal{D}_m$ and $\mathcal{T}$ are drawn from the same distribution. A box plot in Fig. 3 displays the range of results with some '+' indicating extreme values. Table 1 numbers $f_m^{Best}$, the best of the 34 error rates.
- `Oracle` : We also applied the standard SVM with the true labels of $\mathcal{T}$. Its performance is computed by 5-fold cross validation. This quantity is what we can expect when a sufficient amount of labeled data are available for the target task.

For each dataset, we repeated these and reported their results in Fig. 3 and Table 1.

As can be seen in the figure and table, applying one of the $f_m$ to the target task result in a wide range of accuracy. A classifier performing well on a dataset can perform poorly on other datasets due to relative difficulty of a task, dataset shift, or dissimilarity between tasks.

The `Pooling` performs poorly on many datasets. The merging step of the `Pooling` makes the classification problem more difficult. Even if classes are well-separated in each dataset, the separation will be lost in the merged dataset. Additionally, the classifier from `Pooling` can be biased toward larger source data. The optimization algorithm might also terminate prematurely before it converges to an optimal solution because the merged dataset is very large. Therefore, the `Pooling` can perform poorly, sometimes even worse than the worst $f_m$.

Because `Transductive` merges all the labeled and unlabeled datasets for training, it shares the similar problems as the `Pooling`. Moreover, the objective function is non-convex, and the solutions from an optimization algorithm are often suboptimal. The obtained results are very high error rates and low gating quality as shown in the table.

The baseline classifier $f_0$ typically improves when we adapt $f_0$ by changing the bias variable in most cases except Case 14 and Case 23. They further improve by adaptively varying both the direction and the bias. The differences among the $f_m^{Best}$, `Oracle` and $f_t$ are very small. This reveals that our strategy can successfully replicate what experts do in the field without labeled training set for the target task.

## 6. Conclusion

We cast flow cytometry auto-gating as a novel kind of transfer learning problem. By combining existing ideas from transfer learning, together with a low-density separation criterion for class separation, our approach can leverage expert-gated datasets for the automatic gating of a new unlabeled dataset.

Although linear classifiers are sufficient to gate lymphocytes in peripheral blood, nonlinear classifiers may be necessary for other kinds of auto-gating. For example, a bone marrow sample contains cells of whole range of developmental stages and is known to be more difficult to gate. Depending on diseases being screened for, other types of cells need to be separated or the separated lymphocytes need to be further gated. Our approach accommodates the incorporation of inner-product kernels, which may offer a solution to such problems. It is also quite likely that several other strategies from the transfer learning literature can be adapted to this problem.

Biological and technical variation pose challenges for the analysis of many types of biomedical data. Typically one or both types of variation is accounted for by performing task-independent "normalization" prior to analysis. Our approach to flow cytometry auto-gating can be viewed as a task-dependent approach. The application of transfer learning to overcome biological and/or technical variations in other kinds of biomedical data is an interesting problem for future work.

Table 1: The error rates (%) of various classifiers on each flow cytometry dataset, with the other 34 treated as labeled datasets. The results from $f_t$ adapted to the unlabeled target data are comparable to Oracle trained on labeled target data, and make less errors than Pooling.

| Case | Trans | Pool | $f_0$ | $f_b$ | $f_t$ | $f_m^B$ | Oracle |
|---|---|---|---|---|---|---|---|
| 1 | 44.64 | 38.65 | 2.91 | 3.05 | 3.17 | 2.87 | 2.79 |
| 2 | 70.66 | 20.27 | 5.44 | 2.09 | 2.10 | 2.06 | 1.71 |
| 3 | 18.46 | 2.05 | 1.66 | 1.00 | 0.94 | 0.91 | 0.74 |
| 4 | 19.27 | 2.93 | 2.62 | 2.54 | 2.67 | 2.44 | 2.56 |
| 5 | 34.10 | 5.06 | 1.50 | 1.40 | 1.44 | 1.41 | 1.58 |
| 6 | 31.90 | 1.60 | 1.84 | 1.60 | 1.80 | 1.62 | 1.56 |
| 7 | 90.82 | 7.00 | 0.91 | 0.82 | 0.77 | 0.80 | 0.79 |
| 8 | 89.28 | 2.44 | 0.65 | 0.60 | 0.50 | 0.52 | 0.47 |
| 9 | 16.92 | 8.31 | 2.19 | 1.91 | 1.83 | 1.78 | 1.71 |
| 10 | 37.14 | 26.65 | 2.16 | 1.09 | 1.09 | 1.03 | 1.03 |
| 11 | 73.47 | 2.67 | 5.11 | 1.86 | 1.86 | 1.77 | 1.79 |
| 12 | 65.48 | 21.89 | 6.69 | 1.60 | 1.63 | 1.78 | 1.54 |
| 13 | 58.44 | 39.44 | 1.69 | 1.63 | 1.65 | 1.59 | 1.64 |
| 14 | 75.60 | 3.67 | 2.29 | 3.55 | 0.87 | 0.71 | 0.81 |
| 15 | 32.68 | 5.90 | 1.78 | 1.16 | 1.22 | 1.11 | 1.11 |
| 16 | 64.41 | 4.34 | 3.79 | 3.19 | 3.23 | 2.82 | 2.83 |
| 17 | 56.92 | 7.70 | 2.75 | 3.49 | 3.51 | 2.47 | 2.44 |
| 18 | 63.88 | 2.53 | 1.86 | 1.64 | 1.67 | 1.59 | 1.60 |
| 19 | 84.48 | 8.25 | 3.44 | 3.45 | 3.14 | 2.46 | 2.29 |
| 20 | 31.01 | 3.03 | 4.48 | 2.39 | 2.37 | 2.56 | 2.45 |
| 21 | 63.61 | 10.14 | 7.71 | 6.28 | 6.30 | 5.64 | 5.08 |
| 22 | 18.69 | 4.16 | 1.60 | 1.81 | 1.82 | 1.54 | 1.42 |
| 23 | 75.95 | 21.73 | 2.89 | 7.51 | 1.58 | 1.61 | 1.43 |
| 24 | 29.89 | 2.79 | 2.41 | 2.06 | 2.06 | 1.91 | 1.89 |
| 25 | 6.57 | 1.98 | 2.22 | 2.25 | 2.32 | 2.04 | 1.47 |
| 26 | 56.89 | 1.55 | 2.13 | 1.82 | 1.83 | 1.42 | 1.39 |
| 27 | 56.83 | 11.34 | 11.22 | 9.02 | 9.18 | 8.17 | 8.72 |
| 28 | 53.73 | 2.21 | 1.68 | 2.23 | 2.17 | 1.56 | 1.48 |
| 29 | 80.58 | 9.19 | 1.06 | 0.96 | 0.97 | 0.77 | 0.73 |
| 30 | 14.20 | 7.80 | 1.25 | 1.24 | 1.25 | 1.25 | 1.24 |
| 31 | 61.83 | 16.08 | 13.46 | 4.57 | 4.59 | 4.80 | 4.45 |
| 32 | 80.56 | 20.39 | 12.66 | 2.62 | 2.62 | 2.72 | 2.21 |
| 33 | 75.87 | 5.57 | 4.58 | 5.74 | 5.74 | 2.28 | 1.77 |
| 34 | 20.82 | 4.66 | 2.10 | 1.90 | 1.93 | 1.79 | 1.80 |
| 35 | 55.84 | 9.33 | 6.68 | 5.46 | 5.49 | 5.56 | 5.59 |
| avg | 51.76 | 9.80 | 3.70 | 2.73 | 2.49 | 2.21 | 2.12 |
| std err | 4.12 | 1.68 | 0.54 | 0.33 | 0.30 | 0.26 | 0.27 |

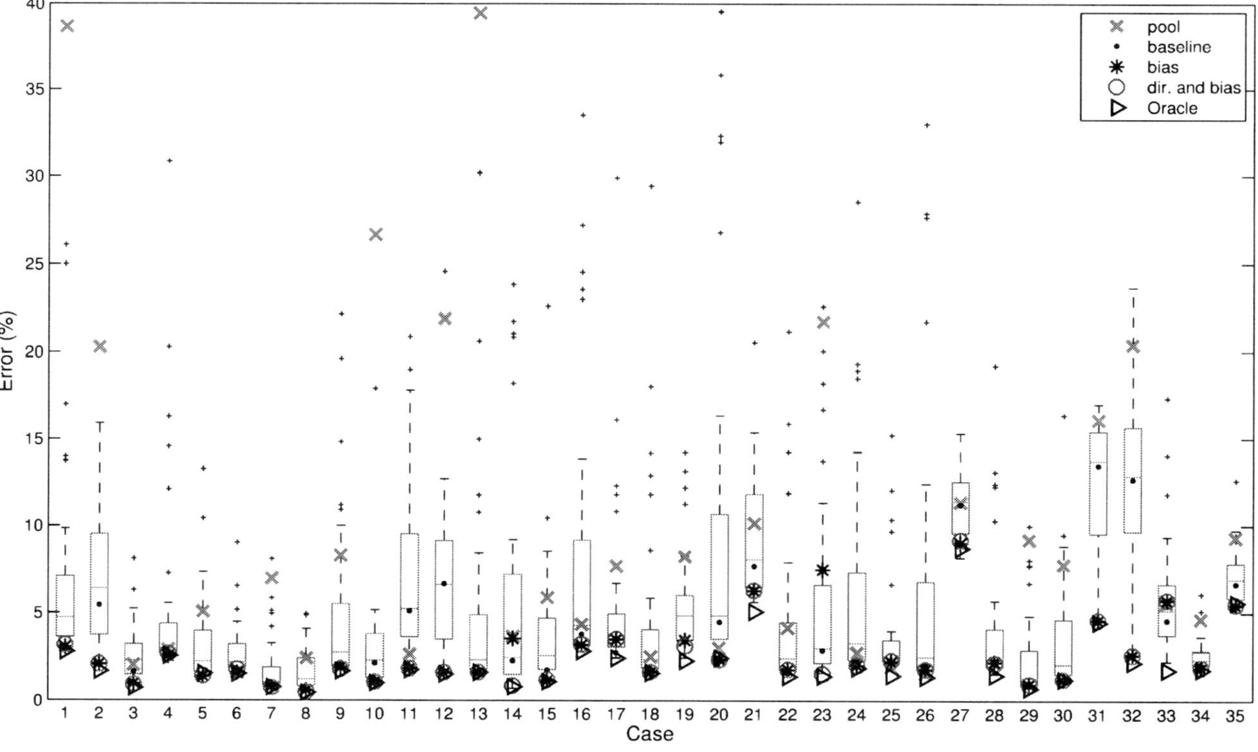


Figure 3: The error rates of various classifiers on each unlabeled dataset. The box plot corresponds to the range of results when one of $f_m$ is applied to $\mathcal{T}$. '+' marks an extreme value that deviates from the others. The results from $f_t$, `Oracle` and the best of $f_m$ are usually indistinguishable.

## References


R. K. Ando and T. Zhang. A high-performance semi-supervised learning method for text chunking. *Proceedings of the 43rd Annual Meeting on Association for Computational Linguistics (ACL 05)*, pages 1–9, 2005.

A. Arnold, R. Nallapati, and W.W. Cohen. A comparative study of methods for transductive transfer learning. *Seventh IEEE International Conference on Data Mining Workshops*, pages 77–82, 2007.

A. Bashashati and R. R. Brinkman. A survey of flow cytometry data analysis methods. *Advances in Bioinformatics*, 2009:Article ID 584603, 2009. doi: 10.1155/2009/584603.

N. A. Campbell. Robust procedures in multivariate analysis I: Robust covariance estimation. *Journal of the Royal Statistical Society. Series C (Applied Statistics)*, 29: 231–237, 1980.

R. Caruana. Multitask learning. *Machine Learning*, 28:41–75, 1997.

C. Chan, F. Feng, J. Ottinger, D. Foster, M. West, and T.B. Kepler. Statistical mixture modeling for cell subtype identification in flow cytometry. *Cytometry Part A*, 73: 693–701, 2008.

C. Chang and C. Lin. *LIBSVM: a library for support vector machines*, 2001. Software available at http://www.csie.ntu.edu.tw/~cjlin/libsvm.

T. Evgeniou and M. Pontil. Regularized multi–task learning. *Proceedings of the tenth ACM SIGKDD international conference on Knowledge Discovery and Data mining (KDD 04)*, pages 109–117, 2004.

G. H. Golub and C. F. Van Loan. *Matrix computations*. Johns Hopkins University Press, 1996.

T. Joachims. Making large-scale SVM learning practical. In B. Schölkopf, C. Burges, and A. Smola, editors, *Advances in Kernel Methods - Support Vector Learning*, chapter 11, pages 169–184. MIT Press, Cambridge, MA, 1999.

K. Lo, R. R. Brinkman, and R. Gottardo. Automated gating of flow cytometry data via robust model-based clustering. *Cytometry Part A*, 73:321 – 332, 2008.

S. J. Pan and Q. Yang. A survey on transfer learning. *IEEE Transactions on Knowledge and Data Engineering*, 22:1345–1359, 2010.

S. Pyne, X. Hu, K. Wang, E. Rossin, T. Lin, L. M. Maier, C. Baecher-Allan, G. J. McLachlan, P. Tamayo, D. A. Hafler, P. L. De Jager, and J. P. Mesirov. Automated high-dimensional flow cytometric data analysis. *PNAS*, 106:8519–8524, 2009.

A. Rettinger, M. Zinkevich, and M. Bowling. Boosting expert ensembles for rapid concept recall. *Proceedings of the 21st National Conference on Artificial Intelligence (AAAI 06)*, 1:464–469, 2006.

B. Schölkopf and A.J. Smola. *Learning with Kernels*. MIT Press, Cambridge, MA, 2002.

H. Shapiro. *Practical Flow Cytometry*. Wiley-Liss, 3rd edition, 1994.

B. W. Silverman. *Density Estimation for Statistics and Data Analysis.* Chapman and Hall, London, 1986.

J. Toedling, P. Rhein, R. Ratei, L. Karawajew, and R. Spang. Automated in-silico detection of cell populations in flow cytometry readouts and its application to leukemia disease monitoring. *BMC Bioinformatics*, 7:282, 2006.

# Inductive Transfer for Bayesian Network Structure Learning*

**Alexandru Niculescu-Mizil**                                ALEX@NEC-LABS.COM
*NEC Laboratories America, 4 Independence Way, Princeton, NJ 08540*

**Rich Caruana**                                    RCARUANA@MICROSOFT.COM
*Microsoft Research, One Microsoft Way, Redmond, WA 98052-6399*

**Editor:** I. Guyon, G. Dror, V. Lemaire, G. Taylor, and D. Silver

## Abstract

We study the multi-task Bayesian Network structure learning problem: given data for multiple related problems, learn a Bayesian Network structure for each of them, sharing information among the problems to boost performance. We learn the structures for all the problems *simultaneously* using a *score and search* approach that encourages the learned Bayes Net structures to be similar. Encouraging similarity promotes information sharing and prioritizes learning structural features that explain the data from all problems over features that only seem relevant to a single one. This leads to a significant increase in the accuracy of the learned structures, especially when training data is scarce.

## 1. Introduction

Bayes Nets (Pearl, 1988) provide a compact description of the dependency structure of a domain by using a directed acyclic graph to encode probabilistic dependencies between variables. The ability to learn this structure from data makes Bayes Nets an appealing data analysis tool, as the learned structure can convey, in an intuitive manner, a wealth of information about the domain at hand.

Until now, Bayes Net structure learning research has focused on learning *a single* structure for a single problem (task) in isolation (e.g. learn the structure for only one species of yeast from the gene expression data from that one species alone) (e.g. Cooper and Hersovits, 1992; Heckerman, 1999; Spirtes et al., 2000; Teyssier and Koller, 2005). In many situations, however, we are faced with multiple problems (tasks) that are related in some way (e.g. learn about the gene regulatory structure of several species of yeast, not just one). In these cases, rather than learning the Bayes Net structure for each problem in isolation, and ignoring the relationships with the other tasks, it would be beneficial to learn all the structure for all the problems jointly. Indeed, the transfer learning literature (e.g. Caruana, 1997; Baxter, 1997; Thrun, 1996) suggests that significant benefits can be obtained by *transferring* relevant information among the related problems.

---

* A version of this paper appeared in (Niculescu-Mizil and Caruana, 2007). The work was done while both authors were at Cornell University.

In this paper we present a transfer learning approach that jointly learns multiple Bayesian Network structures from multiple related datasets. We follow a score and search approach, where the search is performed over *sets of DAGs* rather than over single DAGs as in case of traditional structure learning. We derive a principled measure of the quality of a *set of structures* that rewards both a good fit of the training data as well as high similarity between the structures in the set. This score is then used to guide a greedy hill climbing procedure in a properly defined search space to find a high quality set of Bayes Net structures.

We evaluate the proposed technique on problems generated from the benchmark ALARM (Beinlich et al., 1989) and INSURANCE (Binder et al., 1997) networks, as well as on a real bird ecology problem. The results of the empirical evaluation show that learning the Bayes Net structures jointly in a multi-task manner does indeed yield a boost in performance and leads to learning significantly more accurate structures than when learning each structure independently. As with other transfer learning techniques, the benefit is especially large when the training data is scarce.

## 2. Background: Learning the Bayes Net Structure for a Single Problem

A Bayesian Network $\mathcal{B} = \{G, \theta\}$ compactly encodes the joint probability distribution of a set of $n$ random variables $X = \{X_1, X_2, \ldots, X_n\}$. It is specified by a directed acyclic graph (DAG) $G$ and a set of conditional probability functions parametrized by $\theta$ (Pearl, 1988). The Bayes Net *structure*, $G$, encodes the probabilistic dependencies in the data: the presence of an edge between two variables means that there exists a direct dependency between them. An appealing feature of Bayes Nets is that the dependency graph $G$ is easy to interpret and can be used to aid understanding the problem domain.

Given a dataset $D = \{x^1, \ldots, x^m\}$ where each $x^i$ is a complete assignment of variables $X_1, \ldots, X_n$, it is possible to learn both the structure $G$ and the parameters $\theta$ (Cooper and Hersovits, 1992; Heckerman, 1999; Spirtes et al., 2000). In this paper we will focus on structure learning, and more specifically on the score and search approach to it.

Following the Bayesian paradigm, the posterior probability of the structure given the data is estimated via Bayes rule:

$$P(G \mid D) \propto P(G)P(D \mid G) \tag{1}$$

The prior, $P(G)$, indicates the belief, before seeing any data, that the structure $G$ is correct. If there is no reason to prefer one structure over another, one should assign the same probability to all structures. If there exists a known ordering on the nodes in $G$ such that all the parents of a node precede it in the ordering, a prior can be assessed by specifying the probability that each of the $n(n-1)/2$ possible arcs is present in the correct structure (Buntine, 1991). Alternately, when there is access to a structure believed to be close to the correct one (e.g. from an expert), $P(G)$ can be specified by penalizing each difference between $G$ and the given structure by a constant factor (Heckerman et al., 1995).

The marginal likelihood, $P(D \mid G)$, is computed by integrating over all parameter values:

$$P(D \mid G) = \int P(D \mid G, \theta) P(\theta \mid G) d\theta \tag{2}$$

When the local conditional probability distributions are from the exponential family, the parameters $\theta_i$ are mutually independent, we have conjugate priors for these parameters, and the data is complete, $P(D \mid G)$ can be computed in closed form (Heckerman, 1999).

Treating the posterior, $P(G \mid D)$, as a score, one can search for a high scoring network using heuristic search (Heckerman, 1999). Greedy search, for example, starts from an initial structure, evaluates the score of all the *neighbors* of that structure and moves to the neighbor with the highest score. A common definition of the *neighbor* of a structure G is a DAG obtained by removing or reversing an existing arc in G, or by adding an arc that is not present in G. The search terminates when the current structure is better than all it's neighbors. Because it is possible to get stuck in a local minima, this procedure is usually repeated a number of times starting from different initial structures.

## 3. Learning Bayes Net Structures for Multiple Related Problems

In the previous section we reviewed how to learn the Bayes Net structure for a single problem. What if we have data for a number of related problems (e.g., gene expression data for several species) and we want to jointly learn Bayes Net structures for each of them?

Given $k$ data-sets, $D_1, \ldots, D_k$, defined on overlapping but not necessarily identical sets of variables, we want to learn the structures of the Bayes Nets $\mathcal{B}_1 = \{G_1, \theta_1\}, \ldots, \mathcal{B}_k = \{G_k, \theta_k\}$. In what follows, we will use the term *configuration* to refer to a set of structures $(G_1, \ldots, G_k)$.

From Bayes rule, the posterior probability of a configuration given the data is:

$$P(G_1, \ldots, G_k \mid D_1, \ldots, D_k) \propto P(G_1, \ldots, G_k) P(D_1, \ldots, D_k \mid G_1, \ldots, G_k) \tag{3}$$

The marginal likelihood $P(D_1, \ldots, D_k \mid G_1, \ldots, G_k)$ is computed by integrating over all parameter values for all the $k$ networks:

$$P(D_1, \ldots, D_k \mid G_1, \ldots, G_k) =$$

$$\int P(D_1, \ldots, D_k \mid G_1, \ldots, G_k, \theta_1, \ldots, \theta_k) \cdot P(\theta_1, \ldots, \theta_k \mid G_1, \ldots, G_k) d\theta_1 \ldots d\theta_k$$

$$= \int P(\theta_1, \ldots, \theta_k \mid G_1, \ldots, G_k) \prod_{p=1}^{k} P(D_p \mid G_p, \theta_p) d\theta_1 \ldots d\theta_k \tag{4}$$

If we make the parameters of different networks independent *a priori* (i.e. $P(\theta_1, \ldots, \theta_k \mid G_1, \ldots, G_k) = P(\theta_1 \mid G_1) \ldots P(\theta_k \mid G_k)$ ), the marginal likelihood factorizes into the product of the marginal likelihoods of each data set given its network structure. In this case the posterior probability of a configuration is:

$$P(G_1,..,G_k \mid D_1,..,D_k) \propto P(G_1,..,G_k) \prod_{p=1}^{k} P(D_p \mid G_p) \tag{5}$$

Making the parameters independent *a priori* is unfortunate, and contradicts the intuition that related problems should have similar parameters, but it is needed in order to make the learning efficient (see Section 3.3). Note that this is not a restriction on the model. Unlike Naive Bayes for instance, where the attribute independence assumption restricts the class of models that can be learned, here the learned parameters will be similar if the data supports it. The only downside of making the parameters independent *a priori* is that it prevents multi-task structure learning from taking advantage of the similarities between the parameters during the structure learning phase. After the structures have been learned, however, such similarities could be leveraged to learn more accurate parameters. Finding ways to allow for some *a priori* parameter dependence while still maintaining computational efficiency is an interesting direction for future work.

## 3.1. The Prior

The prior knowledge of how related the different problems are and how similar their structures should be is encoded in the prior $P(G_1,\ldots,G_k)$. If there is no reason to believe that the structures for each task should be related, then $G_1,\ldots,G_k$ should be made independent *a priori* (i.e. $P(G_1,\ldots,G_k) = P(G_1) \cdot \ldots \cdot P(G_k)$). In this case the structure-learning can be performed independently on each problem.

At the other extreme, if the structures for all the different tasks should be identical, the prior $P(G_1,\ldots,G_k)$ should put zero probability on any configuration that contains nonidentical structures. In this case one can efficiently learn the same structure for all tasks by creating a new data set with attributes $X_1,\ldots,X_n,TSK$, where $TSK$ encodes the problem each case is coming from.[1] Then learn the structure for this new data set under the restriction that $TSK$ is always the parent of all the other nodes. The common structure for all the problems is exactly the learned structure, with the node $TSK$ and all the arcs connected to it removed. This approach, however, does not easily generalize to the case where the problems have only partial overlap in their attributes.

Between these two extremes, the prior should encourage configurations with similar network structures. One way to generate such a prior for two structures is to penalize each arc $(X_i, X_j)$ that is present in one structure but not in the other by a constant $\delta \in [0, 1]$:

$$P(G_1, G_2) = Z_\delta \cdot (P(G_1)P(G_2))^{\frac{1}{1+\delta}} \prod_{\substack{(X_i,X_j)\in \\ G_1 \Delta G_2}} (1 - \delta) \tag{6}$$

where $Z_\delta$ is a normalization constant and $G_1 \Delta G_2$ represents the symmetric difference between the edge sets of the two DAGs (in case some variables are only present in one of the tasks, arcs connected to such variables are not counted).

---

1. This is different from pooling the data, which would mean that not only the structures, but also the parameters will be identical for all problems.

If $\delta = 0$ then $P(G_1,G_2) = P(G_1)P(G_2)$, so the structures are learned independently. If $\delta = 1$ then $P(G_1,G_2) = \sqrt{P(G)P(G)} = P(G)$ for $G_1 = G_2 = G$ and $P(G_1,G_2) = 0$ for $G_1 \neq G_2$, leading to learning identical structures for all problems. For $\delta$ between 0 and 1, the higher the penalty, the higher the probability of more similar structures. The advantage of this prior is that $P(G_1)$ and $P(G_2)$ can be any structure priors that are appropriate for the task at hand.

One way to interpret the above prior is that it penalizes by $\delta$ each *edit* (i.e. arc addition, arc removal or arc reversal) that is necessary to make the two structures identical (arc reversals can count as one or two edits). This leads to a natural extension to more than two tasks: penalizes each edit that is necessary to obtain a set of identical structures:

$$P(G_1,\ldots,G_k) = Z_{\delta,k} \cdot \prod_{1 \leq s \leq k} P(G_s)^{\frac{1}{1+(k-1)\delta}} \times \prod_{i,j} (1-\delta)^{edits_{i,j}} \tag{7}$$

where $edits_{i,j}$ is the minimum number of edits necessary to make the arc between $X_i$ and $X_j$ the same in all the structures. We will call this prior the *Edit* prior. The exponent $1/(1+(k-1)\delta)$ is used to transition smoothly between the case where structures should be independent (i.e. $P(G_1,\ldots,G_k) = (P(G_1)\ldots P(G_k))^1$ for $\delta = 0$) and the case where structures should be identical (i.e. $P(G,..,G) = (P(G)\ldots P(G))^{1/k}$ for $\delta = 1$). This prior can be easily generalized by using different penalties for different edges, and/or different penalties for different edit operations.

Another way to specify a prior in configurations for more than two tasks is to multiply the penalties incurred between all pairs of structures:

$$P(G_1,\ldots,G_k) = Z_{\delta,k} \cdot \prod_{1 \leq s \leq k} P(G_s)^{\frac{1}{1+(k-1)\delta}} \times \prod_{1 \leq s < t \leq k} \left( \prod_{\substack{(X_i,X_j) \in \\ G_s \Delta G_t}} (1-\delta) \right)^{\frac{1}{k-1}} \tag{8}$$

We will call this prior the *Paired* prior. The exponent $1/(k-1)$ is used because each individual structure is involved in $k-1$ terms (one for each other structure).

One advantage that the Paired prior has over the Edit prior is that it can be generalized by specifying different penalties between different pairs of structures. This can handle situations where there is reason to believe that Task1 is related to Task2, and Task2 is related to Task3, but the relationship to between Task1 and Task3 is weaker.

There are, of course, other priors that encourage finding similar networks for each task in different ways. In particular, if the process that generated the related tasks is know, it might be possible to design a suitable prior.

## 3.2. Greedy Structure Learning

Treating $P(G_1,\ldots,G_k \mid D_1,\ldots,D_k)$ as a score, we can search for a high scoring configuration using an heuristic search algorithm. If we choose to use greedy search for example, we start from an initial configuration, compute the scores of the neighboring configurations, then move to the configuration that has the highest score. The search ends when no neighboring configuration has a higher score than the current one.

One question remains: what do we mean by the neighborhood of a configuration? An intuitive definition of a neighbor is the configuration obtained by modifying a single arc in a single DAG in the configuration, such that the resulting graph is still a DAG. With this definition, the size of the neighborhood of a configuration is $O(k * n^2)$ for $k$ problems and $n$ variables. Unfortunately, this definition introduces a lot of local minima in the search space and leads to significant loss in performance. Consider for example the case where there is a strong belief that the structures should be similar (i.e. the penalty parameter of the prior, $\delta$, is near one resulting in a prior probability near zero when the structures in the configuration differ). In this case it would be difficult to take any steps in the greedy search since modifying a single edge for a single DAG would make it different from the other DAGs, resulting in a very low posterior probability (score).

To correct this problem, we have to allow all structures to change at the same time. Thus, we will define the neighborhood of a configuration to be the set of all configurations obtained by changing the same arc in any subset of the structures. Examples of such changes are removing an existing arc from all the structures, or just removing it from half of the structures, or removing it from one structure, reverse it in another, and leave it unchanged in the rest. This way we avoid creating local minimas in the search space while still ensuring that every possible configuration can be reached. Given this definition, the size of a neighborhood is $O(n^2 3^k)$, which is exponential in the number of problems, but only quadratic in the number of nodes.[2] When setting $\delta = 1$, leading to learning identical structures, multi-task learning with this definition of neighborhood will find the same structures as the specialized algorithm described in Section 3.1.

### 3.3. Searching for the Best Configuration

At each iteration, the greedy procedure described in the previous section must find the best scoring configuration from a set $N$ of neighboring configurations. In a naive implementation, the score of every configuration in $N$ has to be computed to find the best one, which can quickly become computationally infeasible given our definition of neighborhood.

In this section we show how one can use branch-and-bound techniques to find the best scoring configuration without evaluating all configurations in $N$. Let a partial configuration of order $l$, $C_l = (G_1, .., G_l)$, be a configuration where only the *first* $l$ structures are specified and the rest of $k - l$ structures are not specified. We say that a configuration $C$ matches a partial configuration $C_l$ if the first $l$ structures in $C$ are the same as the structures in $C_l$.

A search strategy for finding the best scoring configuration in $N$ can be represented via a search tree of depth $k$ with the following properties: a) each node at level $l$ contains

---

2. The restriction that changes, if any, have to be made to the same arc in all structures could be dropped, but this would lead to a neighborhood that is exponential in both $n$ and $k$. Considering the assumption that the structures should be similar, such a restriction is not inappropriate.

a valid partial configuration of order $l$; b) all nodes in the subtree rooted at node $C_l$ contain only (partial) configurations that match $C_l$ (i.e. the first $l$ structures are the same as in $C_l$.)

If, given a partial configuration, the score of any complete configuration that matches it can be efficiently upper bounded, and the upper bound is lower than the current best score, then the entire subtree rooted at the respective partial configuration can be pruned.

Let $edits_{l,i,j}$ be the minimum number of edits necessary to make the arc between $X_i$ and $X_j$ the same in the *first* $l$ structures, and let $Best_q = \max_{G_q}\{P(G_q)^{\frac{1}{1+(k-1)\delta}} P(D_q \mid G_q)\}$. If the marginal likelihood of a configuration factorizes in the product of the marginal likelihoods of the individual structures, as in equation 5, then the score of any configuration that matches the partial configuration $C_l = (G_1,\ldots,G_l)$ can be upper bounded by:

$$U_N^E(C_l) = \left(\prod_{i,j}(1-\delta)^{edits_{l,i,j}}\right) \cdot \left(\prod_{1 \le p \le l} P(G_p)^{\frac{1}{1+(k-1)\delta}} P(D_p \mid G_p)\right) \cdot \left(\prod_{l+1 \le q \le k} Best_q\right) \quad (9)$$

if using the Edit prior (equation 7), and by

$$U_N^P(C_l) = \left(\prod_{\substack{1 \le s < t \le l \, (X_i,X_j) \in \\ G_s \Delta G_t}}(1-\delta)\right)^{\frac{1}{k-1}} \cdot \left(\prod_{1 \le p \le l} P(G_p)^{\frac{1}{1+(k-1)\delta}} P(D_p \mid G_p)\right) \cdot \left(\prod_{l+1 \le q \le k} Best_q\right) \quad (10)$$

if using the Paired prior (equation 8).

This branch and bound search significantly reduces the number of partial configurations (and consequently complete configurations) that need to be explored. As an example of how much the branch and bound search can help, Figure 1 shows the fraction of configurations that are evaluated by branch and bound as the multi-task penalty parameter $\delta$ is varied, for a problem with five tasks and thirty seven variables. In this case, branch and bound evaluates four orders of magnitude less configurations than a naive search would.

## 4. Experimental Results

We evaluate the performance of multi-task structure learning using multi-task problems generated by perturbing the ALARM (Beinlich et al., 1989) and INSURANCE (Binder et al., 1997) networks, and on a real problem in bird ecology.

Multi-task structure learning is compared to single-task structure learning, and learning identical structures for all tasks. Single-task structure learning uses greedy hill-climbing with 100 restarts and tabu lists to learn the structure of each task independently of the others. The learning of identical structures is performed via the algorithm

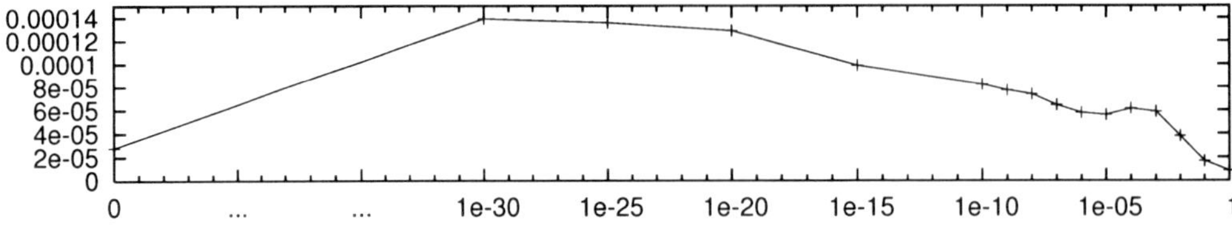

Figure 1: Fraction of partial configurations evaluated as a function of the penalty

presented in Section 3.1 and also uses greedy hillclimbing with 100 restarts and tabu lists.[3] Multi-task structure learning uses the greedy algorithm described in Section 3.2 with the solution found by single-task learning as the starting point.[4] The penalty parameter of the multi-task prior, $\delta$, is selected from a logarithmic grid to maximize the mean log-likelihood of a small validation set. For all methods, the Bayes net parameters are learned using Bayesian updating ((see e.g. Cooper and Hersovits, 1992)) independently for each problem.

The goal is to recover as closely as possible the true Bayes Net structures for all the related tasks, so the main measure of performance we use is average edit distance[5] between the true structures and learned structures. Edit distance directly measures the quality of the learned structures, independently of the parameters of the Bayes Net. We also measure the average empirical KL-divergence (computed on a large test set) between the distributions encoded by the true networks and the learned ones. Since KL-Divergence is also sensitive to the parameters of the Bayes Net it does not measure directly the quality of the learned structures, but, in general, more accurate structures should lead to models with lower KL-Divergence. For the bird ecology problem, where the true networks are unknown, we measure performance in terms of mean log likelihood on a large independent test set.

## 4.1. The ALARM and INSURANCE problems

For the experiments with the ALARM and INSURANCE networks, we generate five related tasks by perturbing the original structures. We use two qualitatively different methods for perturbing the networks: randomly deleting edges with some probability, and changing entire subgraphs. In the first case, we create five related tasks by starting with the original network and deleting arcs with probability $P_{del}$. This way, the structures of the five tasks can be made more or less similar by varying $P_{del}$ (For $P_{del} = 0$

---

3. Learning identical structures and single-task structure learning can be viewed as learning an augmented naive Bayesian network and a Bayesian multi-net (Friedman et al., 1997) respectively, where the "class" of each example is the task it belongs to . Unlike in the usual setting, however, here we are not interested in predicting to which task an example belongs to. We are only interested in recovering accurate network structures for each task.

4. Initializing MTL search with the STL solution does not provide an advantage to MTL, but makes the search more efficient.

5. Edit distance measures how many edits (arc additions, deletions or reversals) are needed to get from one structure to the other.

all the structures are identical. Given the restriction we imposed in Section 3 that parameters for different tasks should be independent *a priori*, we want to investigate the performance of multi-task structure learning in settings where the parameters are indeed independent between tasks (ALARM-IND and INSURANCE-IND), as well as in settings where the parameters are actually correlated between tasks (ALARM and INSURANCE).

We also experiment with a qualitatively different way of generating related tasks (ALARM-COMP). We split the ALARM network in 4 subgraphs with disjoint sets of nodes. For each of the five tasks, we randomly change the structure and parameters of zero, one or two of the subgraphs, while keeping the rest of the Bayes net (including parameters) unchanged. This way parts of the structures are shared between tasks while other parts are completely unrelated.

Figures 2 and 3 show the average percent reduction in loss, in terms of edit distance and KL-divergence, achieved by multi-task learning over single-task learning for a training set of 1000 points on the ALARM and INSURANCE-IND problems. The figures for the ALARM-IND, ALARM-COMP, and INSURANCE problems are similar and are not included. On the x-axis we vary the penalty parameter of the multi-task prior on a log-scale. Note that the x-axis plots $1 - \delta$. The higher the penalty (the lower $1 - \delta$), the more similar the learned structures will be, with all the structures being identical for a penalty of one ($1 - \delta = 0$, left end of graphs). Each line in the figure corresponds to a particular value of $P_{del}$. Error bars are omitted to maintain the figure readable.

The trends in the graphs are exactly as expected. For all values of $P_{del}$, as the penalty increases, the performance increases because the learning algorithm takes into account information from the other tasks when deciding whether to add a new arc or not. If the penalty is too high, however, the algorithm loses the ability to find true differences between tasks and the performance drops. As the tasks become more similar (lower values of $P_{del}$), the best performance is obtained at higher penalties. Also as the tasks become more similar, more information can be extracted from the related tasks, so usually multi-task learning provides more benefit. Multi-task learning provides similar benefits whether the tasks have highly correlated parameters (ALARM and INSURANCE problems) or independent parameters (ALARM-IND and INSURANCE-IND problems). This shows that making the parameters independent *a priori* (see Section 3) does not hurt the performance of multi-task learning.

One thing to note is that multi-task structure learning provides a larger relative improvement in edit distance than in KL-divergence. This happens because multi-task structure learning helps to correctly identify the arcs that encode weaker dependencies (or independences) which have a smaller effect on KL-divergence. The arcs that encode strong dependencies, and have the biggest effect on KL-divergence, can be easily learned without help from the other tasks.

Figure 4 shows the edit distance and KL-Divergence performance for single task learning (STL), learning identical networks (IDENTICAL), and multi-task learning (MTL). The training set has 1000 instances with 50 instances used to select the penalty

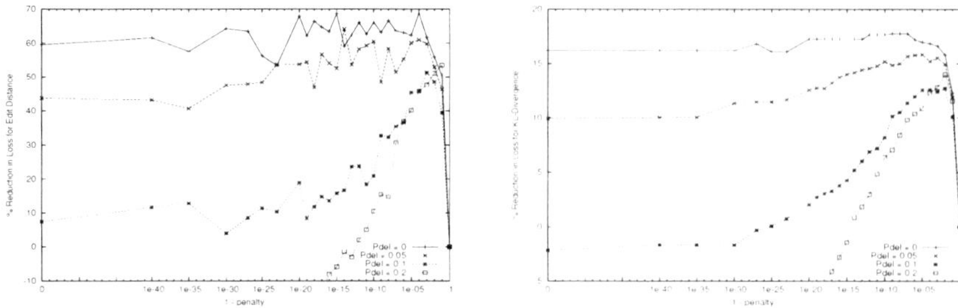

Figure 2: Reduction in edit distance (left) and KL-Divergence (right) for ALARM

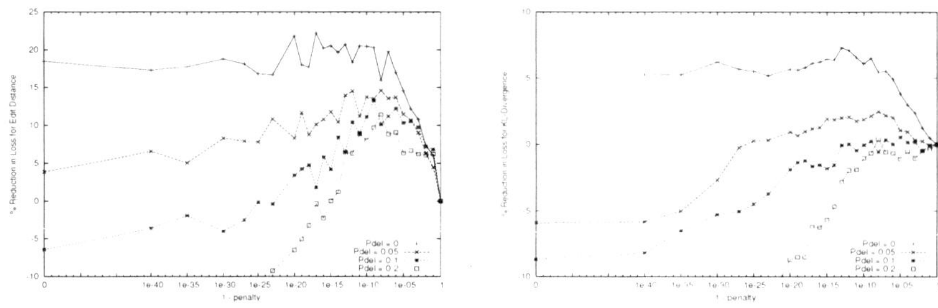

Figure 3: Reduction in edit distance (left) and KL-Divergence (right) for INSURANCE-IND

parameter for the multi-task prior. Single-task learning and identical structure learning use all the data for learning since they do not have free parameters. The figure shows that multi-task learning yields a 10%-54% reduction in edit distance and a 2% - 13% reduction in KL-divergence when compared to single task structure learning. All differences except for KL-divergence on ALARM-IND and INSURANCE-IND problems are .95 significant according to paired T-tests. When compared to learning identical structures, multi-task learning reduces the KL-divergence 7% - 32% and the number of incorrect arcs in the learned structures by 4% - 60%. All differences are .95 significant, except for edit distance on the ALARM-IND problem. Since the five tasks for the ALARM, INSURANCE, and ALARM-COMP problems share a large number of their parameters, one might believe that simply pooling the data would work well. This is, however, not the case. Except for the ALARM problem, where it achieves about the same edit distance as learning identical structures, pooling the data has much worse performance both in terms of edit distance and in terms of KL-divergence.

Figure 5 shows the performance of single and multi-task learning as the train set size varies from 250 to 16000 cases (MTL uses 5% of the training points as a validation set to select the penalty parameter). As expected, the benefit from multi-task learning is

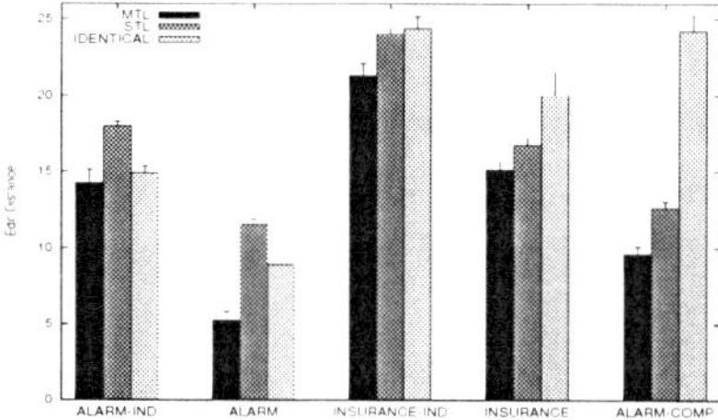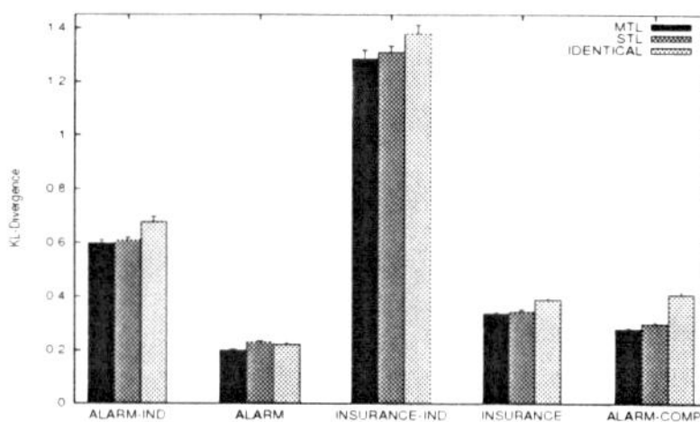

Figure 4: Edit distance (left) and KL-Div (right) for STL, learning identical structures and MTL

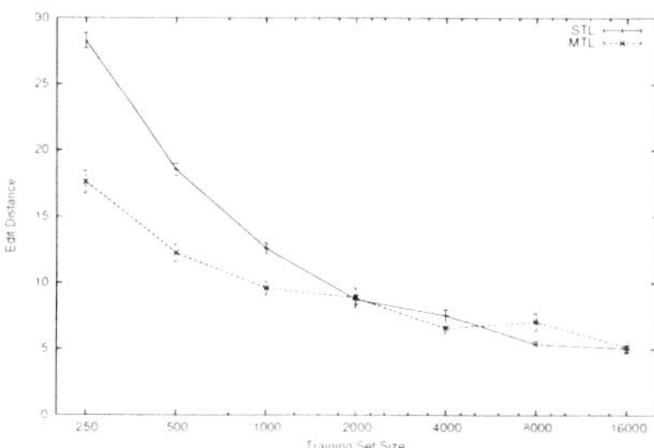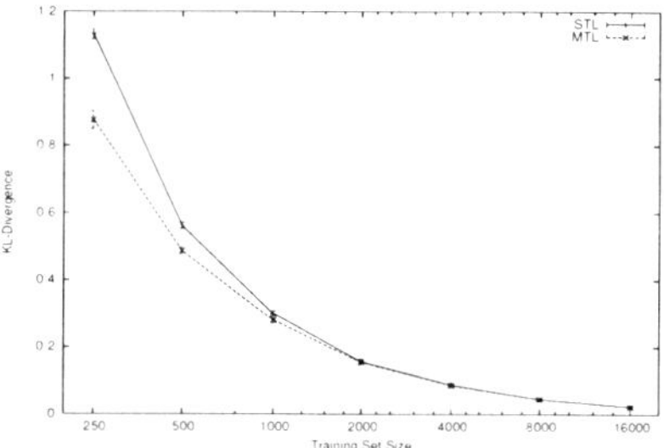

Figure 5: Edit distance (left) and KL-Divergence (right) vs. train set size for ALARM-COMP.

larger when the data is scarce and it diminishes as more training data is available. This is consistent with the behavior of multi-task learning in other learning setting (see e.g. (Caruana, 1997)). For smaller training set sizes multi-task learning needs about half as much data as single-task learning to achieve the same edit distance. In terms of KL-divergence, multi-task learning provides smaller savings in sample size. One reason for this is that, as discussed before, multi-task learning yields lower improvements in KL-divergence than in edit distance. For the most part however, the smaller savings in sample size are due to the fact that more training data leads not only to more accurate structures, but also to more accurate parameters. Since multi-task structure learning only improves the structure and not the parameters, it is not able to *make up* for the loss of large amounts of training data.

## 4.2. The Bird Ecology Problem

We also evaluate the performance of multi-task structure learning on a real bird ecology problem. The data for this problem comes from Project FeederWatch (PFW)[6], a winter-long survey of North American birds observed at bird feeders. Each PFW submission is

---

6. http://birds.cornell.edu/pfw

described by multiple attributes, which can be roughly grouped into features related to which birds have been observed, observer effort, weather during the observation period, and attractiveness of the location and neighborhood area for birds. The goal is to gain a better understanding of the various bird species by identifying environmental factors that attract or detract certain bird species, as well as how different bird species interact with each other.

Ecologists have divided North America into a number of ecologically distinct Bird Conservation Regions (BCRs; see Figure 6). This division naturally splits the data into multiple tasks, one task per BCR. For the results in this section we use six related tasks corresponding to BCRs 30, 29, 28, 22, 13 and 23. Because each bird species lives in some BCRs but not in others, this is an instance of a problem where the different tasks are not defined over identical sets of variables.

The results on the BIRD problem mimic the ones in the previous section. Figure 7 shows the average (across the 6 BCRs/tasks) mean log likelihood on a large independent test set for multi-task structure learning as a function of the penalty parameter of the multi-task prior. Each line corresponds to a different type of multi-task prior. The x-axis plots $1 - \delta$, so the right most point corresponds to no penalty (single task learning) and the leftmost point corresponds to a penalty of one (learning identical structures). Higher mean log likelihood represents better performance. As the penalty parameter increases ($1 - \delta$ decreases), information starts to be transferred between the different tasks and the performance quickly increases. After reaching a peak, the performance starts to decrease slowly as the penalty increases further. Since the tasks are not all defined on the same set of variables, the algorithm for learning identical structures for all tasks from Section 3.1 can not be easily applied. Our algorithm on the other hand can handle this situation and learns a set of identical structures for all tasks that performs reasonably well (left end of the plot). The type of multi-task prior does not have a significant impact on the performance for this problem.

Figure 8 shows the average mean log likelihood performance of multi-task structure learning and single task structure learning as a function of the training set size. Multi-task learning uses 5% of the training data to select the penalty parameter for the multi-task prior. Again, the benefit from multi-task learning is largest for smaller training set sizes. As the training size increases single-task learning catches up and eventually outperforms multi-task learning. Unfortunately, since we do not know the real network structures for this problem, we can not directly asses the quality of the learned structures. The results on the ALARM and INSURANCE problems, however, suggest that the improvement provided by multi-task learning in terms of structural accuracy (edit distance) would probably be even larger than the improvement in terms of average mean log likelihood.

## 5. Conclusions and Discussion

Learning the structure of Bayes Nets from data has received a lot of attention in the literature and numerous techniques have been developed to solve this problem (e.g. (Cooper and Hersovits, 1992; Heckerman, 1999; Buntine, 1996; Spirtes et al., 2000)).

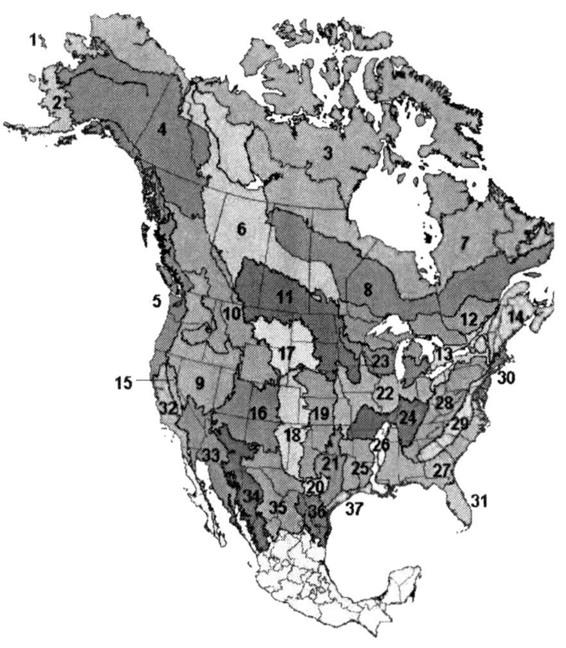

Figure 6: North American Bird Conservation Regions.

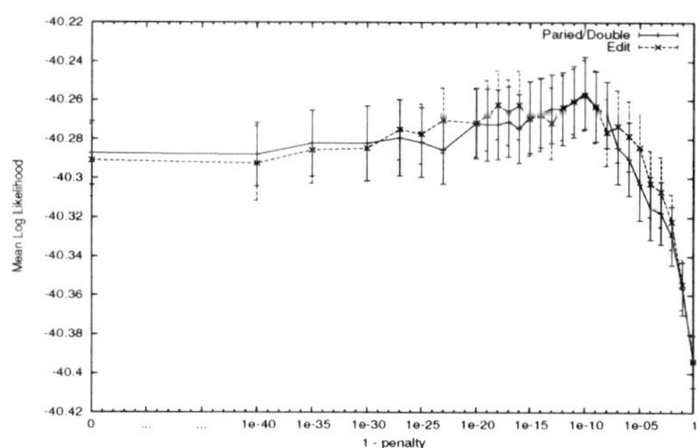


Figure 7: Average mean log likelihood vs. the penalty parameter for multi-task structure learning on the BIRD problem.

In this paper, we have focused on, arguably, the most basic one: score-and-search using
greedy hill-climbing in the space of network structures (DAG-search), and extended this
technique to the multi-task learning scenario. The key ingredients in achieving this have
been: defining a principled scoring function that takes into account the data from all the

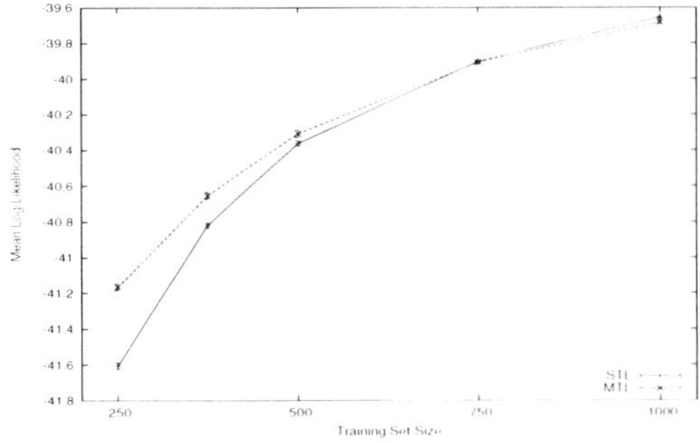

Figure 8: Average mean log likelihood vs. training set size for the BIRD problem.

tasks and encourages the learning of similar structures, defining a suitable search space, and devising a branch and bound procedure that enables efficient moves in this search space. We experimented with perturbed ALARM and INSURANCE networks and a real bird ecology problem, and showed that the multi-task structure learning technique yields significantly more accurate Bayes Net structures, especially when training data is scarce.

Even though in the paper we have focused on DAG-search, one can straightforwardly obtain multi-task Bayes Net structure learning algorithms based on other techniques such as greedy search in the space of equivalence classes (Chickering, 1996), obtaining confidence measures on the structural features of the configurations via bootstrap analysis (Friedman et al., 1999), and structure learning from incomplete datasets via the structural EM algorithm (Friedman, 1998). Other extensions such as obtaining a sample from the posterior distribution via MCMC methods might be more problematic. Because of the larger search space, MCMC methods might not converge in reasonable time. Evaluating different MCMC schemes is a direction for future work.

Another open question is whether we can relax the requirement that the parameters of the Bayes Nets for the different related tasks are independent *a priori*. Relaxing this requirement might further improve the performance of multi-task learning since the task would be able to share not only the structures but also the parameters, thus having more opportunities for inductive transfer. Further improvement is also possible by eliminating the need for the user to specify the penalty parameter $\delta$. At this point, one has to rely on cross-validation to determine a reasonable value for this parameter, which leads to a loss in performance and an increase in computational time. It would be very desirable to find techniques to infer $\delta$ directly from the data, or integrate over it in a bayesian manner.

Multi-task structure learning might also prove useful in learning Bayesian multi-nets (Friedman et al., 1997). In Bayesian multi-nets a special attribute is selected (usually the class attribute), and a separate network is learned for each value of that attribute. To the best of our knowledge, all work in learning Bayesian multi-nets treats each separate network as an independent learning problem, in a single-task manner. Since it is reasonable to assume that the networks for the different values of the class attribute should be similar, learning all the networks jointly using multi-task structure learning might yield improved performance.

## References


J. Baxter. A bayesian/information theoretic model of learning to learn via multiple task sampling. *Mach. Learn.*, 28(1):7–39, 1997. ISSN 0885-6125.

I.A. Beinlich, H.J. Suermondt, R.M. Chavez, and G.F. Cooper. The ALARM monitoring system: A case study with two probabilistic inference techniques for belief networks. In *Proceedings of the Second European Conference on Artificial Intelligence in Medicine*, 1989.

J. Binder, D. Koller, S. Russell, and K. Kanazawa. Adaptive probabilistic networks with hidden variables. *Machine Learning*, 29, 1997.

W. Buntine. Theory refinement on bayesian networks. In *Proc. 7th Conference on Uncertainty in Artificial Intelligence (UAI '91)*, 1991.

W. Buntine. A guide to the literature on learning probabilistic networks from data. *IEEE Trans. On Knowledge and data Engineering*, 8:195–210, 1996. URL `http://citeseer.nj.nec.com/buntine96guide.html`.

R. Caruana. Multitask learning. *Machine Learning*, 28(1):41–75, 1997.

D. Chickering. Learning equivalence classes of Bayesian network structures. In *Proc. 12th Conference on Uncertainty in Artificial Intelligence (UAI'96)*, 1996. ISBN 1-55860-412-X. URL `http://citeseer.nj.nec.com/chickering96learning.html`.

G. Cooper and E. Hersovits. A bayesian method for the induction of probabilistic networks from data. *Maching Learning*, 9:309–347, 1992.

N. Friedman. The Bayesian structural EM algorithm. In *Proc. 14th Conference on Uncertainty in Artificial Intelligence (UAI '98)*, 1998. URL `citeseer.ist.psu.edu/article/friedman98bayesian.html`.

N. Friedman, D. Geiger, and M. Goldszmidt. Bayesian Network Classifiers. *Machine Learning*, 29(2):131–163, 1997.

N. Friedman, M. Goldszmidt, and A. J. Wyner. Data analysis with bayesian networks: A bootstrap approach. In *Proc. 15th Conference on Uncertainty in Artificial Intelligence*, 1999.

D. Heckerman. A tutorial on learning with bayesian networks. *Learning in graphical models*, pages 301–354, 1999.

D. Heckerman, A. Mamdani, and M.P. Wellman. Real-world applications of Bayesian networks. *Communications of the ACM*, 38(3):24–30, 1995.

A. Niculescu-Mizil and R. Caruana. Inductive transfer for bayesian network structure learning. In *Proc. 11th International Conf. on AI and Statistics*, 2007.

J. Pearl. *Probabilistic Reasoning in Intelligent Systems: Networks of Plausible Inference*. Morgan Kaufmann, San Mateo, CA, 1988.

P. Spirtes, C. Glymour, and R. Scheines. *Causation, Prediction, and Search*. The MIT Press, Cambridge, MA, second edition, 2000.

M. Teyssier and D. Koller. Ordering-based search: A simple and effective algorithm for learning bayesian networks. In *Proceedings of the Twenty-first Conference on Uncertainty in AI (UAI)*, pages 584–590, Edinburgh, Scottland, UK, July 2005.

S. Thrun. Is learning the n-th thing any easier than learning the first? In *Advances in Neural Information Processing Systems*, 1996.

# Unsupervised dimensionality reduction via gradient-based matrix factorization with two adaptive learning rates

**Vladimir Nikulin**          VNIKULIN.UQ@GMAIL.COM
**Tian-Hsiang Huang**          HUANGTX@GMAIL.COM
*Department of Mathematics, University of Queensland, Australia*

**Editor:** I. Guyon, G. Dror, V. Lemaire, G. Taylor, and D. Silver

## Abstract

The high dimensionality of the data, the expressions of thousands of features in a much smaller number of samples, presents challenges that affect applicability of the analytical results. In principle, it would be better to describe the data in terms of a small number of meta-features, derived as a result of matrix factorization, which could reduce noise while still capturing the essential features of the data. Three novel and mutually relevant methods are presented in this paper: 1) gradient-based matrix factorization with two adaptive learning rates (in accordance with the number of factor matrices) and their automatic updates; 2) nonparametric criterion for the selection of the number of factors; and 3) nonnegative version of the gradient-based matrix factorization which doesn't require any extra computational costs in difference to the existing methods. We demonstrate effectiveness of the proposed methods to the supervised classification of gene expression data.

**Keywords:** matrix factorization, unsupervised learning, clustering, nonparametric criterion, nonnegativity, bioinformatics, leave-one-out, classification

## 1. Introduction

The analysis of gene expression data using matrix factorization has an important role to play in the discovery, validation, and understanding of various classes and subclasses of cancer. One feature of microarray studies is the fact that the number of samples collected is relatively small compared to the number of genes per sample which are usually in the thousands. In statistical terms this very large number of predictors compared to a small number of samples or observations makes the classification problem difficult.

Many standard classification algorithms have difficulties in handling high dimensional data and due to a relatively low number of training samples, they tend to overfit. Moreover, usually only a small subset of examined genes is relevant in the context of a given task. For these reasons, feature selection methods are an inevitable part of any successful microarray data classification algorithm.

Another approach to reduce the overfitting is to describe the data in terms of meta-genes as a linear combinations of the original genes.

Recently, models such as independent component analysis and nonnegative matrix factorization (NMF) have become popular research topics due to their obvious usefulness (Oja *et al.*, 2010). All these blind latent variable models in the sense that no prior knowledge on the variables is used, except some broad properties like gaussianity, statistical independence, or nonnegativity.

As pointed out by Tamayo *et al.* (2007), the metagene factors are a small number of gene combinations that can distinguish expression patterns of subclasses in a data set. In many cases, these linear combinations of the genes are more useful in capturing the invariant biological features of the data. In general terms, matrix factorization, an unsupervised learning method, is widely used to study the structure of the data when no specific response variable is specified.

Examples of successful matrix factorization methods are singular value decomposition and nonnegative matrix factorization (Lee and Seung, 1999). In addition, we developed a novel and very fast algorithm for gradient-based matrix factorization (GMF), which was introduced in our previous study (Nikulin and McLachlan, 2009).

The main subject of this paper is a more advanced version of the GMF. We call this algorithm as GMF with two learning rates and their automatic updates (A2GMF). Details about this algorithm are given in Section 2.1. The main features of the A2GMF are two flexible (adaptive) learning rates in accordance to the number of factor matrices. By the definition, the learning rates will be updated during learning process. The explicit update formulas are given in Section 2.1.

Clearly, the number of factors is the most important input parameter for any matrix factorization algorithm. In Section 2.2 we are proposing a novel unsupervised method for the selection of the number of factors. This method is absolutely general, and we show in Section 4.5 one possible extension to the clustering methodology, see Figure 3. Using supervised classification with leave-one-out (LOO) evaluation criterion, we can develop another approach to select the numbers of factors. Classification results, which are based on five well-known and publicly available datasets, and presented in Section 4 demonstrate correspondence between the outcomes of both methods. However, speed is the main advantage of the proposed here nonparametric criterion.

The third proposed novelty is a nonnegative version of GMF (NN-GMF), see Section 2.3. Essentially, an implementation of the NN-GMF doesn't require any extra computational costs. Consequently, the NN-GMF is as fast as GMF, and may be particularly useful for a wide range of the professionals who are working with real data directly and who are interested to find out an interpretation of the data in terms of meta-variables.

## 2. Methods

Let $\left(\mathbf{x}_j, y_j\right), j = 1, \ldots, n$, be a training sample of observations where $\mathbf{x}_j \in \mathbb{R}^p$ is $p$-dimensional vector of features, and $y_j$ is a multiclass label, which will not be used in this section. Boldface letters denote vector-columns, whose components are labelled using a normal typeface. Let us denote by $\mathbf{X} = \{x_{ij}, i = 1, \ldots, p, j = 1, \ldots, n\}$ the matrix containing the observed values on the $n$ samples.

For gene expression studies, the number $p$ of genes is typically in the thousands, and the number $n$ of experiments is typically less than 100. The data are represented by an expression matrix $\mathbf{X}$ of size $p \times n$, whose rows contain the expression levels of the $p$ genes in the $n$ samples. Our goal is to find a small number $k \ll p$ of metagenes or factors. We can then approximate the gene expression patterns of samples as a linear combinations of these metagenes. Mathematically, this corresponds to factoring matrix $\mathbf{X}$ into two matrices

$$\mathbf{X} \sim \mathbf{AB}, \tag{1}$$

where weight matrix $\mathbf{A}$ has size $p \times k$, and the matrix of metagenes $\mathbf{B}$ has size $k \times n$, with each of $k$ rows representing the metagene expression pattern of the corresponding sample.

---

**Algorithm 1** A2GMF.

---

1: Input: $\mathbf{X}$ - microarray matrix.
2: Select $m$ - number of global iterations; $k$ - number of factors; $\lambda_a, \lambda_b > 0$ - initial learning rates, $0 < \tau(\ell) \leq 1$ is a scaling function of the global iteration $\ell$.
3: Initial matrices $\mathbf{A}$ and $\mathbf{B}$ are generated randomly.
4: Global cycle: for $\ell = 1$ to $m$ repeat steps 5 - 16:
5: genes-cycle: for $i = 1$ to $p$ repeat steps 6 - 15:
6: samples-cycle: for $j = 1$ to $n$ repeat steps 7 - 15:
7: compute prediction $S_{ij} = \sum_{f=1}^{k} a_{if} b_{fj}$;
8: compute error of prediction: $E_{ij} = x_{ij} - S_{ij}$;
9: internal factors-cycle: for $f = 1$ to $k$ repeat steps 10 - 15:
10: compute $\alpha = a_{if} b_{fj}$;
11: update $a_{if} \Leftarrow a_{if} - \tau(\ell) \cdot \lambda_a \cdot g_{ifj}$;
12: $E_{ij} \Leftarrow E_{ij} + \alpha - a_{if} b_{fj}$;
13: compute $\alpha = a_{if} b_{fj}$;
14: update $b_{fj} \Leftarrow b_{fj} - \tau(\ell) \cdot \lambda_b \cdot h_{ifj}$;
15: $E_{ij} \Leftarrow E_{ij} + \alpha - a_{if} b_{fj}$;
16: update $\lambda_a$ and $\lambda_b$ according to (5) and (7);
17: Output: $\mathbf{A}$ and $\mathbf{B}$ - matrices of metagenes or latent factors.

---

## 2.1. A2GMF algorithm under the stochastic gradient descent framework

Let us consider the following loss function

$$L(\mathbf{A}, \mathbf{B}) = \sum_{i=1}^{p} \sum_{j=1}^{n} E_{ij}^2, \tag{2}$$

where $E_{ij} = x_{ij} - \sum_{f=1}^{k} a_{if} \cdot b_{fj}$.

The above target function (2) includes in total $k(p + n)$ regulation parameters and may be unstable if we minimise it without taking into account the mutual dependence between elements of the matrices $\mathbf{A}$ and $\mathbf{B}$.

Derivatives of the function $E$ are given below:

$$g_{ifj} = \frac{\partial E_{ij}^2}{\partial a_{if}} = -2 \cdot E_{ij} \cdot b_{fj}, \tag{3a}$$

$$h_{ifj} = \frac{\partial E_{ij}^2}{\partial b_{fj}} = -2 \cdot E_{ij} \cdot a_{if}. \tag{3b}$$

Considering step 11 of Algorithm 1, we can replace in (2) $a_{if}$ by its update $a_{if} - \lambda_a g_{ifj}$, assuming that $\tau(\ell) = 1$, where scaling function $\tau$ is defined in Remark 2.1. After that, we can rewrite (2) in the following way

$$L(\mathbf{A}, \mathbf{B}, \lambda_a) = \sum_{i=1}^{p} \sum_{j=1}^{n} (E_{ij} + \lambda_a U_{ij})^2, \tag{4}$$

where

$$U_{ij} = \sum_{f=1}^{k} g_{ifj} \cdot b_{fj} = -2E_{ij} \sum_{f=1}^{k} b_{fj}^2.$$

Minimising (4) as a function $\lambda_a$, we shall find

$$\lambda_a = -\frac{\sum_{i=1}^{p} \sum_{j=1}^{n} E_{ij} U_{ij}}{\sum_{i=1}^{p} \sum_{j=1}^{n} U_{ij}^2}$$

$$= \frac{1}{2} \frac{\sum_{j=1}^{n} \sum_{f=1}^{k} b_{fj}^2 \sum_{i=1}^{p} E_{ij}^2}{\sum_{j=1}^{n} (\sum_{f=1}^{k} b_{fj}^2)^2 \sum_{i=1}^{p} E_{ij}^2} = \frac{1}{2} \frac{\sum_{j=1}^{n} b_j \phi_j}{\sum_{j=1}^{n} b_j^2 \phi_j}, \tag{5}$$

where

$$b_j = \sum_{f=1}^{k} b_{fj}^2, \quad \phi_j = \sum_{i=1}^{p} E_{ij}^2.$$

Considering step 14 of Algorithm 1, we can replace in (2) $b_{fj}$ by its update $b_{fj} - \lambda_b h_{ifj}$, assuming that $\tau(\ell) = 1$. After that, we can re-write (2) in the following way

$$L(\mathbf{A}, \mathbf{B}, \lambda_b) = \sum_{i=1}^{p} \sum_{j=1}^{n} (E_{ij} + \lambda_b V_{ij})^2, \tag{6}$$

where

$$V_{ij} = \sum_{f=1}^{k} h_{ifj} \cdot a_{if} = -2E_{ij} \sum_{f=1}^{k} a_{if}^2.$$

Minimising (6) as a function $\lambda_b$, we shall find

$$\lambda_b = -\frac{\sum_{i=1}^{p} \sum_{j=1}^{n} E_{ij} V_{ij}}{\sum_{i=1}^{p} \sum_{j=1}^{n} V_{ij}^2}$$

$$= \frac{1}{2} \frac{\sum_{i=1}^{p} \sum_{f=1}^{k} a_{if}^2 \sum_{j=1}^{n} E_{ij}^2}{\sum_{i=1}^{p} (\sum_{f=1}^{k} a_{if}^2)^2 \sum_{j=1}^{n} E_{ij}^2} = \frac{1}{2} \frac{\sum_{i=1}^{p} a_i \psi_i}{\sum_{i=1}^{p} a_i^2 \psi_i}, \tag{7}$$

where

$$a_i = \sum_{f=1}^{k} a_{if}^2, \ \psi_i = \sum_{j=1}^{n} E_{ij}^2.$$

Figure 1(a-e) illustrates clear advantage of the A2GMF compared to the GMF algorithm. We did not try to optimize performance of the A2GMF, and used the same regulation parameters against all five datasets described in Section 3.

**Remark 1** *Scaling function $\tau$ is a very important in order to ensure stability of Algorithm 1. Assuming that the initial values of matrices A and B were generated randomly, we have to use smaller values of $\tau$ during the first few global iterations. The following structure of the function $\tau(\ell)$ was used for the illustrations given in Figure 1:*

$$\tau(\ell) = 0.005(1 - r) + 0.2r, r = \sqrt{\frac{\ell}{m}},$$

*where parameters $\ell$ and $m$ are defined in the body of Algorithm 1.*

**Remark 2** *We note an important fact: a significant difference in behavior of the learning rates $\lambda_a$ and $\lambda_b$ as functions of the global iteration, see Figure 1(f). This difference reflects, also, the main conceptual difference between A2GMF and GMF algorithms.*

**Definition 3** *We define GMF algorithm as Algorithm 1 but without the scaling function $\tau$ and with constant and equal learning rates $\lambda_a - \lambda_b - 0.01$ (Nikulin and McLachlan, 2009). The value 0.01 was used in the comparison experiments Figure 1(a-e). However, the optimal settings may be different depending on the particular dataset.*

## 2.2. Nonparametric unsupervised criterion for selection of the number of meta-variables

The basic idea of nonparametric inference is to use data to infer an unknown quantity while making as few assumptions as possible (Wasserman, 2006). The criterion, which we propose in this section is the most general and may attract interest as a self-target in fundamental sense, but not only as an application for the selection of the number of meta-variables. It is very logical to assume that the microarray matrix contains some systematic dependencies, which will be discovered or revealed by the matrix factorization. We can easily destroy those hidden dependencies by randomly re-shuffling the elements within any column. Note, also, that all the columns will be re-shuffled independently.

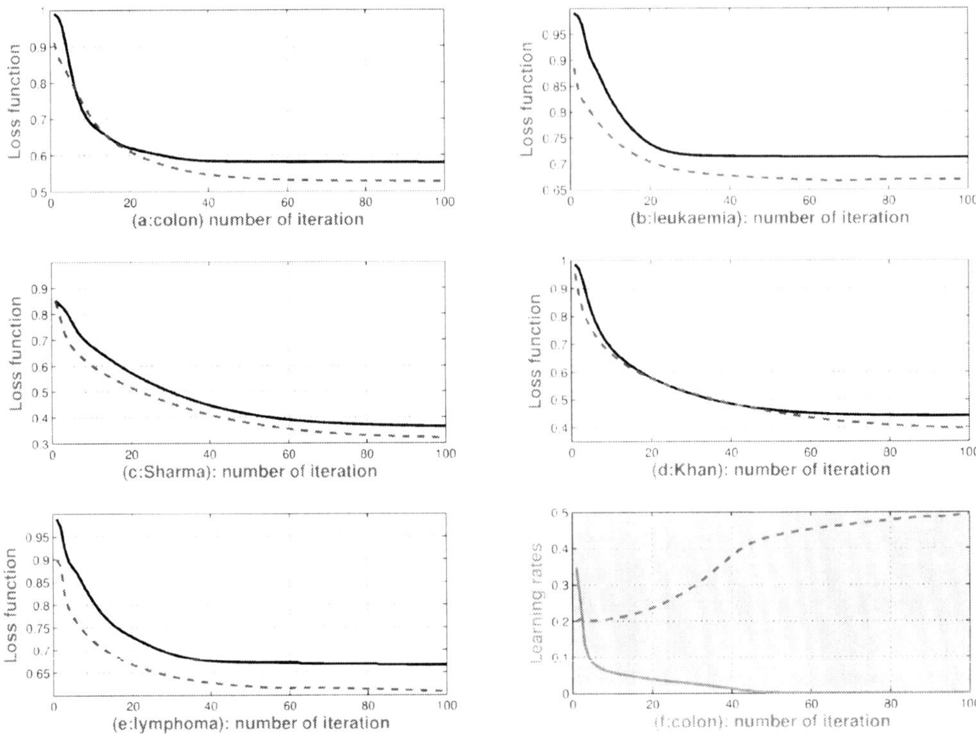

Figure 1: Convergence of the GMF (black line) and A2GMF (dashed blue line, Algorithm 1) algorithms in the cases of (a) colon, (b) leukaemia,(c) Sharma, (d) Khan and (e) lymphoma sets. The last window (f) illustrates the difference in behavior between learning rates $\lambda_a$ (red) and $\lambda_b$ (dashed green), see Algorithm 1.

Let us denote by $X^{(\gamma)}, \gamma = 1, \ldots, 3$ three random re-shuffling of the original matrix $X$. By $A^{(\gamma)}$ and $B^{(\gamma)}$ we shall denote the corresponding factor matrices. Figure 2 illustrates behavior of

$$\mathcal{D}_k = \frac{1}{3} \sum_{\gamma=1}^{3} \Phi_k^{(\gamma)} - \Phi_k, \tag{8}$$

as a function of $k$, where

$$\Phi_k = \sqrt{\frac{L(A_k, B_k)}{pn}}, \Phi_k^{(\gamma)} = \sqrt{\frac{L(A_k^{(\gamma)}, B_k^{(\gamma)})}{pn}}, \gamma = 1, \ldots, 3.$$

As expected, the values of $\mathcal{D}_k$ are always positive. Initially, and because of the small number of factors $k$, the values of $\mathcal{D}_k$ are low. This fact has very simple explanation, as far as the number of metagenes is small, those metagenes are not sufficient to contain the hidden dependencies and relations. Then, the values of $\mathcal{D}_k$ will grow to some

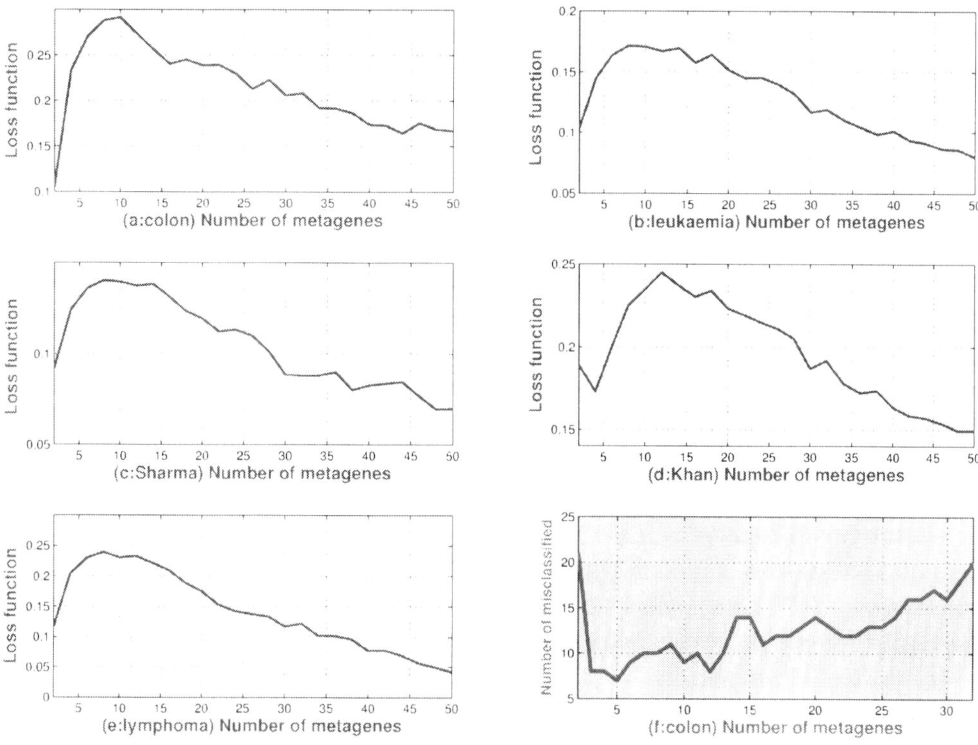

Figure 2: Nonparametric feature selection with criterion (8), where (a) colon, (b) leukaemia, (c) Sharma, (d) Khan and (e) lymphoma datasets. Window (f) illustrates the numbers of misclassified samples as a function of $k$ - number of metagenes, where we used evaluation scheme $Alg.1 + LOO\{lm\}$ applied to the colon dataset.

higher level. The pick point appears to be an appropriate criterion for the selection of the optimal number of meta-variables. After that point, the values of $\mathcal{D}_k$ will decline because of the overfitting.

## 2.3. On the nonnegative modification of the GMF (NN-GMF)

Many real-world data are nonnegative and the corresponding hidden components have a physical meaning only when nonnegative (Cichocki *et al.*, 2009). Suppose that microarray matrix $X$ is nonnegative (that means, all the elements of the matrix are nonnegative). Then, it is very logical to have factor matrices $A$ and $B$ in (1) to be nonnegative as well. In order to implement this, we can easily create a special modification of Algorithm 1.

The main idea behind this modification is a remarkably simple. Let us consider a generalization of (1)

$$\mathbf{X} \sim \xi(\mathbf{A})\xi(\mathbf{B}),\tag{9}$$

where $\xi$ is a differentiable function, and $\xi(A)$ is a matrix with the same dimensions as matrix $A$ and elements $\{\xi(a_{ij}), i = 1,\ldots,p, j = 1,\ldots,k\}$.

Taking into account the fact that an exponential function is always nonnegative, we can consider $\xi(\cdot) = \exp(\cdot)$. Equation (9) represents a flexible framework, where we can apply any suitable function. For example, we can create another modification of the NN-GMF with squared or logistic function.

According to Gao and Church (2005), it is obvious that dimension reduction to a much lower dimension (smaller than the number of observations) is appropriate. Principal component analysis (PCA) or singular value decomposition and partial least squares are two such methods that have been applied to cancer classification with satisfactory results. However, due to the holistic nature of PCA, the resulting components are global interpretations and lack intuitive meaning. To solve this problem, Lee and Seung (1999) demonstrated that NMF is able to learn localized features with obvious interpretation. Their work was applied elegantly to image and text analysis.

According to Fogel *et al.* (2007), singular value decomposition (SVD) attempts to separate mechanisms in an orthogonal way, although nature is all but orthogonal. As a consequence, SVD components are unlikely to match with real mechanisms and so are not easily interpreted. On the contrary, NMF appears to match each real mechanism with a particular component.

Clearly, we can extend above two statements to the NN-GMF algorithm.

As it was pointed by Lin (2007), there are many existing methods for NMF. Most of those method are presented in (Cichocki *et al.*, 2009). However, NN-GMF is an essentially novel algorithm (based on the different platform), and can not be regarded as a modification of NMF (Lee and Seung, 1999).

## 2.4. Sparsity and Regularization

A sparse representation of the data by a limited number of components is an important research problem. In machine learning, sparseness is closely related to feature selection and certain generalizations in learning algorithms, while nonnegativity relates to probability distributions (Cichocki *et al.*, 2009).

The most standard way to achieve sparsity is to include regularization term in (2), which penalises usage of the elements of the matrices **A** and **B**. We shall exploit here flexibility of Algorithm 1. The structure of the regularized GMF will be the same, but we have to make some changes in the formulas for derivatives (3a) and (3b), which may be found using standard techniques (Nikulin and McLachlan, 2010).

## 2.5. Handling missing values

Suppose that some elements of the matrix **X** have low level of confidence or missing. Based on the principles of the stochastic gradient descent technique, we are considering during any global iteration not the whole target function, but particular terms of the target function. Accordingly, we can easily exclude from an update process those terms, which have low level of confidence or missing. Using GMF algorithm we can make possible factorization of a huge matrices, for example, in the case of the well known Netflix Cup (marketing applications) we are dealing with hundreds of thousands of customers and tens of thousands of items, assuming that only about 5% or 10% of true relationships or preferences are available (Koren, 2009).

## 2.6. Boosting with GMF

Algorithm 1 represents a flexible framework, where we can easily insert suitable model as a component. For example, we can use the idea of boosting (Dettling and Buhlmann, 2003) in the following way. In the case if the element $x_{ij}$ has been approximated poorly (step N8), we can increase the value of learning rate. On the other hand, in the case if an absolute value of $E_{ij}$ is small enough, we can overpass the steps NN9-15. As a direct consequence, the algorithm will run faster and faster in line with general improvement of the quality of approximation.

## 3. Data

All experiments were conducted against 5 well known and publicly available microarray datasets named colon, leukaemia, Sharma, Khan and lymphoma. Some basic statistical characteristics of the datasets are given in Table 2, and more details are presented in (Nikulin *et al.*, 2011).

## 4. Experiments

After decomposition of the original matrix $\mathbf{X}$, we used the leave-one-out (LOO) evaluation scheme, applied to the matrix of metagenes $\mathbf{B}$. This means that we set aside the $i$th observation and fit the classifier by considering remaining $(n-1)$ data points. We conducted experiments with *lm* function in R, which is a standard linear regression without any regulation parameters.

### 4.1. Evaluation scheme

We shall denote such a scheme (Nikulin and McLachlan, 2009) as

$$Alg.1 + LOO\{lm\}. \tag{10}$$

Above formula has very simple interpretation as a sequence of two steps: 1) compute the matrix of metagenes $B$ using Algorithm 1 (dimensionality reduction), and 2) run LOO evaluation scheme with *lm* as a base classification model.

As a reference, we mention that similar evaluation model was employed in (Li and Ngom, 2010) to classify samples in NMF space using kNN algorithm. In accordance with the evaluation framework (10), we can express this model in the following way: $NMF + LOO\{kNN\}$.

**Remark 4** *As far as Algorithm 1 is an unsupervised (labels are not used), we do not have to include it into the LOO internal loop in (10).*

Table 1: Classification results (numbers of misclassified samples) in the case of the scheme: $SVD + LOO\{lm\}$.

| k | colon | leukaemia | Sharma | Khan | lymphoma |
|---|---|---|---|---|---|
| 2 | 22 | 8 | 24 | 48 | 6 |
| 4 | 10 | 5 | 22 | 27 | 2 |
| 6 | 9 | 2 | 18 | 15 | 1 |
| 8 | **7** | **1** | 14 | 5 | **0** |
| 10 | 8 | 1 | **9** | 6 | 0 |
| 12 | 8 | 1 | 10 | 2 | 0 |
| 14 | 8 | 2 | 9 | **0** | 0 |
| 16 | 10 | 2 | 10 | 0 | 0 |
| 18 | 9 | 2 | 10 | 0 | 0 |
| 20 | 10 | 2 | 11 | 0 | 0 |
| 22 | 11 | 5 | 10 | 0 | 0 |
| 24 | 13 | 3 | 13 | 0 | 0 |
| 26 | 13 | 1 | 15 | 1 | 1 |
| 28 | 11 | 1 | 16 | 1 | 1 |
| 30 | 14 | 2 | 16 | 1 | 2 |
| 32 | 13 | 3 | 16 | 3 | 1 |
| 34 | 9 | 2 | 18 | 2 | 1 |
| 36 | 12 | 3 | 18 | 3 | 1 |
| 38 | 14 | 4 | 18 | 3 | 2 |
| 40 | 15 | 6 | 20 | 3 | 3 |

Table 2: Comparison between the numbers of factors (metagenes) $k_{NP}$ and $k^*$, where $k_{NP}$ was selected according to the nonparametric criterion of Section 2.2 as a point on the horizontal axis corresponding to the maximum loss, see Figure 2(a-e); and $k^*$ was selected according to the independent scheme $SVD + LOO\{lm\}$, see Table 2 (in supplementary material), where column "NM" indicates the numbers of corresponding misclassified samples.

| Data | n | p | $k^*$ | $k_{NP}$ | NM |
|---|---|---|---|---|---|
| Colon | 62 | 2000 | 8 | 10 | 7 |
| Leukaemia | 72 | 1896 | 8 | 8 | 1 |
| Sharma | 60 | 1368 | 10 | 8 | 9 |
| Khan | 83 | 2308 | 14 | 12 | 0 |
| Lymphoma | 62 | 4026 | 8 | 8 | 0 |

Table 3: Classification results in the case of the scheme $Alg.1 + LOO\{lm\}$, where we used 20 runs against colon dataset with random initial settings, columns *"min, mean, max, std"* represent the corresponding statistical characteristics for the numbers of misclassified samples. In the column "SVD" we presented the numbers of misclassified samples in the case of the scheme $Alg.1 + LOO\{lm\}$, where initial settings were generated using $SVD$ method.

| k | min | mean | max | std | SVD |
|---|---|---|---|---|---|
| 2 | 16 | 21.89 | 28 | 4.50 | 21 |
| 4 | 7 | 15.22 | 19 | 3.53 | 10 |
| 6 | 7 | 10.56 | 18 | 3.39 | 9 |
| 8 | 5 | 8.50 | 11 | 2.09 | 7 |
| 10 | 6 | 8.92 | 12 | 1.72 | 6 |
| 12 | 6 | 9.33 | 12 | 2 | 7 |
| 14 | 8 | 9.22 | 11 | 1.09 | 9 |
| 16 | 7 | 9.89 | 12 | 1.54 | 10 |
| 18 | 8 | 10 | 11 | 1.12 | 10 |
| 20 | 8 | 10.33 | 13 | 1.58 | 10 |
| 22 | 11 | 12.56 | 15 | 1.59 | 10 |
| 24 | 11 | 12.78 | 15 | 1.20 | 13 |
| 26 | 12 | 12.56 | 14 | 0.73 | 12 |
| 28 | 10 | 12.67 | 16 | 2.06 | 12 |
| 30 | 10 | 13.22 | 16 | 1.64 | 14 |
| 32 | 12 | 14.33 | 17 | 1.58 | 14 |
| 34 | 12 | 15.11 | 18 | 1.90 | 12 |
| 36 | 11 | 14 | 17 | 2.18 | 14 |
| 38 | 13 | 14.78 | 17 | 1.39 | 14 |
| 40 | 13 | 15 | 17 | 1.22 | 14 |

## 4.2. Singular value decomposition (SVD)

An alternative factorisation may be created using *svd* function in Matlab or R:

$$X = UDV,$$

where $U$ is $p \times n$ matrix of orthonormal eigenvectors, $D$ is $n \times n$ diagonal matrix of nonnegative eigenvalues, which are sorted in a decreasing order, and $V$ is $n \times n$ matrix of orthonormal eigenvectors.

The absolute value of an eigenvalue indicates the significance of the corresponding eigenvector relative to the others. Accordingly, we shall use the matrix $V_k$ with $k$ first eigenvectors (columns) taken from the matrix $V$ as a replacement to the matrix $X$. According to (10), we shall denote above model by $SVD + LOO\{lm\}$, where classification results are presented in Table 1.

It is a well known fact that principal components, or eigenvectors corresponding to the biggest eigenvalues, contain the most important information. However, the other eigenvectors, which may contain some valuable information will be ignored. In contrast, nothing will be missed by definition in the case of the GMF, because the required number of metavariables will be computed using the whole microarray matrix. The main weakness of the GMF algorithm is its dependence on the initial settings, see Table 3. By combining SVD and GMF algorithms together, we can easily overcome this weakness: it appears to be logical to use matrices $V_k$ and $U_k$ as an initial for the GMF algorithm, where $U_k$ is defined according to $U$ using the same method as in the case of $V_k$ and $V$.

In the column "SVD", Table 3, we presented classification results, which were calculated using the model $Alg.1 + LOO\{lm\}$ with initial setting produced by the SVD method. These results demonstrate some improvement compared to the "colon" column of Table 1. Similar improvements were observed in application to four remaining datasets.

## 4.3. Partial least squares and selection bias

Partial least squares (PLS) (Nguyen and Rocke, 2002) is an essentially different compared to GMF or to PCA, because it is a supervised method. Therefore, the evaluation scheme (10) with PLS as a first component will not be free of a selection bias. As an alternative, we can apply evaluation model

$$LOO\{PLS + lm\}, \tag{11}$$

which is about $n$ times more expensive computationally compared to (10). According to (11), we have to exclude from the training process features of the test sample (that means we have to run PLS within any LOO evaluation loop), and this may lead to the overpessimistic results, because those features are normally available and may be used in the real life.

Very reasonably (Cawley and Talbot, 2006), the absence of the selection bias is regarded as a very significant advantage of the model, because, otherwise, we have to deal with nested cross-validation, which is regarded as an impractical (Jelizarow *et al.*, 2010, p.1996) in general terms.

Based on our own experience with the nested CV, this tool should not be used until it is absolutely necessary, because nested CV may generate secondary serious problems as a consequence of 1) the dealing with an intense computations, and 2) very complex software (and, consequently, high level of probability to make some mistakes) used for the implementation of the nested CV. Moreover, we do believe that in most of the cases scientific results produced with the nested CV are not reproducible (in the sense of an absolutely fresh data, which were not used prior). In any case, "low skills bias" could be much more damaging compared to the selection bias.

We fully agree with Jelizarow *et al.* (2010), p.1990, that the superiority of new algorithms should always be demonstrated on an independent validation data. In this sense, an importance of the data mining contests is unquestionable. The rapid popularity growth of the data mining challenges demonstrates with confidence that it is the best known way to evaluate different models and systems.

## 4.4. Classification results

As it was noticed in (Nikulin and McLachlan, 2009), the number of factors/metagenes must not be too large (in order to prevent overfitting), and must not be too small. In the latter case, the model will suffer because of the over-smoothing and loss of essential information as a consequence, see Figure 2(f), Tables 2 and 3 (in supplementary material).

As we reported in our previous studies (Nikulin and McLachlan, 2009, Nikulin *et al.*, 2010a), classification results observed with the model $GMF + LOO\{lm\}$ are competitive with those in (Dettling and Buhlmann, 2003, Hennig, 2007). Further improvements may be achieved using $A2GMF$ algorithm, because an adaptive learning rates make the factorization model more flexible. In addition, the results of Tables 2 and 3 (in supplementary material) may be easily improved if we shall apply instead of function $lm$ more advanced classifier with parameter tuning depending on the particular dataset.

The numbers of factors $k_{NP}$ and $k^*$, reported in Table 2, were computed using absolutely different methods, and we can see quite close correspondence between $k_{NP}$ and $k^*$. In two cases (leukaemia and lymphoma), when $NM = 0$, we observed an exact correspondence. This fact indicates that the nonparametric method of Section 2.2 is capable to discover some statistical fundamentals in microarray datasets.

**Remark 5** *Additionally, we conducted extensive experiments with NMF and SVD matrix factorizations. The observed classification results were similar in terms of the numbers of misclassified. However, misclassified samples were not the same. This fact may be exploited by the suitable ensembling technique.*

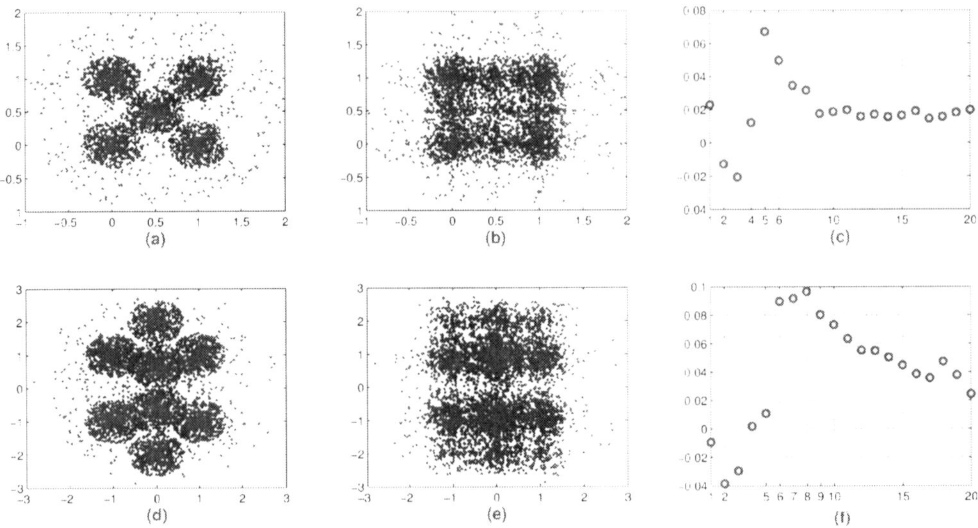

Figure 3: Illustrative examples based on two simulated datasets with known solutions (numbers of clusters) in support to the nonparametric criterion for the selection of the numbers of factors, Section 2.2. See for more details Section 4.5.

## 4.5. Illustrative example

The goal of cluster analysis is to partition the observations into groups (clusters) so that the pairwise dissimilarities between those assigned to the same cluster tend to be smaller than those in different clusters (Hastie *et al..*, 2008).

Figure 3 illustrates two experiments with *kmeans* algorithm, which were conducted against simulated 2D data with known numbers of clusters, see Figure 3(a) - five clusters ($n = 10000$); Figure 3(d) - eight clusters ($n = 20000$).

As a next step we re-shuffled independently the coordinates (columns) of the data, see Figure 3(b); Figure 3(e).

Figures 3(c) and (f) illustrate behavior of

$$\widehat{\mathcal{D}}_k = \widehat{\mathcal{R}}_k - \mathcal{R}_k, \tag{12}$$

as a function of the number of the clusters $k$, where 1) $\widehat{\mathcal{R}}$ and 2) $\mathcal{R}$ are two averaged dissimilarity measures with squared loss functions corresponding to 1) the re-shuffled and to 2) the original data.

In both cases, we can see correct detection: Figures 3(c) - $k = 5$, and (f) - $k = 8$.

## 4.6. Computation time

A multiprocessor Linux workstation with speed 3.2GHz was used for most of the computations. The time for 100 global iterations (see, for example, trajectory of Figure 1) against colon set in the cases of the GMF and A2GMF algorithm was 6 and 8 sec. (expenses for the other sets were similar).

## 5. Concluding remarks

It seems natural to use different learning rates applied to two factor matrices. Also, the values of the learning rates must not be fixed and it is proposed to update them after any global iteration according to the given formulas. Based on our experimental results, the A2GMF algorithm presented in this paper is significantly faster. That means, less number of global iterations will be required in order to achieve the same quality of factorization. Besides, the final results have smaller value of the loss function. This fact indicates that the A2GMF algorithm is better not only technically in the sense of speed, but is better in principle: to achieve essentially better quality of factorization for the given number of factors.

The proposed criterion for the selection of the number of factors/metagenes is non-parameteric and general. By application of such criterion (as an alternative to an extensive LOO experiments) it will be possible to save valuable computational time. Besides, we are confident that similar method is applicable elsewhere. An example with *kmeans* algorithm in given in Section 4.5.

An implementation of the third proposed novelty, a nonnegative modification of GMF, doesn't require any extra computational time. As an outcome, the NN-GMF is as fast as GMF algorithm itself. We can expect that practitioners in many fields will find this algorithm to be a competitive alternative to the well-known NMF.

### Acknowledgments

This work was supported by a grant from the Australian Research Council. Also, we are grateful to three anonymous reviewers for the very helpful and constructive comments.

## References

A. Cichocki, R. Zdunek, A. Phan and S. Amari. Nonnegative Matrix and Tensor Factorizations. *Wiley*, 2009.

M. Dettling and P. Buhlmann. Boosting for tumor classification with gene expression data. *Bioinformatics*, **19**(9), 1061-1069, 2003.

Q. Dong, X. Wang and L. Lin. Application of latent semantic analysis to protein remote homology detection, *Bioinformatics*, **22**, 285-290, 2006.

S. Dudoit, J. Fridlyand and T. Speed. Comparison of discrimination methods for the classification of tumors using gene expression data. *Journal of American Statistical Association*, **97**(457), 77-87, 2002.

P. Fogel, S. Young, D. Hawkins and N. Ledirac. Inferential, robust nonnegative matrix factorization analysis of microarray data, *Bioinformatics*, **23**(1), 44-49, 2007.

Y. Gao and G. Church. Improving molecular cancer class discovery through sparse nonnegative matrix factorization, *Bioinformatics*, **21**(21), 3970-3975, 2005.

G. Cawley and N. Talbot Gene selection in cancer classification using sparse logistic regression with Bayesian regularization. *Bioinformatics*, **22**(19), 2348-2355.

M. Jelizarow, V. Guillemot, A. Tenenhaus, K. Strimmer and A.-L. Boulesteix Over-optimism in bioinformatics: an illustration, *Bioinformatics*, **26**(16), 1990-1998.

T. Hastie, R. Tibshirani and J. Friedman. The Elements of Statistical Learning, *Springer-Verlag*, 2008.

C. Hennig. Cluster-wise assessment of cluster stability. *Computational Statistics and Data Analysis*, **52**, 258-271, 2007.

Y. Koren. Collaborative filtering with temporal dynamics. *KDD, Paris*, 447-455, 2009.

D. Lee and H. Seung. Learning the parts of objects by nonnegative matrix factorization. *Nature*, **401**, 788-791, 1999.

Y. Li and A. Ngom. Nonnegative matrix and tensor factorization based classification of clinical microarray gene expression data. *In Proceedings of 2010 IEEE International Conference on Bioinformatics and Biomedicine*, Hong Kong, 438-443, 2010.

C.-J. Lin. Projected gradient method for nonnegative matrix factorization. *Neural Computation*, **19**, 2756-2779, 2007.

D. Nguyen and D. Rocke Tumor classification by partial least squares using microarray gene expression data, *Bioinformatics*, **18**(1), 39-50.

V. Nikulin and G. J. McLachlan. Classification of imbalanced marketing data with balanced random sets. *JMLR: Workshop and Conference Proceedings*, 7, pp. 89-100, 2009.

V. Nikulin and G. J. McLachlan. On a general method for matrix factorization applied to supervised classification. *Proceedings of the 2009 IEEE International Conference on Bioinformatics and Biomedicine Workshops*, Washington D.C., 44-49, 2009.

V. Nikulin and G. J. McLachlan. On the gradient-based algorithm for matrix factorisation applied to dimensionality reduction. *Proceedings of BIOINFORMATICS 2010, Edited by Ana Fred, Joaquim Filipe and Hugo Gamboa*, Valencia, Spain, 147-152, 2010.

V. Nikulin, T.-H. Huang and G.J. McLachlan. A Comparative Study of Two Matrix Factorization Methods Applied to the Classification of Gene Expression Data. *In Proceedings of 2010 IEEE International Conference on Bioinformatics and Biomedicine*, Hong Kong, 618-621, 2010.

V. Nikulin, T.-H. Huangb, S.-K. Ng, S. Rathnayake, G.J. McLachlan. A very fast algorithm for matrix factorization. *Statistics and Probability Letters*, **81**, 773-782, 2011.

E. Oja, A. Ilin, J. Luttinen and Z. Yang. Linear expansions with nonlinear cost functions: modelling, representation, and partitioning. *Plenary and Invited Lectures, WCCI 2010, Edited by Joan Aranda and Sebastia Xambo*, Barcelona, Spain, 105-123, 2010.

P. Tamayo et al. Metagene projection for cross-platform, cross-species characterization of global transcriptional states. *Proceedings of the National Academy of Sciences USA*, **104**(14) 5959-5964, 2007.

L. Wasserman. All of Nonparametric Statistics. *Springer*, 2006.

# One-Shot Learning with a Hierarchical Nonparametric Bayesian Model


**Ruslan Salakhutdinov**            RSALAKHU@MIT.EDU
*Department of Statistics, University of Toronto*
*Toronto, Ontario, Canada*

**Josh Tenenbaum**            JBT@MIT.EDU
*Department of Brain and Cognitive Sciences, MIT*
*Cambridge, MA, USA*

**Antonio Torralba**            TORRALBA@MIT.EDU
*CSAIL, MIT*
*Cambridge, MA, USA*


**Editor:** I. Guyon, G. Dror, V. Lemaire, G. Taylor, and D. Silver

## Abstract


We develop a hierarchical Bayesian model that learns categories from single training examples. The model transfers acquired knowledge from previously learned categories to a novel category, in the form of a prior over category means and variances. The model discovers how to group categories into meaningful super-categories that express different priors for new classes. Given a single example of a novel category, we can efficiently infer which super-category the novel category belongs to, and thereby estimate not only the new category's mean but also an appropriate similarity metric based on parameters inherited from the super-category. On MNIST and MSR Cambridge image datasets the model learns useful representations of novel categories based on just a single training example, and performs significantly better than simpler hierarchical Bayesian approaches. It can also discover new categories in a completely unsupervised fashion, given just one or a few examples.


## 1. Introduction

In typical applications of machine classification algorithms, learning curves are measured in tens, hundreds or thousands of training examples. For human learners, however, the most interesting regime occurs when the training data are very sparse. Just a single example is often sufficient for people to grasp a new category and make meaningful generalizations to novel instances, if not to classify perfectly (Pinker, 1999). Human categorization often asymptotes after just three or four examples (Xu and Tenenbaum, 2007; Smith et al., 2002; Kemp et al., 2006; Perfors and Tenenbaum, 2009). To illustrate, consider learning entirely novel "alien" objects, as shown in Fig. 1, left panel. Given just three examples of a novel "tufa" concept (boxed in red), almost all human learners select just the objects boxed in gray (Schmidt, 2009). Clearly this requires very strong but also appropriately tuned inductive biases. A hierarchical Bayesian model we describe here takes a step towards this "one-shot learning" ability by learning abstract knowledge that support transfer of useful inductive biases from previously learned concepts to novel ones.

At a minimum, categorizing an object requires information about the category's mean and variance along each dimension in an appropriate feature space. This is a similarity-based approach, where the mean represents the category prototype, and the inverse variances (or precisions) correspond to the dimensional weights in a category-specific similarity metric. One-shot learning may seem impossible because a single example provides information about the mean or prototype of the category, but not about the variances or the similarity metric. Giving equal weight to every dimension in a large a priori-defined feature space, or using the wrong similarity metric, is likely to be disastrous.

Our model leverages higher-order knowledge abstracted from previously learned categories to estimate the new category's prototype as well as an appropriate similarity metric from just one example. These estimates are also improved as more examples are observed. To illustrate, consider how human learners seeing one example of an unfamiliar animal, such as a wildebeest (or gnu), can draw on experience with many examples of "horse", "cows", "sheep", and more familiar related categories. These similar categories have similar prototypes – horses, cows, and sheep look more like each other than like furniture or vehicles – but they also have similar variability in their feature-space representations, or similar similarity metrics: The ways in which horses vary from the "horse" prototype are similar to the ways in which sheep vary from the "sheep" prototype. We may group these similar basic-level categories into an "animal" super-category, which captures these classes' similar prototypes as well as their similar modes of variation about their respective prototypes, as show in Fig. 1, right panel. If we can identify the new example of "wildebeest" as belonging to this "animal" super-category, we can transfer an appropriate similarity metric and thereby generalize informatively even from a single example.

Learning similarity metric over the high-dimensional input spaces has become an important task in machine learning as well. A number of recent approaches (Weinberger and Saul, 2009; Babenko et al., 2009; Singh-Miller and Collins, 2009; Goldberger et al., 2004; Salakhutdinov and Hinton, 2007; Chopra et al., 2005) have demonstrated that learning a class-specific similarity metric can provide some insights into how high-dimensional data is organized and it can significantly improve the performance of algorithms like K-nearest neighbours that are based on computing distances. Most this work, however, focused on learning similarity metrics when many labeled examples are available, and did not attempt to address the one-shot learning problem.

Although inspired by human learning, our approach is intended to be broadly useful for machine classification and AI tasks. To equip a robot with human-like object categorization abilities, we must be able to learn tens of thousands of different categories, building on (and not disrupting) representations of old ones (Bart and Ullman, 2005; Biederman, 1995). In these settings, learning from one or a few labeled examples and performing efficient inference will be crucial. Our method is designed to scale up

**Learning from Three Examples**

**Learning Class-specific Similarity Metric from One Example**

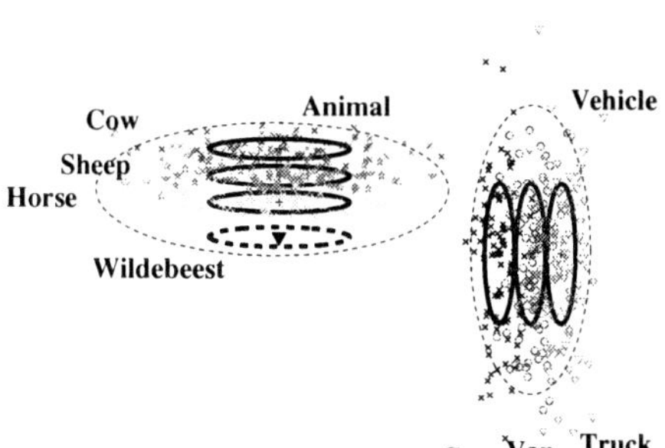


Figure 1: **Left:** Given only three examples (boxed in red) of a novel "tufa" object, which other objects are tufas? Most human learners select just the objects boxed in gray, as shown by Schmidt (2009). **Right:** Learning a similarity metric for a novel "wildebeest" class based on one example. The goal is to identify that the new "wildebeest" belongs to the "animal" super-category, which would allow to transfer an appropriate similarity metric and thereby generalize informatively from a single example.

in precisely these ways: a nonparametric prior allows new categories to be formed at any time in either supervised or unsupervised modes, and conjugate distributions allow most parameters to be integrated out analytically for very fast inference.

## 2. Related Prior Work

Hierarchical Bayesian models have previously been proposed (Kemp et al. (2006); Heller et al. (2009)) to describe how people learn to learn categories from one or a few examples, or learn similarity metrics, but these approaches were not focused on machine learning settings – large-scale problems with many categories and high-dimensional natural image data. A large class of models based on hierarchical Dirichlet processes (Teh et al. (2006)) have also been used for transfer learning (Sudderth et al. (2008); Canini and Griffiths (2009)). There are two key difference: First, HDPs typically assume a fixed hierarchy of classes for sharing parameters, while we learn the hierarchy in an unsupervised fashion. Second, HDPs are typically given many examples for each category rather than the one-shot learning cases we consider here. Recently introduced nested Dirichlet processes can also be used for transfer learning (Rodriguez and Vuppala (2009); Rodriguez et al. (2008)). However, this work assumes a fixed number of classes (or groups) and did not attempt to address one-shot learning problem. A recent hierarchical model of Adams et al. (2011) could also be used for transfer learning tasks. However, this model does not learn hierarchical priors over covariances, which is crucial for transferring an appropriate similarity metric to new basic-level categories in order to support learning from few examples. These recently introduced models are complementary to our approach, and can be combined productively, although we leave that as a subject for future work.

There are several related approaches in the computer vision community. A hierarchical topic model for image features (Bart et al. (2008); Sivic et al. (2008)) can discover visual taxonomies in an unsupervised fashion from large datasets but was not designed for one-shot learning of new categories. Perhaps closest to our work, Fei-Fei et al. (2006) also gave a hierarchical Bayesian model for visual categories with a prior on the parameters of new categories that was induced from other categories. However, they learned a single prior shared across all categories and the prior was learned only from three categories, chosen by hand.

More generally, our goal contrasts with and complements that of computer vision efforts on one-shot learning. We have attempted to minimize any tuning of our approach to specifically visual applications. We seek a general-purpose hierarchical Bayesian model that depends minimally on domain-specific representations but instead learns to perform one-shot learning by finding more intelligent representations tuned to specific sub-domains of a task (our "super-categories").

## 3. Hierarchical Bayesian Model

Consider observing a set of $N$ $i.i.d$ input feature vectors $\{\mathbf{x}^1, \ldots, \mathbf{x}^N\}$, $\mathbf{x}^n \in R^D$. In general, features will be derived from high-dimensional, highly structured data, such as images of natural scenes, in which case the feature dimensionality $D$ can be quite large (e.g. 50,000). For clarity of presentation, let us first assume that our model is presented with a fixed two-level category hierarchy. In particular, suppose that $N$ objects are partitioned into $C$ basic-level (or level-1) categories. We represent such partition by a vector $\mathbf{z}^b$ of length $N$, each entry of which is $z_n^b \in \{1, \ldots, C\}$. We also assume that our $C$ basic-level categories are partitioned into $K$ super-categories (level-2 categories), which we represent by $\mathbf{z}^s$ of length $C$, with $z_c^s \in \{1, \ldots, K\}$.

For any basic-level category $c$, the distribution over the observed feature vectors is assumed to be Gaussian with a category-specific mean $\mu^c$ and a category-specific *diagonal* precision matrix, whose entries are $\{\tau_d^c\}_{d=1}^D$. The distribution takes the following product form:

$$P(\mathbf{x}^n \mid z_n^b = c, \theta^1) = \prod_{d=1}^{D} \mathcal{N}(x_d^n \mid \mu_d^c, 1/\tau_d^c), \tag{1}$$

where $\mathcal{N}(x \mid \mu, 1/\tau)$ denotes a Gaussian distribution with mean $\mu$ and precision $\tau$ and $\theta^1 = \{\mu^c, \tau^c\}_{c=1}^C$ denotes the level-1 category parameters. We next place a conjugate Normal-Gamma prior over $\{\mu^c, \tau^c\}$. Let $k = z_c^s$, i.e. let the level-1 category $c$ belong to level-2 category $k$, where $\theta^2 = \{\mu^k, \tau^k, \alpha^k\}_{k=1}^K$ denote the level-2 parameters. Then: $P(\mu^c, \tau^c \mid \theta^2, \mathbf{z}^s) = \prod_{d=1}^D P(\mu_d^c, \tau_d^c \mid \theta^2, \mathbf{z}^s)$, where for each dimension $d$ we have:

$$P(\mu_d^c, \tau_d^c \mid \theta^2) = P(\mu_d^c \mid \tau_d^c, \theta^2) P(\tau_d^c \mid \theta^2) = \mathcal{N}(\mu_d^c \mid \mu_d^k, 1/(\nu\tau_d^c)) \Gamma(\tau_d^c \mid \alpha_d^k, \alpha_d^k/\tau_d^k). \tag{2}$$

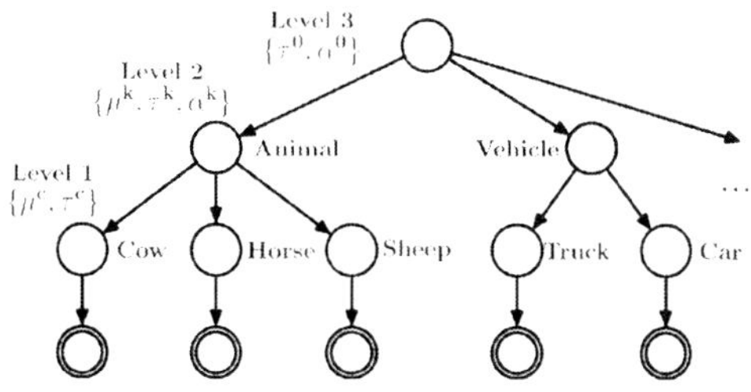



- For each super-category $k = 1,..,\infty$:
  draw $\theta^2$ using Eq. 4.

- For each basic category $c^k = 1,...,\infty$, placed under each super-category $k$:
  draw $\theta^1$ using Eq. 2.

- For each observation $n = 1,\ldots,N$
  draw $\mathbf{z}_n \sim \text{nCRP}(\gamma)$
  draw $\mathbf{x}^n \sim \mathcal{N}(\mathbf{x}^n \mid \theta^1, \mathbf{z}_n)$ using Eq. 1

Figure 2: **Left:** Hierarchical Bayesian model that assumes a fixed tree hierarchy for sharing parameters. **Right:** Generative process of the corresponding non-parametric model.

Our parameterization of the Gamma density is in terms of its shape $\alpha^k$ and mean $\tau^k$ parameters:

$$\Gamma(\tau \mid \alpha^k, \alpha^k/\tau^k) = \frac{(\alpha^k/\tau^k)^{\alpha^k}}{\Gamma(\alpha^k)} \tau^{\alpha^k-1} \exp\left(-\tau\frac{\alpha^k}{\tau^k}\right). \tag{3}$$

Such a parameterization is more interpretable, since $E[\tau] = \tau^k$. In particular, from Eq. 2, we can easily derive that $E[\mu^c] = \mu^k$ and $E[\tau^c] = \tau^k$. This gives our model a very intuitive interpretation: the expected values of the basic level-1 parameters $\theta^1$ are given by the corresponding level-2 parameters $\theta^2$. The parameter $\alpha^k$ further controls the variability of $\tau^c$ around its mean, i.e. $\text{Var}[\tau^c] = (\tau^k)^2/\alpha^k$. For the level-2 parameters $\theta^2$, we shall assume the following conjugate priors:

$$P(\mu_d^k) = \mathcal{N}(\mu_d^k \mid 0, 1/\tau^0), \quad P(\alpha_d^k \mid \alpha^0) = \text{Exp}(\alpha_d^k \mid \alpha^0), \quad P(\tau_d^k \mid \theta^0) = \text{IG}(\tau_d^k \mid a^0, b^0), \tag{4}$$

where $\text{Exp}(x \mid \alpha)$ denotes an exponential distribution with rate parameter $\alpha$, and $\text{IG}(x \mid \alpha, \beta)$ denotes an inverse-gamma distribution with shape parameter $\alpha$ and scale parameter $\beta$. We further place a diffuse Gamma prior $\Gamma(1,1)$ over the level-3 parameters $\theta^3 = \{\alpha^0, \tau^0\}$. Throughout our experimental results, we also set $a^0 = 1$ and $b^0 = 1$.

### 3.1. Modelling the number of super-categories

So far we have assumed that our model is presented with a two-level partition $\mathbf{z} = \{\mathbf{z}^s, \mathbf{z}^b\}$. If, however, we are not given any level-1 or level-2 category labels, we need to infer the distribution over the possible category structures. We place a nonparametric two-level nested Chinese Restaurant Prior (CRP) (Blei et al. (2003, 2010)) over $\mathbf{z}$, which defines a prior over tree structures and is flexible enough to learn arbitrary hierarchies. The main building block of the nested CRP is the Chinese restaurant process, a distribution on partition of integers. Imagine a process by which customers enter a restaurant with an unbounded number of tables, where the $n^{th}$ customer occupies a table $k$ drawn from:

$$P(z_n = k \mid z_1, \ldots, z_{n-1}) = \begin{cases} \frac{n^k}{n-1+\gamma} & n^k > 0 \\ \frac{\gamma}{n-1+\gamma} & k \text{ is new} \end{cases}, \tag{5}$$

where $n^k$ is the number of previous customers at table $k$ and $\gamma$ is the concentration parameter.

The Nested CRP, nCRP($\gamma$), extends CRP to nested sequence of partitions, one for each level of the tree. In this case each observation $n$ is first assigned to the super-category $z_n^s$ using Eq. 5. Its assignment to the basic-level category $z_n^b$, that is placed under a super-category $z_n^s$, is again recursively drawn from Eq. 5 (for details see Blei et al. (2010)). For our model, a two-level nested CRP allows flexibility of having a potentially unbounded number of super-categories as well as an unbounded number of basic-level categories placed under each super-category. Finally, we also place a Gamma prior $\Gamma(1,1)$ over $\gamma$. The full generative model is given in Fig. 2, right panel. Unlike in many conventional hierarchical Bayesian models, here we infer both the model parameters as well as the hierarchy for sharing those parameters.

Our model can be readily used in unsupervised or semi-supervised modes, with varying amounts of label information. Here we focus on two settings. First, we assume basic-level category labels have been given for all examples in a training set, but no super-category labels are available. We must infer how to cluster basic categories into super-categories at the same time as we infer parameter values at all levels of the hierarchy. The training set includes many examples of familiar basic categories but only one (or few) example for a novel class. The challenge is to generalize the new class intelligently from this one example by inferring which super-category the new class comes from and exploiting that super-category's implied priors to estimate the new class's prototype and similarity metric most accurately. This training regime reflects natural language acquisition, where spontaneous category labeling is frequent, almost all spontaneous labeling is at the basic level (Rosch et al., 1976) yet children's generalizations are sensitive to higher superordinate structure (Mandler, 2004), and where new basic-level categories are typically learned with high accuracy from just one or a few labeled examples. Second, we consider a similar labeled training set but now the test set consists of many unlabeled examples from an unknown number of basic-level classes – including both familiar and novel classes. This reflects the problem of "unsupervised category learning" a child or robot faces in discovering when they have encountered novel categories, and how to break up new instances into categories in an intelligent way that exploits knowledge abstracted from a hierarchy of more familiar categories.

## 4. Inference

Inferences about model parameters at all levels of hierarchy can be performed by MCMC. When the tree structure $\mathbf{z}$ of the model is not given, the inference process will alternate between fixing $\mathbf{z}$ while sampling the space of model parameters $\theta$ and fixing $\theta$ while sampling category assignments.

**Sampling level-1 and level-2 parameters:** Given level-2 parameters $\theta^2$ and $\mathbf{z}$, the conditional distribution $P(\mu^c, \tau^c \mid \theta^2, \mathbf{z}, \mathbf{x})$ is Normal-Gamma (Eq. 2), which allows us to easily sample level-1 parameters $\{\mu^c, \tau^c\}$. Given $\mathbf{z}$, $\theta^1$, and $\theta^3$, the conditional distributions over the mean $\mu^k$ and precision $\tau^k$ take Gaussian and Inverse-Gamma forms.

The only complicated step involves sampling $\alpha^k$ that control the variation of the precision term $\tau^c$ around its mean (Eq. 3). The conditional distribution over $\alpha^k$ cannot be computed in closed form and is proportional to:

$$p(\alpha^k \mid \mathbf{z}, \theta^1, \theta^3, \tau^k) \propto \frac{(\alpha^k/\tau^k)^{\alpha^k n_k}}{\Gamma(\alpha^k)^{n_k}} \exp\left(-\alpha^k\left(\alpha^0 + S^k/\tau^k - T^k\right)\right), \tag{6}$$

where $S^k = \sum_{c:z(c)=k} \tau^c$ and $T^k = \sum_{c:z(c)=k} \log(\tau^c)$. For large values of $\alpha^k$ the density, specified by Eq. 6, is similar to a Gamma density (Wiper et al. (2001)). We therefore use Metropolis-Hastings with a proposal distribution given by the Gamma density. In particular, we generate a new candidate

$$\alpha^* \sim Q(\alpha^* \mid \alpha^k) \quad \text{with} \quad Q(\alpha^* \mid \alpha^k) = \Gamma(\alpha^* \mid t, t/\alpha^k)$$

and accept it with M-H rule. In all of our experiments we use $t = 3$, which gave an acceptance probability of about 0.6. Sampling level-3 parameters is similar to sampling level-2 parameters.

**Sampling assignments z:** Given model parameters $\theta = \{\theta^1, \theta^2\}$, combining the likelihood term with the nCRP($\gamma$) prior, the posterior over the assignment $\mathbf{z}_n$ can be calculated as follows:

$$p(\mathbf{z}_n \mid \theta, \mathbf{z}_{-n}, \mathbf{x}^n) \propto p(\mathbf{x}^n \mid \theta, \mathbf{z}_n) p(\mathbf{z}_n \mid \mathbf{z}_{-n}), \tag{7}$$

where $\mathbf{z}_{-n}$ denotes variables $\mathbf{z}$ for all observations other than $n$. We can further exploit the conjugacy in our hierarchical model when computing the probability of creating a new basic-level category. Using the fact that Normal-Gamma prior $p(\mu^c, \tau^c)$ is the conjugate prior of a normal distribution, we can easily compute the following marginal likelihood:

$$p(\mathbf{x}^n \mid \theta^2, \mathbf{z}_n) = \int_{\mu^c, \tau^c} p(\mathbf{x}^n, \mu^c, \tau^c \mid \theta^2, \mathbf{z}_n) = \int_{\mu^c, \tau^c} p(\mathbf{x}^n \mid \mu^c, \tau^c) p(\mu^c, \tau^c \mid \theta^2, \mathbf{z}_n).$$

Integrating out basic-level parameters $\theta^1$ lets us more efficiently sample over the tree structures[1]. When computing the probability of placing $\mathbf{x}^n$ under a newly created super-category, its parameters are sampled from the prior.

## 5. One-shot Learning

One of the key goals of our work is to develop a model that has the ability to generalize from a single example. Consider observing a single new instance $\mathbf{x}^*$ of a *novel category* $c^*$[2]. Conditioned on the current setting of the level-2 parameters $\theta^2$ and our current tree structure $\mathbf{z}$, we can first infer which super-category the novel category should belong to, i.e. we can compute the posterior distribution over the assignments $\mathbf{z}_c^*$ using Eq. 7.

---

1. In the supervised case, inference in simplified by only considering which super-category each basic-level category is assigned to.
2. Observing several examples of a new category is treated similarly.

We note that our new category can either be placed under one of the existing super-categories, or create its own super-category, if it is sufficiently different from all of the existing super-categories.

Given an inferred assignment $\mathbf{z}_c^*$ and using Eq. 2, we can infer the posterior mean and precision terms (or similarity metric) $\{\mu^*, \tau^*\}$ for our novel category. We can now test the ability of the HB model to generalize to new instances of a novel category by computing the conditional probability that a new test input $\mathbf{x}^l$ belongs to a novel category $c^*$:

$$p(c^* \mid \mathbf{x}^l) = \frac{p(\mathbf{x}^l \mid \mathbf{z}_c^*)p(\mathbf{z}_c^*)}{\sum_{\mathbf{z}} p(\mathbf{x}^l \mid \mathbf{z})p(\mathbf{z})}, \tag{8}$$

where the prior is given by the nCRP($\gamma$) and the log-likelihood takes form:

$$\log p(\mathbf{x}^l \mid c^*) = \frac{1}{2} \sum_d \log(\tau_d^*) - \frac{1}{2} \sum_d \tau_d^* (x_d^l - \mu_d^*)^2 + C,$$

where $C$ is a constant that does not depend on the parameters. Observe that the relative importance of each feature in determining the similarity is proportional to the category-specific precision of that feature. Features that are salient, or have higher precision, within the corresponding category contribute more to the overall similarity of an input.

## 6. Experimental results

We now present experimental results on the MNIST handwritten digit and MSR Cambridge object recognition image datasets. During the inference step, we run our hierarchical Bayesian (HB) model for 200 full Gibbs sweeps, which was sufficient to reach convergence and obtain good performance. We normalize input vectors to zero mean and scale the entire input by a single number to make the average feature variance be one.

In all of our experiments, we compare performance of the HB model to the following four alternative methods for one-shot learning. The first model, "Euclidean", uses a Euclidean metric, i.e. all precision terms are set to one and are never updated. The second model, that we call "HB-Flat", always uses a single super-category. When presented with a single example of a new category, HB-Flat will inherit a similarity metric that is shared by all existing categories, as done in Fei-Fei et al. (2006). Our third model, called "HB-Var", is similar in spirit to the approach of Heller et al. (2009) and is based on clustering only covariance matrices without taking into account the means of the super-categories. Our last model, "MLE", ignores hierarchical Bayes altogether and estimates a category-specific mean and precision from sample averages. If a category contains only one example, the model uses the Euclidean metric. Finally, we also compare to the "Oracle" model that is the same as our HB model, but always uses the correct, instead of inferred, similarity metric.

## 6.1. MNIST dataset

The MNIST dataset contains 60,000 training and 10,000 test images of ten handwritten digits (zero to nine), with 28×28 pixels. For our experiments, we randomly choose 1000 training and 1000 test images (100 images per class). We work directly in the pixel space because all handwritten digits were already properly aligned. In addition, working in the pixel space allows us to better visualize the kind of transfer of similarity metrics our model is performing. Fig. 3, left panel, shows a typical partition over the basic level categories, along with corresponding mean and similarity metrics, that our model discovers.

We first study the ability of the HB model to generalize from a single training example of handwritten digit "nine". To this end, we trained the HB model on 900 images (100 images of each of zero-to-eight categories), while withholding all images that belong to category "nine". Given a single new instance of a novel "nine" category our model is able to discover that the new category is more like categories that contain images of seven and four, and hence this novel category can inherit the mean and the similarity metric, shared by categories "seven" and "four".

Table 1 further quantifies performance using the area under the ROC curve (AUROC) for classifying 1000 test images as belonging to the "nine" vs. all other categories. (an area of 0.5 corresponds to the classifier that makes random predictions). The HB model achieves an AUROC of 0.81, considerably outperforming HB-Flat, HB-Var, Euclidean, and MLE that achieve an AUROC of 0.71, 0.72, 0.70, and 0.69 respectively. Moreover, with just four examples, the HB model is able to achieve performance close to that of the Oracle model. This is in sharp contrast to HB-Flat, MLE and Euclidean models, that even with four examples perform far worse.

## 6.2. MSR Cambridge Dataset

We now present results on a considerably more difficult MSR Cambridge dataset[3], that contains images of 24 different categories. Fig. 3, right panel, shows 24 basic-level categories along with a typical partition that our model discovers. We use a simple "texture-of-textures" framework for constructing image features (DeBonet and Viola (1997)).

We first tested the ability of our model to generalize from a single image of a cow. Similar to the experiments on the MNIST dataset, we first train the HB model on images corresponding to 23 categories, while withholding all images of cows. In general, our model is able to discover that the new "cow" category is more like the "sheep" category, as opposed to categories that contain images of cars, or forks, or buildings. This allows the new "cow" category inherit sheep's similarity metric.

Table 2 show that the HB model, based on a single example of cow, achieves an AUROC of 0.77. This is compared to an AUROC of only 0.62, 0.61, 0.59, and 0.58 achieved by the HB-Flat, HB-Var, Euclidean, and MLE models. Similar to the results

---

3. Available at `http://research.microsoft.com/en-us/projects/objectclassrecognition/`

Table 1: Performance results using the area under the ROC curve (AUROC) on the MNIST dataset. The Average panel shows results averaged over all 10 categories, using leave-one-out test format.

| Model | Category: Digit 9 | | | | Category: Digit 6 | | | | Average | | | |
|---|---|---|---|---|---|---|---|---|---|---|---|---|
| | 1 ex | 2 ex | 4 ex | 20 ex | 1 ex | 2 ex | 4 ex | 20 ex | 1 ex | 2 ex | 4 ex | 20 ex |
| HB | 0.81 | 0.85 | 0.88 | 0.90 | 0.85 | 0.89 | 0.92 | 0.97 | 0.85 | 0.88 | 0.90 | 0.93 |
| HB-Flat | 0.71 | 0.77 | 0.84 | 0.90 | 0.73 | 0.79 | 0.88 | 0.97 | 0.74 | 0.79 | 0.86 | 0.93 |
| HB-Var | 0.72 | 0.81 | 0.86 | 0.90 | 0.72 | 0.83 | 0.90 | 0.97 | 0.75 | 0.82 | 0.89 | 0.93 |
| Euclidean | 0.70 | 0.73 | 0.76 | 0.80 | 0.74 | 0.77 | 0.82 | 0.86 | 0.72 | 0.76 | 0.80 | 0.83 |
| Oracle | 0.87 | 0.89 | 0.90 | 0.90 | 0.95 | 0.96 | 0.96 | 0.97 | 0.90 | 0.92 | 0.92 | 0.93 |
| MLE | 0.69 | 0.75 | 0.83 | 0.90 | 0.72 | 0.78 | 0.87 | 0.97 | 0.71 | 0.77 | 0.84 | 0.93 |

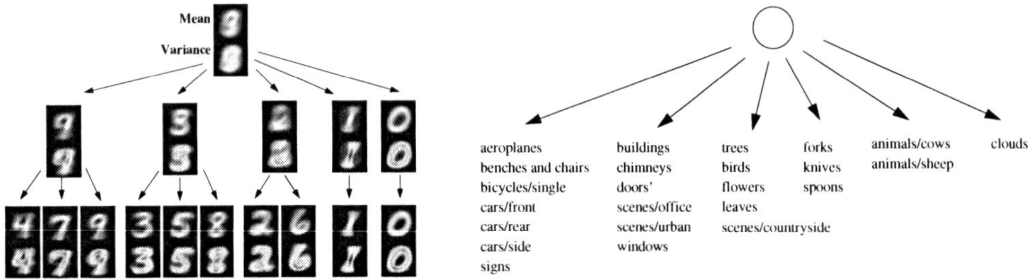


Figure 3: **Left:** MNIST Dataset: A typical partition over the 10 categories discovered by the HB model. Top panels display means and bottom panels display variances (white encodes larger values). **Right:** MSR Dataset: A typical partition over the 24 categories discovered by the HB model.

on the MNIST dataset, the HB model with just one example performs comparably the HB-Flat and MLE models that make use of four examples. Fig. 4 further displays retrieval results based on a single image of a cow. As expected, the HB model performs much better compared to the Euclidean model that does not learn a similarity metric.

### 6.3. Unsupervised Category Discovery

Another key advantage of the hierarchical nonparametric Bayesian model is its ability to infer category structure in an unsupervised fashion, discovering novel categories at both levels 1 and 2 of the hierarchy. We explored the HB model's category discovery ability by training on labeled examples of 21 basic-level MSR categories, leaving out clouds, trees, and chimneys. We then provided six test images: one in each of the three unseen categories and one in each of three familiar basic-level categories (car, airplane, bench). For each test image, using Eq. 8, we can easily compute the posterior probability of forming a new basic-level category. Figure 5, left panel, shows six representative test images, sorted by the posterior probability of forming a novel category. The model correctly identifies the car, the airplane and the bench as belonging to familiar cate-

Table 2: Performance results using the area under the ROC curve (AUROC) on the MSR dataset. The Average panel shows results averaged over all 24 categories, using leave-one-out test format.

| Model | Category: Cow | | | | Category: Flower | | | | Average | | | |
|---|---|---|---|---|---|---|---|---|---|---|---|---|
| | 1 ex | 2 ex | 4 ex | 20 ex | 1 ex | 2 ex | 4 ex | 20 ex | 1 ex | 2 ex | 4 ex | 20 ex |
| HB | 0.77 | 0.81 | 0.84 | 0.89 | 0.71 | 0.75 | 0.78 | 0.81 | 0.76 | 0.80 | 0.84 | 0.87 |
| HB-Flat | 0.62 | 0.69 | 0.80 | 0.89 | 0.59 | 0.64 | 0.75 | 0.81 | 0.65 | 0.71 | 0.78 | 0.87 |
| HB-Var | 0.61 | 0.73 | 0.83 | 0.89 | 0.60 | 0.68 | 0.77 | 0.81 | 0.64 | 0.74 | 0.81 | 0.87 |
| Euclidean | 0.59 | 0.61 | 0.63 | 0.66 | 0.55 | 0.59 | 0.61 | 0.64 | 0.63 | 0.66 | 0.69 | 0.71 |
| Oracle | 0.83 | 0.84 | 0.87 | 0.89 | 0.77 | 0.79 | 0.80 | 0.81 | 0.82 | 0.84 | 0.86 | 0.87 |
| MLE | 0.58 | 0.64 | 0.78 | 0.89 | 0.55 | 0.62 | 0.72 | 0.81 | 0.62 | 0.67 | 0.77 | 0.87 |

Figure 4: Retrieval results based on observing a single example of cow. Top five most similar images were retrieved from the test set, containing 360 images corresponding to 24 categories.

gories, and places much higher probability on forming novel categories for the other images. With only one unlabeled example of these novel classes, the model still prefers two of them in familiar categories: the "tree" is interpreted as an atypical example of "countryside" while the "chimney" is classified as an atypical "building".

The model, however, can correctly discover novel categories given only a little more unlabeled data. With 18 unlabeled test images (see Fig. 5), after running a Gibbs sampler for 100 steps, the model correctly places nine "familiar" images in nine different basic-level categories, while also correctly forming three novel basic-level categories with three examples each. Most interestingly, these new basic-level categories are placed at the appropriate level of the category hierarchy: the novel "tree" category is correctly placed under the super-category containing "leaves" and "countrysides"; the novel "chimney" category is placed together with "buildings" and "doors"; while "clouds" category is placed in its own super-category – all consistent with the hierarchy we originally found from a fully labeled training set (see Fig. 3). Other models we tried for this unsupervised task perform much worse; they confuse "chimneys" with "cows" and "trees" with "countrysides".

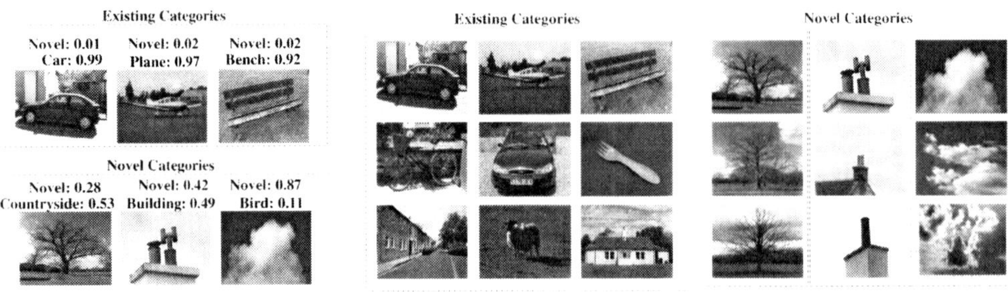


Figure 5: Unsupervised category discovery. **Left:** Six representative test images, sorted by the posterior probability of forming a novel category. **Right:** When presented with 18 unlabeled test images, the model correctly places nine "familiar" images in nine different basic-level categories, while also correctly forming three novel basic-level categories with three examples each.

## 7. Conclusions

In this paper we developed a hierarchical nonparametric Bayesian model for learning a novel category based on a single training example. Our experimental results further demonstrate that our model is able to effectively transfer appropriate similarity metric from the previously learned categories to a novel category based on observing a single example. There are several key advantages to our model. First, due to efficient Gibbs moves that can exploit conjugacy, the model can be efficiently trained. Many of the Gibbs updates can be run in parallel, which will allow our model to potentially handle a large number of basic-level categories. Second, the model is able to discover meaningful super-categories and be able to form coherent novel categories. Finally, given a single example of a novel category, the model is able to quickly infer which super-category the new basic-level category should belong to. This in turns allows us to efficiently infer the appropriate similarity metric for this novel category.

## References

R. Adams, Z. Ghahramani, and M. Jordan. Tree-structured stick breaking processes for hierarchical data. In *To appear in NIPS*. MIT Press, 2011.

B. Babenko, S. Branson, and S. J. Belongie. Similarity functions for categorization: from monolithic to category specific. In *ICCV*, 2009.

E. Bart and S. Ullman. Cross-generalization: Learning novel classes from a single example by feature replacement. In *CVPR*, pages 672–679, 2005.

E. Bart, I. Porteous, P. Perona, and M. Welling. Unsupervised learning of visual taxonomies. In *CVPR*, pages 1–8, 2008.

I. Biederman. Visual object recognition. *An Invitation to Cognitive Science*, 2:152–165, 1995.

D. M. Blei, T. L. Griffiths, M. I. Jordan, and Joshua B. Tenenbaum. Hierarchical topic models and the nested Chinese restaurant process. In *NIPS*. MIT Press, 2003.

D. M. Blei, T. L. Griffiths, and M. I. Jordan. The nested Chinese restaurant process and Bayesian nonparametric inference of topic hierarchies. *J. ACM*, 57(2), 2010.

K. R. Canini and T. L. Griffiths. Modeling human transfer learning with the hierarchical Dirichlet process. In *NIPS 2009 workshop: Nonparametric Bayes*, 2009.

S. Chopra, R. Hadsell, and Y. LeCun. Learning a similarity metric discriminatively, with application to face verification. In *IEEE Computer Vision and Pattern Recognition or CVPR*, pages I: 539–546, 2005.

J. S. DeBonet and P. A. Viola. Structure driven image database retrieval. In Michael I. Jordan, Michael J. Kearns, and Sara A. Solla, editors, *NIPS*. The MIT Press, 1997.

Li Fei-Fei, R. Fergus, and P. Perona. One-shot learning of object categories. *IEEE Trans. Pattern Analysis and Machine Intelligence*, 28(4):594–611, April 2006.

J. Goldberger, S. T. Roweis, G. E. Hinton, and R. R. Salakhutdinov. Neighbourhood components analysis. In *Advances in Neural Information Processing Systems*, 2004.

K. Heller, A. Sanborn, and N. Chater. Hierarchical learning of dimensional biases in human categorization. In *NIPS*, 2009.

C. Kemp, A. Perfors, and J. Tenenbaum. Learning overhypotheses with hierarchical Bayesian models. *Developmental Science*, 10(3):307–321, 2006.

J.M. Mandler. *The foundations of mind: Origins of conceptual thought.* Oxford University Press, USA, 2004.

A. Perfors and J.B. Tenenbaum. Learning to learn categories. In *31st Annual Conference of the Cognitive Science Society*, pages 136–141, 2009.

S. Pinker. *How the Mind Works.* W.W. Norton, 1999.

A. Rodriguez and R. Vuppala. Probabilistic classification using Bayesian nonparametric mixture models. Technical Report, 2009.

A. Rodriguez, D. Dunson, and A. Gelfand. The nested Dirichlet process. *Journal of the American Statistical Association*, 103:1131–1144, 2008.

E. Rosch, C.B. Mervis, W.D. Gray, D.M. Johnson, and P. Boyes-Braem. Basic objects in natural categories* 1. *Cognitive psychology*, 8(3):382–439, 1976.

R. R. Salakhutdinov and G. E. Hinton. Learning a nonlinear embedding by preserving class neighbourhood structure. In *Proceedings of the International Conference on Artificial Intelligence and Statistics*, volume 11, 2007.

L. Schmidt. Meaning and compositionality as statistical induction of categories and constraints. Ph.D Thesis, Massachusetts Institute of Technology, 2009.

N. Singh-Miller and M. Collins. Learning label embeddings for nearest-neighbor multi-class classification with an application to speech recognition. In *Advances in Neural Information Processing Systems*. MIT Press, 2009.

J. Sivic, B. C. Russell, A. Zisserman, W. T. Freeman, and A. A. Efros. Unsupervised discovery of visual object class hierarchies. In *CVPR*, pages 1–8, 2008.

L.B. Smith, S.S. Jones, B. Landau, L. Gershkoff-Stowe, and L. Samuelson. Object name learning provides on-the-job training for attention. *Psychological Science*, pages 13–19, 2002.

E. B. Sudderth, A. Torralba, W. T. Freeman, and A. S. Willsky. Describing visual scenes using transformed objects and parts. *International Journal of Computer Vision*, 77 (1-3):291–330, 2008.

Y. W. Teh, M. I. Jordan, M. J. Beal, and D. M. Blei. Hierarchical Dirichlet processes. *Journal of the American Statistical Association*, 101(476):1566–1581, 2006.

K. Q. Weinberger and L. K. Saul. Distance metric learning for large margin nearest neighbor classification. *Journal of Machine Learning Research*, 3, 2009.

M. Wiper, D. R. Insua, and F. Ruggeri. Mixtures of Gamma distributions with applications. *Journal of Computational and Graphical Statistics*, 10(3), September 2001.

F. Xu and J. B. Tenenbaum. Word learning as Bayesian inference. *Psychological Review*, 114(2), 2007.

# Multitask Learning in Computational Biology

**Christian Widmer**    CWIDMER@TUE.MPG.DE and **Gunnar Rätsch**    RAETSCH@TUE.MPG.DE
*FML, Max Planck Society, Spemannstr. 39, 72076 Tübingen, Germany*

**Editor:** I. Guyon, G. Dror, V. Lemaire, G. Taylor, and D. Silver

## Abstract

Computational Biology provides a wide range of applications for Multitask Learning (MTL) methods. As the generation of labels often is very costly in the biomedical domain, combining data from different related problems or tasks is a promising strategy to reduce label cost. In this paper, we present two problems from sequence biology, where MTL was successfully applied. For this, we use regularization-based MTL methods, with a special focus on the case of a hierarchical relationship between tasks. Furthermore, we propose strategies to refine the measure of task relatedness, which is of central importance in MTL and finally give some practical guidelines, when MTL strategies are likely to pay off.

**Keywords:** Transfer Learning, Multitask Learning, Domain Adaptation, Computational Biology, Bioinformatics, Sequences, Support Vector Machines, Kernel Methods

## 1. Introduction

In Computational Biology, supervised learning methods are often used to model biological mechanisms in order to describe and ultimately understand them. These models have to be rich enough to capture the often considerable complexity of these mechanisms. Having a complex model in turn requires a reasonably sized training set, which is often not available for many applications. Especially in the biomedical domain, obtaining additional labeled training examples can be very costly in terms of time and money. Thus, it may often pay off to combine information from several related tasks as a way of obtaining more accurate models and reducing label cost.

In this paper, we will present several problems where MTL was successfully applied. We will take a closer look at a common special case, where different tasks correspond to different organisms. Examples of such a scenario are splice-site recognition and promoter prediction for different organisms. Because these basic mechanisms tend to be relatively well conserved throughout evolution, we can benefit from combining data from several species. This special case is interesting because we are given a tree structure that relates the tasks at hand. For most parts of the *tree of life*, we have a good understanding of how closely related two organisms are and can use this information in the context of Multitask Learning. To illustrate that the relevance of MTL to Computational Biology goes beyond the *organisms-as-tasks* scenario, we present a problem from a setting, where different tasks correspond to different related protein variants, namely Major Histocompatibility Complex (MHC)-I binding prediction.

Clearly, there exists an abundance of problems in Bioformatics that could be approached from the MTL point of view (e.g. Qi et al., 2010). For instance, different tasks could correspond to different cell lines, pathways (Park et al., 2010), tissues, genes (Mordelet and Vert, 2011), proteins (Heckerman et al., 2007), technology platforms such as microarrays, experimental conditions, individuals, tumor subtypes, just to name a few. This illustrates that MTL methods are of interest to many applications in Computational Biology.

## 2. Methods

Our methods strongly rely on regularization based supervised learning methods, such as the Support Vector Machine (SVM) (Boser et al., 1992) or Logistic Regression. In its most general form, it consists of a loss-term that captures the error with respect to the training data and a regularizer that penalizes model complexity

$$J(\Theta) = L(\Theta|X,Y) + R(\Theta).$$

This formulation can easily be generalized to the MTL setting, where we are interested in obtaining several models parameterized by $\Theta_1,...,\Theta_M$, where $M$ is the number of tasks. The above formulation can be extended by introducing an additional regularization term $R_{MTL}$ that penalizes the discrepancy between the individual models (Evgeniou et al., 2005; Agarwal et al., 2010).

$$J(\Theta_1,...,\Theta_M) = \sum_{i=1}^{M} L(\Theta_i|X,Y) + \sum_{i=1}^{M} R(\Theta_i) + R_{MTL}(\Theta_1,...,\Theta_M).$$

### 2.1. Taxonomy-based transfer learning: Top-down

In this section, we present a regularization-based method that is applicable when tasks are related by a taxonomy (Widmer et al., 2010a). Hierarchical relations between tasks are particularly relevant to Computational Biology where different tasks may correspond to different organisms. In this context, we expect that the longer the common evolutionary history between two organisms, the more beneficial it is to share information between these organisms. In the given taxonomy, tasks correspond to leaves (i.e. taxons) and are related by its inner nodes. The basic idea of the training procedure is to mimic biological evolution by obtaining a more specialized classifier with each step from root to leaf. This specialization is achieved by minimizing the training error with respect to training examples from the current subtree (i.e. the tasks below the current node), while similarity to parent classifier is enforced through regularization.

The primal of the Top-down formulation is given by

$$\min_{\mathbf{w},b} \frac{B}{2}\|\mathbf{w}\|^2 + \frac{1-B}{2}\|\mathbf{w} - \mathbf{w}_p\|^2 + C \sum_{(\mathbf{x},y)\in S} \ell(\langle \mathbf{x},\mathbf{w}\rangle + b, y), \tag{1}$$

where $\ell$ is the hinge loss $\ell(z,y) = \max\{1 - yz, 0\}$, $\mathbf{w}_p$ is the parameter vector from the parent model and $B \in [0,1]$ is the hyper-parameter that controls the degree of MTL regularization. The dual of the above formulation is given by Widmer et al. (2010a):

$$\max_{\alpha} -\frac{1}{2}\sum_{i=1}^{n}\sum_{j=1}^{n}\alpha_i\alpha_j y_i y_j k(\mathbf{x}_i,\mathbf{x}_j) - \sum_{i=1}^{n}\alpha_i\left( B\left(\underbrace{\sum_{j=1}^{m}\alpha'_j y_i y'_j k(\mathbf{x}_i,\mathbf{x}'_j)}_{p_i}\right) - 1\right)$$

$$\text{s.t. } \alpha^T \mathbf{y} = 0, \qquad 0 \le \alpha_i \le C \ \forall i \in \{1,n\},$$

where $n$ and $m$ are the number of training examples at the current level and the level of the parent model, respectively.

The $\alpha_i$ represent the dual variables of the current learning problem, whereas the $\alpha'_j$ represent the dual variables obtained from the parent model $\mathbf{w}_p$; in the case of the linear kernel, it is described as $\mathbf{w}_p = \sum_{j=1}^{m}\alpha'_j y'_j \mathbf{x}'_j$. The resulting predictor is given by

$$f(\mathbf{x}) = \sum_{i=1}^{n}\alpha_i y_i k(\mathbf{x},\mathbf{x}_i) + \sum_{j=1}^{m}\alpha'_j y'_j k(\mathbf{x},\mathbf{x}_j) + b.$$

To clarify the algorithm, an illustration of the training procedure is given in Figure 1. If the nodes in the taxonomy represent species, then interior nodes correspond to extant species. Thus, interior nodes are trained on the union of training data from descendant leaves (see Figure 1). In the dual of the standard SVM, the term $\mathbf{p}$ (corresponding to the $p_i$ in the equation above) is set to $\mathbf{p} = (-1,\ldots,-1)^T$ (i.e. there is no parent model). In our extended formulation, the $p_i$ is pre-computed and passed to the SVM-solver as the linear term of the corresponding quadratic program (QP). Therefore, the solution of the above optimization problem can be efficiently computed using well established SVM solvers. We have extended LibSVM, SVMLight and LibLinear to handle a custom linear term $p$ and provide the implementations as part of the SHOGUN-toolbox (Sonnenburg et al., 2010).

## 2.2. Graph-based transfer learning

Following Evgeniou et al. (2005) we use the following formulation

$$\min_{\mathbf{w}_1,\ldots,\mathbf{w}_M} \frac{1}{2}\sum_{s=1}^{M}\sum_{t=1}^{M}A_{st}\|\mathbf{w}_s - \mathbf{w}_t\|^2 + C\sum_{s=1}^{M}\sum_{(\mathbf{x},y)\in S}\ell(\langle \mathbf{x},\mathbf{w}_s\rangle + b, y), \tag{2}$$

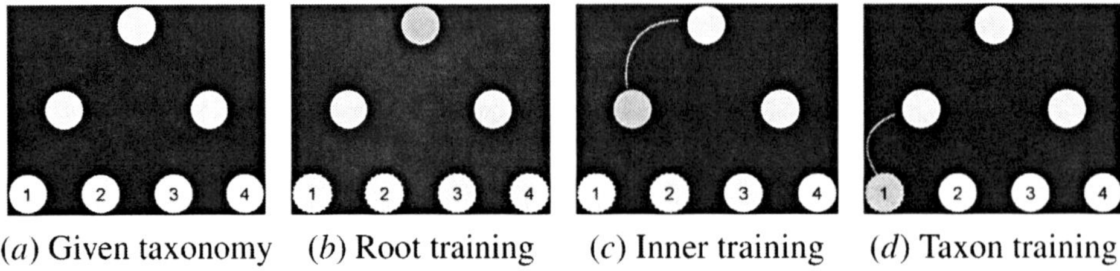

(a) Given taxonomy    (b) Root training    (c) Inner training    (d) Taxon training

Figure 1: Illustration of the Top-down training procedure. In this example, we consider four tasks related via the tree structure in 1(a). Each leaf is associated with a task $t$ and its training sample $S_t = \{(x_1, y_1), ..., (x_n, y_n)\}$. In 1(b), training begins by obtaining the predictor at the root node. Next, we move down one level and train a classifier at an inner node, as shown in 1(c). Here, the training error is computed with respect to $S_1 \cup S_2$, while the classifier *pulled* towards to the parent model $\mathbf{w}_p$ using the newly introduced regularization term, as indicated by the red arrow. Lastly, in 1(d), we obtain the final classifier of task 1 by only taking into account $S_1$ to compute the loss, while again regularizing against the parent predictor. The procedure is applied in a top-down manner to the remaining nodes until we have obtained a predictor for each leaf.

where $\ell$ is the hinge loss $\ell(z, y) = \max\{1 - yz, 0\}$ and $A_{st}$ captures the externally provided task similarity between task $s$ and task $t$. Intuitively speaking, we want to penalize differences between parameter vectors more strongly, if we know a-priori that tasks are closely related. Following Evgeniou et al. (2005), we rewrite the above formulation using the graph Laplacian $L$

$$\min_{\mathbf{w}_1,...,\mathbf{w}_M} \sum_{s=1}^{M} \sum_{t=1}^{M} L_{st} \mathbf{w}_s^T \mathbf{w}_t + C \sum_{s=1}^{M} \sum_{(x,y) \in S} \ell(\langle x, \mathbf{w}_s \rangle + b, y), \tag{3}$$

where $L = D - A$, where $D_{s,t} = \delta_{s,t} \sum_k A_{s,k}$. Finally, it can be shown that this gives rise to the following *multitask* kernel to be used in the corresponding dual (Evgeniou et al., 2005):

$$K((x, s), (z, t)) = L_{st}^+ \cdot K_B(x, z), \tag{4}$$

where $K_B$ is a kernel defined on examples. A closely related formulation was successfully used in the context of Computational Biology by Jacob and Vert (2008), where a kernel on tasks $K_T$ was used instead of the pseudo-inverse of the graph Laplacian $L^+$, giving rise to

$$K((x, s), (z, t)) = K_T(s, t) \cdot K_B(x, z). \tag{5}$$

The corresponding joint feature space between task $t$ and feature vector $x$ can be written as a tensor product $\phi(t, x) = \phi_T(t) \cdot \phi_B(x)$ (Jacob and Vert, 2008). Note that a special case of this was presented by Daumé (2007) in the context of Domain Adaptation, where $\phi_T(t) = (1,1,0)$ was used as the source task descriptor and $\phi_T(t) = (1,0,1)$ for the target task, corresponding to $K_T(s,t) = (1 + \delta_{s,t})$.

## 2.3. Learning Task Similarity

An essential component of MTL is finding measure of task similarity that represents the improvement we expect from the information transfer between two tasks. Often, we are provided with external information on task relatedness (e.g. evolutionary history of organisms). However, the given similarity often only roughly corresponds to the task similarity relevant for the MTL algorithm and therefore one may need to find a suitable transformation to convert one into the other. For this, we propose several approaches, starting with a very simple parameterization of task similarities up to a sophisticated Multiple Kernel Learning (MKL) based formulation that learns task similarity along with the classifiers.

### 2.3.1. A simple approach: Cross-validation

The most general approach to refine a given task-similarity matrix $A_{s,t}$ is to introduce a mapping function $\psi$, parameterized by a vector $\Theta$. The goal is to choose $\Theta$ such that the empirical loss is minimized. For example, we could choose a linear transformation $\psi_\Theta(s,t) = \Theta_1 + A_{s,t}\Theta_2$ or non-linear transformation $\psi_\Theta(s,t) = \Theta_1 \cdot exp(\frac{A_{s,t}}{\Theta_2})$. A straightforward way of finding a good mapping is using cross-validation to choose an optimal value for $\Theta$.

A cross-validation based strategy may also be used in the context of the Top-down method. In principle, each edge $e$ in the taxonomy could carry its own weight $B_e$. Because tuning a grid of all $B_e$ in cross-validation would be quite costly, we propose to perform a *local* cross-validation and optimize the current $B_e$ at each step from top to bottom independently. This can be interpreted as using the taxonomy to reduce the parameter search space.

### 2.3.2. Multitask Multiple Kernel Learning

In the following section, we present a more sophisticated technique to learn task similarity based on MKL. To be able to use MKL for Multitask Learning, we need to reformulate the Multitask Learning problem as a weighted linear combination of kernels (Widmer et al., 2010c). In contrast to Equation 5, the basic idea of our decomposition is to define task-based block masks on the kernel matrix of $K_B$. Given a list of tasks $\mathcal{T} = \{t_1,...,t_T\}$, we define a kernel $K_S$ on a subset of tasks $S \subseteq \mathcal{T}$ as follows:

$$K_S(x,y) = \begin{cases} K_B(x,y), & \text{if } t(x) \in S \wedge t(y) \in S \\ 0, & \text{else} \end{cases}$$

where $t(x)$ denotes the task of example $x$. Here, each $S$ corresponds to a subset of tasks. In the most general formulation, we define a collection $\mathcal{I} = \{S_1,...,S_p\}$ of an arbitrary number $p$ of task sets $S_i$, which defines the latent structure over tasks.

This collection leads to the following linear combination of kernels, which is positive semi-definite for $\beta_i \geq 0$:

$$\hat{K}(x,z) = \sum_{S_i \in I} \beta_i K_{S_i}(x,z).$$

Using $\hat{K}$, we can readily utilize existing MKL methods to learn the weights $\beta_i$. This corresponds to identifying the groups of tasks $S_i$ for which sharing information leads to improved performance. We use the following formulation of $q$-norm MKL by Kloft et al. (2011):

$$\min_{\beta} \max_{\alpha} \quad \mathbf{1}^T \alpha - \frac{1}{2} \sum_{i,j} \alpha_i \alpha_j y_i y_j \sum_{t=1}^{|I|} \beta_t K_{S_t}(x_i, x_j)$$

$$\text{s.t.} \quad \|\beta\|_q^q \leq 1, \beta \geq 0$$

$$\mathbf{Y}^T \alpha = 0, 0 \leq \alpha \leq C$$

To solve the above optimization problem, we follow ideas presented by Kloft et al. (2011) to iteratively solve a convex optimization problem involving only the $\beta$'s and then to solve for $\alpha$ only. This method is known to converge fast even for a relatively large number of kernels (Kloft et al., 2011).

After training using MKL, a classifier $f_t$ for each task $t$ is given by:

$$f_t(z) = \sum_{i=0}^{N} \alpha_i y_i \sum_{S_j \in I : t \in S_j} \beta_j K_{S_j}(x_i, z),$$

where $N$ is the total number of training examples of all tasks combined. What remains to be discussed is how to define the collection $I$ of candidate subsets $S_i$, which are subsequently to be weighted by MKL.

**Powerset MT-MKL**    With no prior information given, a natural choice is to take into account all possible subsets of tasks. Given a set of tasks $\mathcal{T}$, this corresponds to the power set $\mathcal{P}$ of $\mathcal{T}$ (excluding the empty set) $I_{\mathcal{P}} = \{S | S \in \mathcal{P}(\mathcal{T}) \wedge S \neq \emptyset\}$. Clearly, this gives us an exponential number (i.e. $2^T$) of task sets $S_i$ of which only a few will be relevant. To identify the relevant task sets, we propose to use an L1-regularized MKL approach to obtain a sparse solution. Most subset weights will be set to zero, leaving us with only a few relevant subsets with non-zero weights. By summing the weights of the groups containing a particular pair of tasks $(s,t)$, we can recover the similarity $A_{s,t}$. As we do not assume prior information on task structure, this approach may be used to learn the task similarity matrix *de novo* (only for a small number of tasks due to computational complexity of the naïve implementation).

**Hierarchical MT-MKL**    Next, we assume that we are given a tree structure $\mathcal{G}$ that relates our tasks at hand (see Figure 2). In this context, a task $t_i$ corresponds to a leaf in $\mathcal{G}$. We can exploit the hierarchical structure $\mathcal{G}$ to determine which subsets potentially play a role for Multitask Learning. In other words, we use the hierarchy to restrict the space of possible task sets. Let $leaves(n) = \{l|l$ is descendant of $n\}$ be the set of leaves below the sub-tree rooted at node $n$. Then, we can give the following definition for the hierarchically decomposed kernel function

$$\hat{K}(x,y) = \sum_{n_i \in \mathcal{G}} \beta_i K_{leaves(n_i)}.$$

As an example, consider the kernel defined by a hierarchical decomposition according to Figure 2. In contrast to Power-set MT-MKL, we seek a non-sparse weighting of the task sets defined by the hierarchy and will therefore use $L2$-norm MKL (Kloft et al., 2011).

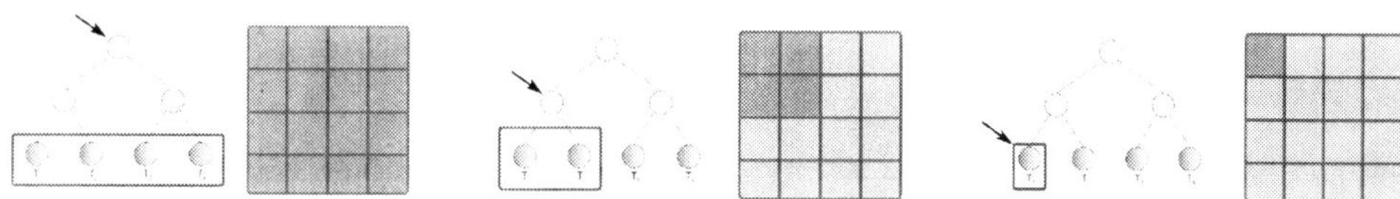

Figure 2: Example of a hierarchical decomposition. According to a simple binary tree, it is shown that each node defines a subset of tasks (a block in the corresponding kernel matrix on the left). Here, the decomposition is shown for only one path: The subset defined by the root node contains all tasks, the subset defined by the left inner node contains $t_1$ and $t_2$ and the subset defined by the leftmost leaf only contains $t_1$. Accordingly, each of the remaining nodes defines a subset $S_i$ that contains all descendant tasks.

Alternatively, one can use a variation of the approach provided above and use boosting algorithms instead of MKL, as similarly done by Gehler and Nowozin (2009) and Widmer et al. (2010b).

## 2.4. When does MTL pay off?

In this section, we give some practical guide lines about when it is promising to use MTL algorithms. First, the tasks should neither be too similar nor too different (Widmer et al., 2010a). If the tasks are too different one will not be able to transfer information, or even end up with *negative transfer* (Pan and Yang, 2009). On the other hand, if tasks are almost identical, it might suffice to pool the training examples and train a single combined classifier. Another integral factor that needs to be considered is whether the problem is *easy* or *hard* with respect to the available training data. In this context, the problem can be considered *easy* if the performance of a classifier saturates as a function of the available training data. In that case using more out-of-domain information will not improve classification performance, assuming that none of the out-of-domain datasets are a better match for the in-domain test set than the in-domain training set.

In order to control problem difficulty in the sense defined above, we can compute a learning curve (i.e. number of examples vs. auROC) and check whether we observe saturation. If this is the case, we do not expect transfer learning to be beneficial as model performance is most likely limited only by label noise. The same idea can be applied to empirically checking similarity between two tasks. For this, we compute pair-wise saturation curves (i.e. train on data from one task, test on the other) between tasks, giving us a useful measure to check whether transferring information between two tasks may be beneficial.

## 3. Applications

In this section, we give a two examples of applications in Computational Biology, where we have successfully employed Multitask Learning.

### 3.1. Splice-site prediction

The recognition of splice sites is an important problem in genome biology. By now it is fairly well understood and there exist experimentally confirmed labels for a broad range of organisms. In previous work, we have investigated how well information can be transfered between source and target organisms in different evolutionary distances (i.e. one-to-many) and training set sizes (Schweikert et al., 2009). We identified TL algorithms that are particularly well suited for this task. In a follow-up project we investigated how our results generalize to the MTL scenario (i.e. many-to-many) and showed that exploiting prior information about task similarity provided by taxonomy can be very valuable (Widmer et al., 2010a). An example how MTL can improve performance compared to baseline methods *plain* (i.e. learn a classifier for each task independently) and *union* (i.e. pool examples from all tasks and obtain a global classifier) is given in Figure 3(*a*). For an elaborate discussion of our experiments with splice-site prediction, please consider the original publications (Schweikert et al., 2009; Widmer et al., 2010a).

### 3.2. MHC-I binding prediction

The second application we want to mention is MHC-I binding prediction. MHC class I molecules are key players in the human immune system. They bind small peptides derived from intracellular proteins and present them on the cell surface for surveillance by the immune system. Prediction of such MHC class I binding peptides is a vital step in the design of peptide-based vaccines and therefore one of the major problems in computational immunology. Thousands of different types of MHC class I molecules exist, each displaying a distinct binding specificity. We consider these different MHC types to be different tasks in a MTL setting. Clearly, we expect information sharing between tasks to be fruitful, if the binding pockets of the molecules exhibit similar properties. These properties are encoded in the protein sequence of the MHC proteins. Thus, we can use the externally provided similarity between MHC proteins to estimate task relatedness. In agreement with previous reports (Jacob and Vert, 2008), we observed that using the approach provided in Equation 5 with $K_T$ defined on the externally provided sequences considerably boosts prediction accuracy compared to baseline methods, as

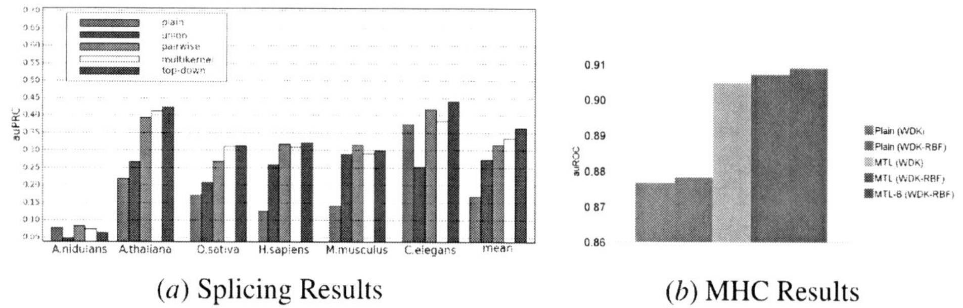

(a) Splicing Results

(b) MHC Results

Figure 3: In 3(a), we show results for 6 out of 15 organisms for baseline methods *plain* and *union* and three multitask learning algorithms. The mean performance is shown in the last column. For each task, we obtained 10000 training examples and an additional test set of 5000 examples. We normalized the data sets such that there are 100 negative examples per positive example. We report the area under the precision recall curve (auPRC), which is an appropriate measure for unbalanced classification problems (i.e. detection problems). In 3(b), we report the average performance over the MHC-I alleles (tasks) as auROC for two baseline methods and three variants of MTL algorithms. Here, au-ROC was used to keep the results comparable to previously published results (Jacob and Vert, 2008).

can be seen in Figure 3(b). Lastly, we present an experiment, where we investigated

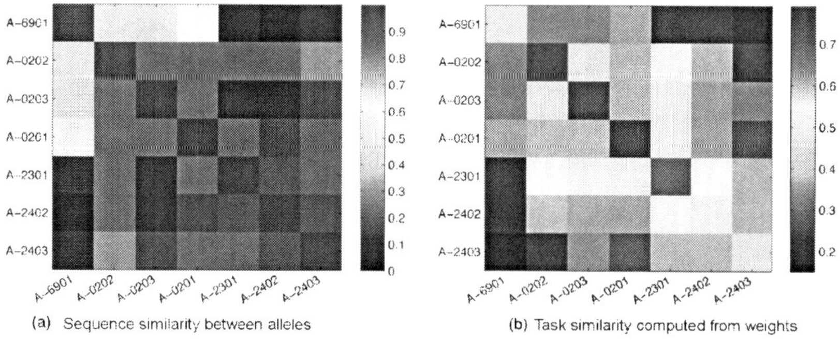

(a) Sequence similarity between alleles

(b) Task similarity computed from weights

Figure 4: Comparison of the task structure as identified by Powerset MKL with an externally provided one.

how well a *learned* task-similarity (using Powerset MT-MKL) matches an externally provided one (i.e. based on sequence similarity). This externally provided task similarity is based on the Hamming-distance between the amino acids contained in the binding pockets of the MHC-I molecules. As can be seen in Figure 4, there is considerable agreement between the two similarity measures (i.e. an overall Spearman correlation coefficient of 0.67). From this, we conclude that MT-MKL may be of value in MTL settings, where relevant external information on the relatedness of tasks is absent.

## 4. Conclusion

We have presented an overview of applications for MTL methods in the field of Computational Biology. Especially in the context of biomedical data, where generating training labels can be very expensive, MTL can be viewed as a means of obtaining more cost-effective predictors. We have presented several regularization-based MTL methods centered around the SVM, amongst them an approach for the special case of hierarchical task structure. Furthermore, we have outlined several strategies how to learn or refine the degree of task relatedness, which is of central importance when applying MTL methods. Lastly, we gave some practical guidelines to quickly check for a given dataset if one can expect MTL strategies to boost performance over straightforward baseline methods.

## Acknowledgments

We would like to acknowledge Jose Leiva, Yasemin Altun, Nora Toussaint, Sören Sonnenburg, Klaus-Robert Müller, Bernhard Schölkopf, Oliver Stegle, Alexander Zien and Jonas Behr.

## References

A. Agarwal, H. Daumé III, and S. Gerber. Learning Multiple Tasks using Manifold Regularization. In *NIPS Proceedings*. NIPS, 2010.

B.E. Boser, I.M. Guyon, and V.N. Vapnik. A training algorithm for optimal margin classifiers. In *Proceedings of the fifth annual workshop on Computational learning theory*, pages 144–152. ACM, 1992.

H. Daumé. Frustratingly easy domain adaptation. In *Annual meeting-association for computational linguistics*, volume 45:1, page 256, 2007.

T. Evgeniou, C.A. Micchelli, and M. Pontil. Learning multiple tasks with kernel methods. *Journal of Machine Learning Research*, 6(1):615–637, 2005. ISSN 1532-4435.

P. Gehler and S. Nowozin. On feature combination for multiclass object classification. In *Proc. ICCV*, volume 1:2, page 6, 2009.

D. Heckerman, C. Kadie, and J. Listgarten. Leveraging information across HLA alleles/supertypes improves epitope prediction. *Journal of Computational Biology*, 14 (6):736–746, 2007. ISSN 1066-5277.

L. Jacob and J. Vert. Efficient peptide-MHC-I binding prediction for alleles with few known binders. *Bioinformatics (Oxford, England)*, 24(3):358–66, February 2008.

M. Kloft, U. Brefeld, S. Sonnenburg, and A. Zien. lp-Norm Multiple Kernel Learning. *Journal of Machine Learning Research*, 12:953–997, 2011.

F. Mordelet and J.P. Vert. ProDiGe: PRioritization Of Disease Genes with multitask machine learning from positive and unlabeled examples. *Arxiv preprint*, 2011.

S.J. Pan and Q. Yang. A survey on transfer learning. *IEEE Transactions on Knowledge and Data Engineering*, pages 1345–1359, 2009.

C.Y. Park, D.C. Hess, C. Huttenhower, and O.G. Troyanskaya. Simultaneous genome-wide inference of physical, genetic, regulatory, and functional pathway components. *PLoS computational biology*, 6(11):e1001009, 2010.

Y. Qi, O. Tastan, J.G. Carbonell, J. Klein-Seetharaman, and J. Weston. Semi-supervised multi-task learning for predicting interactions between hiv-1 and human proteins. *Bioinformatics*, 26(18):i645, 2010.

G. Schweikert, C. Widmer, B. Schölkopf, and G. Rätsch. An Empirical Analysis of Domain Adaptation Algorithms for Genomic Sequence Analysis. In D Koller, D Schuurmans, Y Bengio, and L Bottou, editors, *Advances in Neural Information Processing Systems 21*, pages 1433–1440. NIPS, 2009.

S. Sonnenburg, G. Rätsch, S. Henschel, C. Widmer, A. Zien, F. de Bona, C. Gehl, A. Binder, and V. Franc. The SHOGUN machine learning toolbox. *Journal of Machine Learning Research*, 2010.

C. Widmer, J. Leiva, Y. Altun, and G. Rätsch. Leveraging Sequence Classification by Taxonomy-based Multitask Learning. In B. Berger, editor, *Research in Computational Molecular Biology*, pages 522–534. Springer, 2010a.

C. Widmer, N.C. Toussaint, Y. Altun, O. Kohlbacher, and G. Rätsch. Novel machine learning methods for MHC Class I binding prediction. In *Pattern Recognition in Bioinformatics*, pages 98–109. Springer, 2010b.

C. Widmer, N.C. Toussaint, Y. Altun, and G. Rätsch. Inferring latent task structure for Multitask Learning by Multiple Kernel Learning. *BMC bioinformatics*, 11 Suppl 8 (Suppl 8):S5, January 2010c. doi: 10.1186/1471-2105-11-S8-S5.

# Transfer Learning in Sequential Decision Problems:
# A Hierarchical Bayesian Approach

**Aaron Wilson**      WILSONAA@EECS.OREGONSTATE.EDU
**Alan Fern**      AFERN@EECS.OREGONSTATE.EDU
**Prasad Tadepalli**      TADEPALL@EECS.OREGONSTATE.EDU
*School of Electrical Engineering and Computer Science*
*Oregon State University*
*Corvallis, OR 97331*

**Editor:** I. Guyon, G. Dror, V. Lemaire, G. Taylor, and D. Silver

## Abstract

Transfer learning is one way to close the gap between the apparent speed of human learning and the relatively slow pace of learning by machines. Transfer is doubly beneficial in reinforcement learning where the agent not only needs to generalize from sparse experience, but also needs to efficiently explore. In this paper, we show that the hierarchical Bayesian framework can be readily adapted to sequential decision problems and provides a natural formalization of transfer learning. Using our framework, we produce empirical results in a simple colored maze domain and a complex real-time strategy game. The results show that our Hierarchical Bayesian Transfer framework significantly improves learning speed when tasks are hierarchically related.

**Keywords:** Reinforcement Learning, Markov Decision Processes, Transfer Learning, Hierarchical Bayesian Framework

## 1. Introduction

The goal of the current paper is to explain transfer in sequential decision-making domains using the hierarchical Bayesian framework. Unlike the standard supervised setting, in sequential decision making, the training examples are not independently identically distributed. When learning behaviors from demonstrations, examples consist of states and desired action sequences, which is similar to the supervised setting. However, in the more typical reinforcement learning setting, the target action sequences are not provided. The learner must explore action sequences to determine which are are the most rewarding. Prior knowledge informing the agent's exploration strategy would considerably enhance the speed of convergence. Transfer is especially attractive in reinforcement learning since both the discovery of good policies and their generalization would be potentially impacted by learning from related domains.

There are opportunities to learn different kinds of knowledge through transfer in reinforcement learning. We describe two kinds of knowledge transfer: transfer of domain model knowledge, and transfer of policy knowledge. In the *model-based* approach, model learning is improved by incorporating an informative hierarchical Bayesian prior learned from experience in early tasks. The prior represents knowledge about the world. In the *policy search* approach, a prior is placed on the policy parameters themselves. The prior represents direct knowledge of what the agent should do. In each approach, it is assumed that there is a small but unknown number of component types which can be learned and transferred to new tasks. When the discovered components can be reused Transfer is successful.

## 2. Sequential Decision Problems and Hierarchical Transfer Hypothesis

We formulate the transfer learning problem in sequential decision making domains using the framework of Markov Decision Processes (MDPs). An MDP $M$ is defined by a 6-tuple $(S, A, C, T, I, F)$, where $S$ is a set of states and $A$ is a set of actions. $C(s, a)$ is the immediate cost of executing action $a$ in state $s$. Function $T(s, a, s')$ defines the probability of reaching state $s'$ given that action $a$ is taken in state $s$. $I$ is a distribution of initial states and $F$ is a set of final or terminating states. A policy $\pi$ for $M$ is a mapping from $S$ to $A$. The expected cost of a policy $\pi$ starting from a state $s$ is the cumulative sum of action costs starting from state $s$, and ending when the agent reaches a terminating state. A solution to an MDP is an optimal policy that minimizes the total expected cost from all states. The minimum expected cost from state $s$ is the unique solution to the following Bellman equations:

$$V^*(s) = \begin{cases} 0 \text{ if } s \text{ is a terminal state} \\ \min_a C(s, a) + \sum_{s'}(T(s, a, s')V^*(s')), \text{ else} \end{cases}$$

The actions that minimize the right hand side of this equation constitute the optimal policy.

While most work in RL is concerned with solving a single unknown MDP (task), here we are concerned with transfer across MDPs. Our objective is to develop a life-long learning algorithm that is able to leverage experience in previous MDPs $M_1, \ldots M_n$ to more quickly learn an optimal policy in a newly drawn MDP $M_{n+1}$. If each task faced by an agent is arbitrarily different from the previous tasks, the agent has no way to benefit from having solved the previous tasks. To explain the apparently fast learning of people, and duplicate this learning speed in machines, we explicitly formulate the *hierarchical transfer hypothesis*. It is the claim that tasks have a latent hierarchical relationship, which can be explicated and exploited to transfer knowledge between tasks. We propose a hierarchical Bayesian framework to capture the task similarities. We apply this framework to both model-based RL and to model-free policy search. In model-based RL, the hierarchy represents similarities between MDP models. In the policy search approach, the hierarchy represents similarities between policies. Below we discuss each of these approaches in more detail.

## 3. Hierarchical Model-Based Transfer

In this section, we outline our hierarchical Bayesian approach to model-based transfer for RL problems. Figure 1($a$)(bottom) illustrates our fundamental assumptions of the task generation process. Classes of MDPs are generated from a task prior distribution, and each task generates an MDP. The hierarchical structure assumes that each MDP $M_i$ is an iid sample from one of the unobserved classes. We propose a hierarchical prior distribution, based on the Dirichlet Process (DP), in order to capture this hierarchical structure. We aim to exploit this structure when exploring new tasks.

We describe the generative process for MDP models using Figure 1($b$). The generative model is shown in plate notation. Rectangles indicate probability distributions that are replicated a certain number of times. The number of replications is shown in the bottom-right corner of each plate. The figure shows a distribution over $N$ MDPs. In each MDP the agent has made $R$ observations. Each class $C = c$ is associated with a vector of parameters $\theta_c$. These parameter vectors define distributions from which individual MDPs will be drawn. The parameters $c_i$ indicate the class of MDP $M_i$. $G_0$ is a distribution over classes that corresponds to the Task Class Prior in Figure 1($a$)(bottom). Finally the parameter $\alpha$ is the concentration parameter of the Dirichlet Process (DP) model. $\Psi$ refers to the tuple of parameters $\Psi = (\theta, c, \alpha)$. The DP generalizes the case of a known finite number of MDP classes to a infinite number of classes. This non-parametric model allows us to (a) model a potentially unbounded number of classes, (b) adapt to and make inferences about classes we have not seen before, and (c) learn the shared structure that constitutes each class, rather than predefining class specifications.

The key steps in our approach to exploiting this model are outlined in Algorithm 1. The Hierarchical DP model is used as a prior for Bayesian RL in a new MDP. Before any MDPs are experienced, the hierarchical model parameters $\Psi$ are initialized to uninformed values. The Sample function (line 5) uses the DP model to generate a sample of several MDPs from $\Pr(M_i \mid O_i, \Psi)$. From this sample, the $\hat{M}_i$ with the highest probability is selected. To select an action we solve $\hat{M}_i$ using value iteration (line 6) and then follow the greedy policy for $k$ steps (line 7). Observations gathered while executing the greedy policy are added to the collection of observation tuples $O_i$ for the current MDP. Next, the posterior distribution is updated, and a new sample of MDPs is taken from the updated posterior distribution. This loop is repeated for MDP $M_i$ until the policy has converged. After convergence, we update the hierarchical model parameters $\Psi$ based on the parameter estimates $\hat{M}_i$. This is done using the SampleMap procedure. SampleMap uses the machinery of the DP to assign (possibly new) classes to each observed MDP $\hat{M}_1, \ldots, \hat{M}_i$. Given the assignments, we compute the parameters associated with each class distribution. This process is iterated until convergence. Intuitively, as the agent gains experience with more tasks, the samples generated by this procedure will tend to represent the hierarchical class structures generating the MDPs. Learning speed will improve if a new task is usefully described by the sampled hierarchical structure.

---

**Algorithm 1** Hierarchical Model-Based Transfer Algorithm HMBT

---

1:  Initialize the hierarchical model parameters $\Psi$
2:  **for** each MDP $M_i$ from $i = 1, 2, \ldots$ **do**
3:   $O_i = \emptyset$ //Initialize the set of observation tuples (s,a,r,s',a') for $M_i$
4:   **while** policy $\pi$ has not converged **do**
5:    $\hat{M}_i \leftarrow \text{Sample}(P(M \mid O_i, \Psi))$
6:    $\pi = \text{Solve}(\hat{M}_i)$ //Solve the sampled MDP using value iteration.
7:    Run $\pi$ in $M_i$ for $k$ steps.
8:    $O_i = O_i \cup \{\text{observations from } k \text{ steps}\}$
9:   **end while**
10:   $\Psi \leftarrow \text{Sample}(P(\Psi \mid \hat{M}_1, \ldots, \hat{M}_i))$ //Given fixed estimates of the MDP model parameters generate samples from the posterior distribution and select the best.
11: **end for**

---

Details for this algorithm, including specifications of the posterior sampling algorithm, are found in Wilson et al. (2007).

**Experimental Results.** To test our hypotheses, we consider a simple colored maze domain. An example maze is shown in Figure 2. The goal of the agent is to traverse the least cost path from the start location to the final goal location. To do so, the agent may navigate to adjacent squares by executing one of four directional actions. The agent's reward function is a linear function combining the costs of all squares adjacent to the agent's location on the map. The contributed cost of each adjacent square is a function of its color. To do well, the agent must determine which colors have low cost and identify a path minimizing the total cost from start to finish. In the multi-task setting the parameters of the reward function, governing the penalty for traversing certain colors, characterize each task, and are assumed to be generated from a hierarchical structure. To generate a task, the world first generates a class and then generates a vector of color costs from the sampled class. Each map is randomly colored before the agent begins its interaction.

In the first instance of this domain the goal location is fixed. The agent must traverse from the top left to the bottom right of the map. The objective is to maximize the total reward per episode. Episodes end only after the goal is reached. Map squares have 8 different colors and reward functions are generated from 4 distinct classes. Figure 3 shows results for this setting. Good performance is possible with a small number of steps in each sampled MDP. Even so the benefit of transfer is obvious. Given experience in 16 MDPs the number of exploratory moves is reduced from 2500 to 100 steps – an order of magnitude difference. The curves also indicate that the algorithm avoids negative transfer. Exploring according to the learned prior results in consistently improved performance even when early tasks are not related to the current task.

In order to better illustrate the benefit of transfer, we consider a more difficult problem. We no longer assume that goal locations are fixed. Instead, goal locations are sampled from a hierarchical distribution. This confounds the exploration problem for uninformed agents, which must now identify the goal location and the color costs. Termination is achieved when the agent finds the goal location. We present results for larger maps, 30x30 squares, with 4 underlying classes of goal locations (each class has goal locations distributed around a fixed map square). Results are shown in Figure 4. The graph depicts the average performance in the first episode given a fixed number of prior tasks. By comparison, we show the performance of the algorithm with zero prior tasks. Improvements indicate a reduction in the number of steps needed to find the goal. The observed improvement in average total reward is dramatic. After ten tasks the agent has inferred a reasonable representation of the hierarchical structure that generates the goal locations. Knowledge of this hierarchical structure reduces the exploration problem. The agent only checks regions where goals are likely to exist. Exploiting this hierarchical knowledge reduces the cost of exploration by an order of magnitude.

## 4. Hierarchical Bayesian Policy Transfer

The premise behind the hierarchical transfer hypothesis, as it applies to policy search, is that policies are hierarchically composed of shared components. Our Bayesian Policy Search (BPS) algorithm learns these components in simple tasks and recombines them into a joint policy in more difficult tasks. In some cases, tasks that are impossible to tackle with no prior knowledge become trivial if the necessary component set is learned from simpler tasks.

We pursue this basic idea in the context of a real time strategy game called Wargus. The tactical conflicts we consider include groups of heterogeneous units fighting to destroy one another. An example Wargus map is shown in Figure 5(a). Units controlled by the agent include a ballista, a knight, and several archers. Each unit has a set of physical characteristics including armor, speed, location, and so on. These characteristics naturally predispose units to particular roles in confrontations. For instance, due to its long range the ballista is useful for destroying structures, while units close to key pieces (such as the opposing ballista) may be best employed targeting those pieces. Therefore, we view a policy for Wargus conflicts as a composition of roles (components) and a means of assigning roles to units (combining components). In practice, the agent does not know the number of roles, the type of roles, or the correct method of assigning roles to units. Therefore, the goal of our lifelong learning agents will be to learn (or be taught by example) a collection of roles useful for a variety of confrontations and to adapt its method of assigning units to roles as is appropriate for novel tasks.

Our approach to the role discovery and assignment problem is to define a hierarchical prior distribution for the set of individual agent roles, and then to search this structure using stochastic simulation algorithms adapted to the policy search problem. We first define a set of policy components in the form of a parametric class of policies, then define a hierarchical Bayesian prior in the form of a modified DP model, and finally, we define a means of searching the policy space for solutions.

It is assumed that each unit in the game acts independently of the others given their assigned roles. The distribution of the joint action for a set of agents, conditioned on the assigned roles, is the product of the individual action choices, $P(A|s) = \prod_u P(a_u|s, \theta_{c_u})$, where $A$ is the joint action, $c_u$ is a role index for unit $u$, and each $\theta_{c_u}$ represents the vector of role parameters assigned to unit $u$.

The hierarchical prior distribution for the role parameters, which describes a joint policy for all agents, is shown in Figure 5($b$). The prior illustrates how roles are combined and associated with units in the game. There are $N$ tasks. Each task is composed of $U$ units. Each unit is assigned a vector of role parameters $\theta_{c_u}$ indexed by $c_u$. The variable $\xi_u$ represents a set of trajectories, a sequence of state-action observations, made by agent $u$ when executing its policy. Assignment of units to roles depends on unit-specific features $f_u$ and a vector of role assignment parameters $\phi$. The vector of assignment parameters $\phi$ represent a strategy for assigning the collection of units to roles. An instance of the role assignment strategy is associated with each of the $N$ tasks. The set of available roles $\theta_{c_u}$ is shared across all of the $N$ tasks. Taken together, the set of parameters, $\Psi = (\theta, c, \phi, \alpha)$, represents a joint policy for all observed tasks.

For purposes of searching the joint policy space we build on the work by Hoffman et al. (2007). We define an artificial probability distribution,

$$q(\Psi) \propto U(\Psi)P(\Psi) = P(\Psi) \int R(\xi)P(\xi|\Psi)d\xi,$$

proportional to the prior distribution and a utility function. For purposes of applying this idea to the policy search problem the utility function is defined to be the expected return for a policy executed in an episodic environment, $V(\Psi) = E(\sum_{t=1}^{T} R(s_t, a_t)|\Psi)$. This formulation allows us to adapt advanced methods of stochastic simulation, developed for estimation of probability distributions, to the problem of policy search. Essentially, the prior distribution acts as a bias. It guides the search to regions of the policy space believed to have a high return. The utility function plays a similar role. Samples drawn from $q(\Psi)$, will tend to focus around regions of the policy space with high utility.

Given $N$ previously observed tasks, we take the following strategy. We fix the set of previously identified roles and use this role set as prior knowledge in the new task. Discovery of roles is a difficult process. Therefore, role reuse in the new task can dramatically improve learning speed. Within a new task we search for strategies that recombine these existing roles and propose new roles to overcome novel obstacles. Algorithm 2 shown in Figure 2 is designed to implement this strategy by sampling role-based policies given the informed prior.

The algorithm has four basic parts. We assume that trajectories of optimal policies are available for the first N tasks. If no prior task is available the trajectory set will be empty. Trajectories are stored to be used during inference. This brings us to the main body of the sampling procedure, which generates samples from the artificial distribution $q(\Psi)$. On line 5, after initializing the state of the Markov chain, the new role assignments are sampled. Each agent is placed into one of the existing components, or (using the mechanics of the DP) a new component generated from $G_0$. Next, line 6 updates the role parameters using Hybrid Markov Chain Monte-Carlo (Andrieu et al. (2003)). The key to this portion of the algorithm is the use of gradient information to guide the simulation toward roles which yield high expected return. Next, we perform a simple Metropolis-Hastings update for the vector of assignment parameters. Finally, the role-based policy with highest predicted utility is used to generate a new set of trajectories. The concrete implementation of each sampling procedure can be found in Wilson et al. (2010), which includes details of the model-free method of estimating the utility.

**Experimental Results.** Our algorithm is agnostic about the process generating trajectories. Trajectories may be produced by an expert, or automatically generated by the learning agent. Figure 6 illustrates the results of transfer from a simple map, when given examples of expert play in the simple map, to a more complex map. An expert demonstrates 40 episodes of play in the simple map. The expert controls 6 friendly units. Given the expertly generated trajectories, roles and an assignment strategy are learned using the BPS algorithm. The agent attempts to transfer these learned roles to a more difficult Wargus map where it will control 10 friendly units (see Figure 5(a)). We compare the transfer performance, BPS:Expert, to three baselines. The first two baselines (BPS Independent and BPS Single) represent the BPS algorithm with an enforced role structure. In the Independent case, all units are forcibly assigned to their own role. In the Single case, all units share the same role. In principle, BPS with the independent role structure can learn the correct policy and this fixed role structure may be beneficial when roles are not shared. Similarly, if a single role will suffice, then BPS run with all agents assigned to a single policy will be ideal. Independent and Single baselines cannot make use of the role structure, and therefore they cannot benefit from transfer. Instead, they must learn their policies from scratch. Finally, we compare our performance to OLPOMDP where units independently learn a policy (Baxter et al. (2001)). Results are shown in Figure 6. The BPS algorithm benefits substantially from identifying and transferring the expert roles.

Experts are not always available. Unfortunately, learning in our test map is more difficult when examples of expert play are not provided. On the test map used above, which represents a large multi-agent system, it was difficult to learn a good policy using BPS without prior knowledge. To overcome this problem without resorting to expert help we first allow BPS to learn a useful role structure in a simpler map (by observation of its own behavior). Subsequently, we use the learned prior distribution in the larger 10 unit map. The results are shown in Figure 7, BPS:BPS Map 1. The roles learned in the simpler problem make learning the harder task tractable. By comparison, BPS without the prior knowledge takes much longer to identify a policy.


## 5. Conclusion

This work focuses on a hierarchical Bayesian framework for transfer in sequential decision making tasks. We describe means of transferring two basic kinds of knowledge. The HMBT algorithm learns a hierarchical structure representing classes of MDPs. The HMBT algorithm transfers knowledge of domain models to new tasks by estimating a hierarchical generative model of domain model classes. We illustrated the effectiveness of learning the MDP class hierarchy in a simple colored maze domain. Results show that transferring the class structure can dramatically improve the quality of exploration in new tasks. Furthermore, the special properties of the HMBT algorithm

---

**Algorithm 2** Bayesian Policy Search

---

1: Initialize parameters of the role-based policy: $\Psi_0$

2: Initialize the sets of observed trajectories: $\xi = \emptyset$

3: Generate $n$ trajectories from the domain using $\Psi_0$.

4: $\xi \leftarrow \xi \cup \{\xi_i\}_{i=1..n}$

5: $S = \emptyset$

6: **for** $t = 1 : T$ **do**

7:    //Generate a sequence of samples from the artificial distribution.

8:    $\Psi_t \leftarrow Sample(\hat{U}(\Psi_{t-1})P(\Psi_{t-1}))$

9:    Record the samples:

10:    $S \leftarrow S \cup (\Psi_t)$

11: **end for**

12: Set $\Psi_0 = argmax_{(\Psi_t)\in S} U(\Psi_t)$ //Select the role-based policy maximizing the expected return.

13: Return to Line 2.

---

prevent transfer of incorrect knowledge. The BPS algorithm learns a hierarchical prior for multi-agent policies. It assumes that policies for each domain are composed of multiple roles, and the roles are assigned to units in the game via a hierarchical strategy. Our empirical results show that tactical conflicts in Wargus exemplify this hierarchical structure. Distinct roles can be discovered and transferred across tasks. To facilitate this transfer, the BPS algorithm can learn roles from expert examples as well as observing its own play in simple tasks. By learning in simple task instances first the algorithm can tackle increasingly sophisticated problems.

## References


Christophe Andrieu, Nando de Freitas, Arnaud Doucet, and Michael I. Jordan. An introduction to MCMC for machine learning. *Machine Learning*, 50(1):5–43, January 2003. doi: 10.1023/A:1020281327116.

Jonathan Baxter, Peter L. Bartlett, and Lex Weaver. Experiments with infinite-horizon, policy-gradient estimation. *Journal of Artificial Intelligence Research*, 15(1):351–381, 2001. ISSN 1076-9757.

Mathew Hoffman, Arnaud Doucet, Nando de Freitas, and Ajay Jasra. Bayesian policy learning with trans-dimensional MCMC. *NIPS*, 2007.

Aaron Wilson, Alan Fern, Soumya Ray, and Prasad Tadepalli. Multi-task reinforcement learning: a hierarchical bayesian approach. In *ICML*, pages 1015–1022, 2007.

Aaron Wilson, Alan Fern, and Prasad Tadepalli. Bayesian role discovery for multi-agent reinforcement learning. In *AAAI*, 2010.

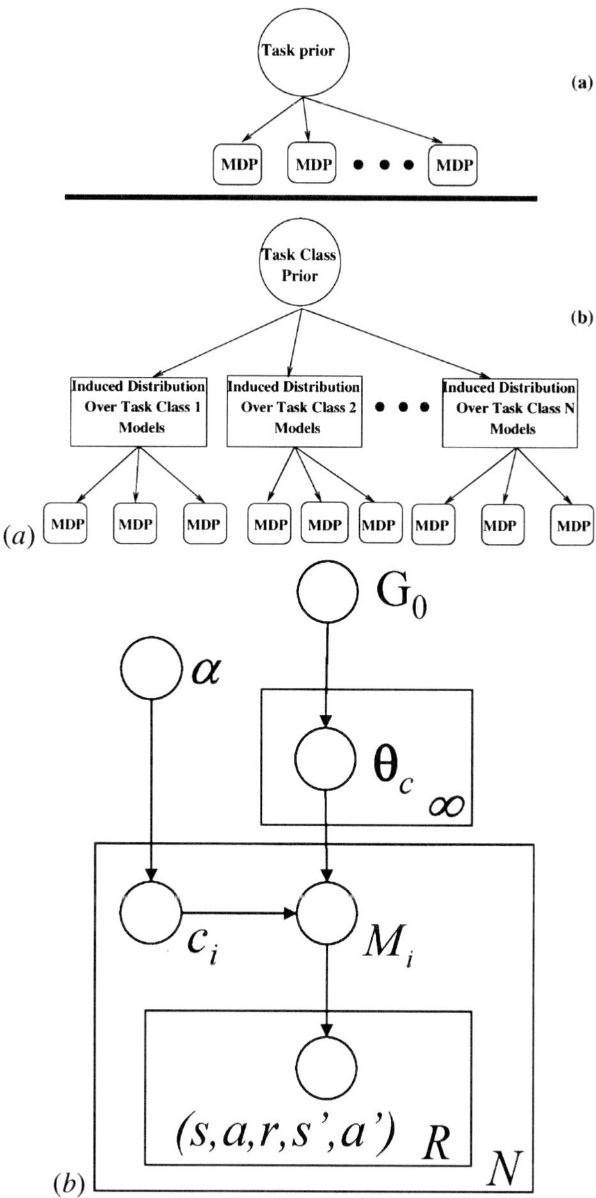


Figure 1: (a) (top) Single task Bayesian RL vs. (bottom) Hierarchical Model-Based Transfer. The number $N$ and parameters of the induced distributions are learned from data. (b) The corresponding Infinite Mixture Model (Parameter set $\Psi$). The task prior is represented by $G_0$, task class distributions have parameters $\theta$, classes are indexed by c, $M_i$ represents the $i^{th}$ MDP model parameters, the tuples $(s,a,r,s',a')$ are the observed transition and reward data, and $\alpha$ is the Dirichlet Process concentration parameter.

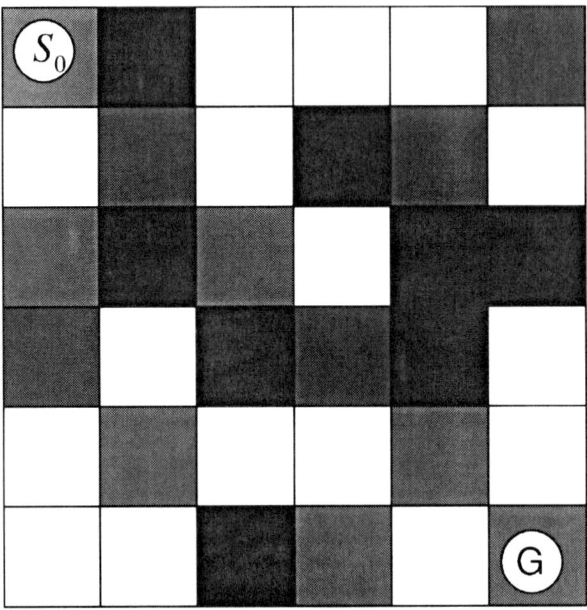



Figure 2: The colored map domain. Tile squares are randomly colored. Each color has a cost. When the agent lands on a square it receives a penalty proportional to the cost of the occupied square and the costs of adjacent squares. The agent's goal is to traverse the least cost path from the top left to the bottom right.

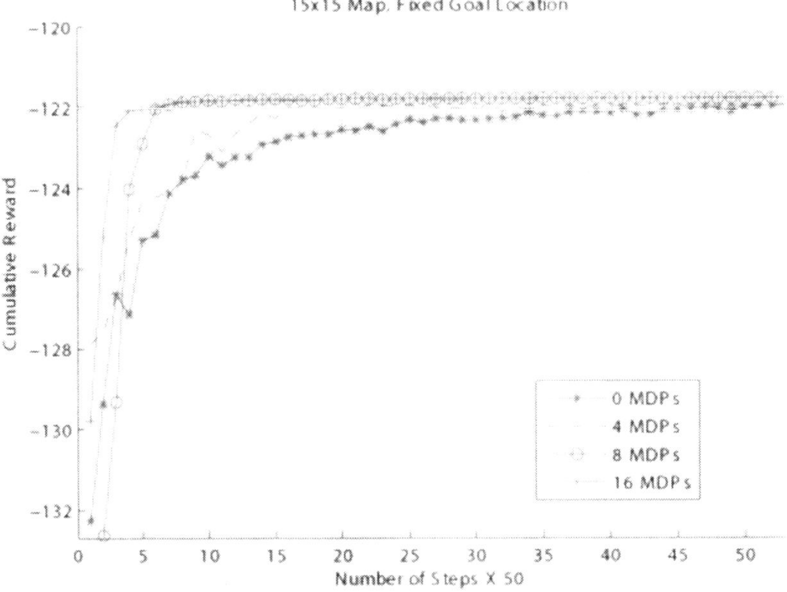



Figure 3: Learning curves for the fixed goal location problem. Performance is evaluated after 0, 4, 8, and 16 tasks have been solved.

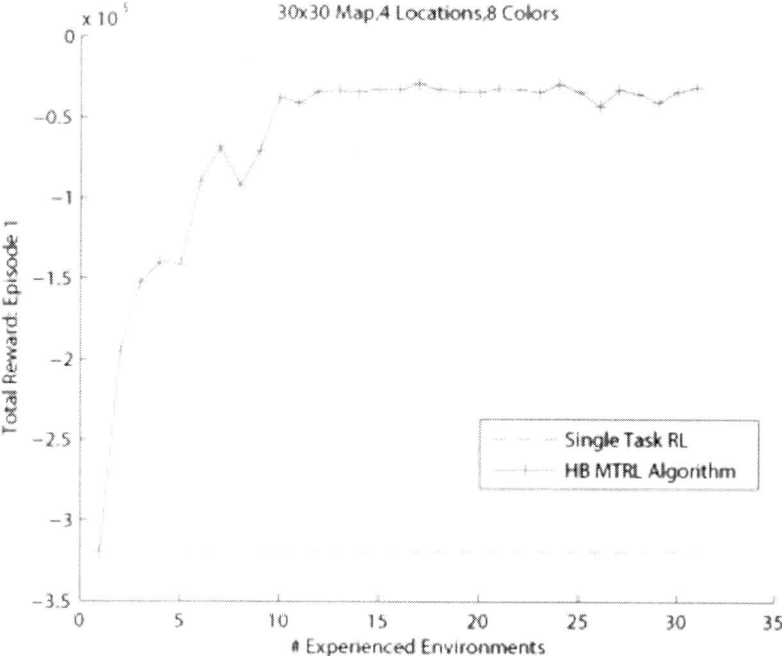



Figure 4: Average cost of finding the hidden goal location in the first episode. Performance is a function of experience in previous tasks.

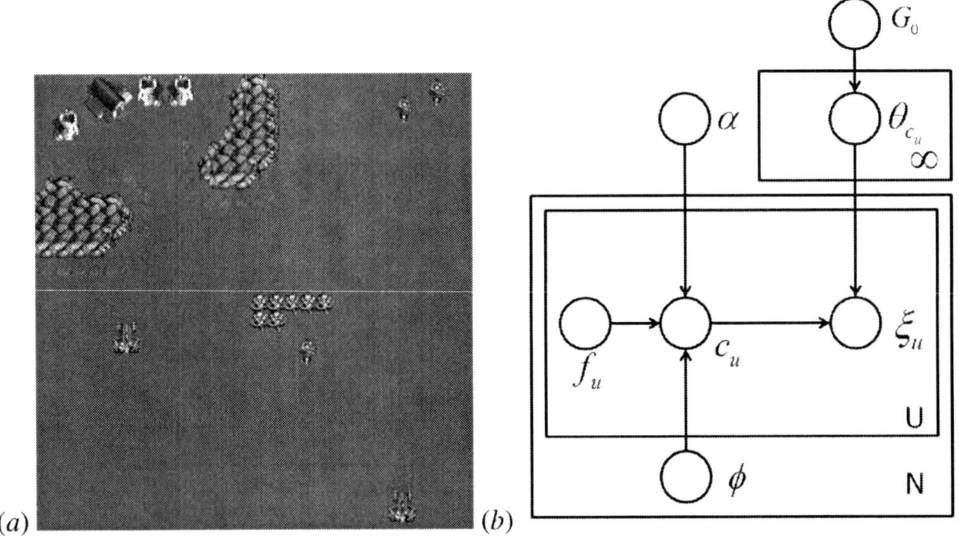



Figure 5: (a) A Wargus map. The friendly (red) team is composed of one ballista, seven archers, and one knight. The agent's objective is to destroy the structures and units on the blue team. The agent must select, at each time step, a target for each unit on its team (targets can include friendly units). (b) The hierarchical prior distribution for unit roles. The set of parameters $\Psi$ includes: The prior distribution on role parameters $G_0$, the set of roles $\theta_c$, the assignment parameters $\phi$, unit features $f_u$, the concentration parameter of the DP model $\alpha$, the assignment variable for each unit $c_u$, and the observed trajectory data $\xi$ generated by executing role based policies. The plate notation indicates that there are $N$ tasks and each task is composed of $U$ units.

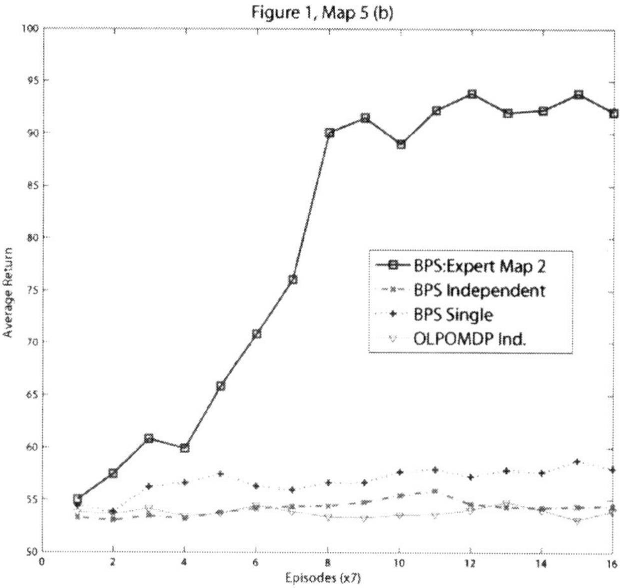


Figure 6: Transfer results of BPS given expert examples of play and tested on a map with 10 units.

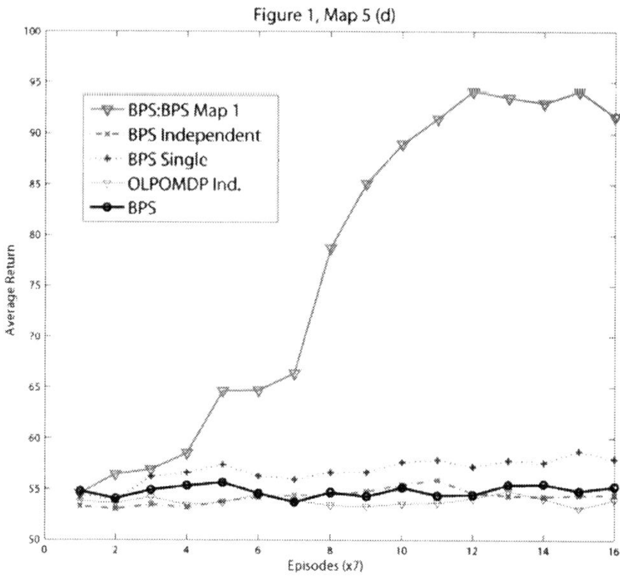


Figure 7: Transfer results of BPS algorithm given training in a simple map with 6 units and tested on a map with 10 units.

# Datasets of the Unsupervised and Transfer Learning Challenge

**Editor:** I. Guyon, G. Dror, V. Lemaire, G. Taylor, and D. Silver

Report prepared by Isabelle Guyon with information from the data donors listed below:

**Handwriting recognition (AVICENNA)** – Reza Farrahi Moghaddam, Mathias Adankon, Kostyantyn Filonenko, Robert Wisnovsky, and Mohamed Chériet (Ecole de technologie supérieure de Montréal, Quebec) contributed the dataset of Arabic manuscripts.

**Human action recognition (HARRY)** – Ivan Laptev and Barbara Caputo collected and made publicly available the KTH human action recognition datasets. Marcin Marszałek, Ivan Laptev and Cordelia Schmid collected and made publicly available the Hollywood 2 dataset of human actions and scenes.

**Object recognition (RITA)** – Antonio Torralba, Rob Fergus, and William T. Freeman, collected and made available publicly the 80 million tiny image dataset. Vinod Nair and Geoffrey Hinton collected and made available publicly the CIFAR datasets. See the techreport Learning Multiple Layers of Features from Tiny Images, by Alex Krizhevsky, 2009, for details.

**Ecology (SYLVESTER)** – Jock A. Blackard, Denis J. Dean, and Charles W. Anderson of the US Forest Service, USA, collected and made available the (Forest cover type) dataset.

**Text processing (TERRY)** – David Lewis formatted and made publicly available the RCV1-v2 Text Categorization Test Collection derived from REUTER news clips.

**The toy example (ULE)** is the MNIST handwritten digit database made available by Yann LeCun and Corinna Costes.

## 1. Data formats

All the data sets are in the same format; xxx should be replaced by one of:

**devel:** development data

**valid:** evaluation data used as validation set

**final:** final evaluation data

Table 1: Datasets of the unsupervised and transfer learning challenge.

| Dataset | Domain | Feat. num. | Sparsity (%) | Development num. | Transfer num. | Validation num. | Final Eval. num. | Data (text) | Data (Matlab) |
|---|---|---|---|---|---|---|---|---|---|
| AVICENNA | Arabic manuscripts | 120 | 0.00 | 150205 | 50000 | 4096 | 4096 | 16 MB | 14 MB |
| HARRY | Human action recognition | 5000 | 98.12 | 69652 | 20000 | 4096 | 4096 | 13 MB | 15 MB |
| RITA | Object recognition | 7200 | 1.19 | 111808 | 24000 | 4096 | 4096 | 1026 MB | 762 MB |
| SYLVESTER | Ecology | 100 | 0.00 | 572820 | 100000 | 4096 | 4096 | 81 MB | 69 MB |
| TERRY | Text recognition | 47236 | 99.84 | 217034 | 40000 | 4096 | 4096 | 73 MB | 56 MB |
| (toy data) | Handwritten digits | 784 | 80.85 | 26808 | 10000 | 4096 | 4096 | 7 MB | 13 MB |

The participants have access only to the files outlined in red:

**dataname.param:** Parameters and statistics about the data

dataname_xxx.data: Unlabeled data (a matrix of space delimited numbers, patterns in lines, features in columns).

dataname_xxx.mat: The same data matrix in Matlab format in a matrix called X_xxx.

dataname_transfer.label: Target values provided for transfer learning only. Multiple labels (1 per column), label values are -1, 0, or 1 (for negative class, unknown, positive class).

**dataname_valid.label:** Target values, not provided to participants.

**dataname_final.label:** Target values, not provided to participants.

**dataname_xxx.dataid:** Identity of the samples (lines of the data matrix).

**dataname_xxx.labelid:** Identity of the labels (variables that are target values, i.e., columns of the label matrix.)

**dataname.classid:** strings representing the names of the classes.

The participants will use the following formats results:

**dataname_valid.prepro:** Preprocessed data send during the development phase.

**dataname_final.prepro:** Preprocessed data for the final submission.

## 2. Metrics

The data representations are assessed automatically by the evaluation platform. To each evaluation set (validation set or final evaluation set) the organizers have assigned several binary classification tasks unknown to the participants. The platform will use the data representations provided by the participants to train a linear classifier (code provided in Appendix) to solve these tasks.

To that end, the evaluation data (validation set or final evaluation set) are partitioned randomly into a training set and a test set. The parameters of the linear classifier are adjusted using the training set. Then, predictions are made on test data using the trained model. The Area Under the ROC curve (AUC) is computed to assess the performance of the linear classifier. The results are averaged over all tasks and over several random splits into a training set and a complementary test set.

The number of training examples is varied and the AUC is plotted against the number of training examples in a log scale (to emphasize the results on small numbers of training examples). The area under the learning curve (ALC) is used as scoring metric to synthesize the results.

The participants are ranked by ALC for each individual dataset. The participants having submitted a **complete experiment** (results on all 5 datasets of the challenge) enter the final ranking. The winner is determined by the best average rank over all datasets for the results of their last complete experiment.

### 2.1. Global Score: The Area under the Learning Curve (ALC)

The prediction performance is evaluated according to the Area under the Learning Curve (ALC). A learning curve plots the Area Under the ROC curve (AUC) averaged over all the binary classification tasks and all evaluation data splits. The AUC is the area of the curve that plots the sensitivity (error rate of the "positive class") vs. the specificity (error rate of the "negative" class).

We consider two baseline learning curves:

1. The ideal learning curve, obtained when perfect predictions are made (AUC=1). It goes up vertically then follows AUC=1 horizontally. It has the maximum area "Amax".

2. The "lazy" learning curve, obtained by making random predictions (expected value of AUC: 0.5). It follows a straight horizontal line. We call its area "Arand".

To obtain our ranking score displayed in Mylab and on the Leaderboard, we normalize the ALC as follows:

$$\text{global\_score} = (\text{ALC} - \text{Arand})/(\text{Amax} - \text{Arand})$$

For simplicity, we call ALC the normalized ALC or global score.

We show in Figure 3 examples of learning curves for the toy example ULE, obtained using the sample code. Note that we interpolate linearly between points. The global score depends on how we scale the x-axis. We use a log2 scaling for all datasets.

# 3. A – ULE



## 3.1. Topic

The task of ULE is handwritten digit recognition.

## 3.2. Sources

### 3.2.1. ORIGINAL OWNERS

The data set was constructed from the MNIST data that is made available by Yann Le-Cun of the NEC Research Institute at `http://yann.lecun.com/exdb/mnist/`.

The digits have been size-normalized and centered in a fixed-size image of dimension $28 \times 28$. We show examples of digits in Figure 1.

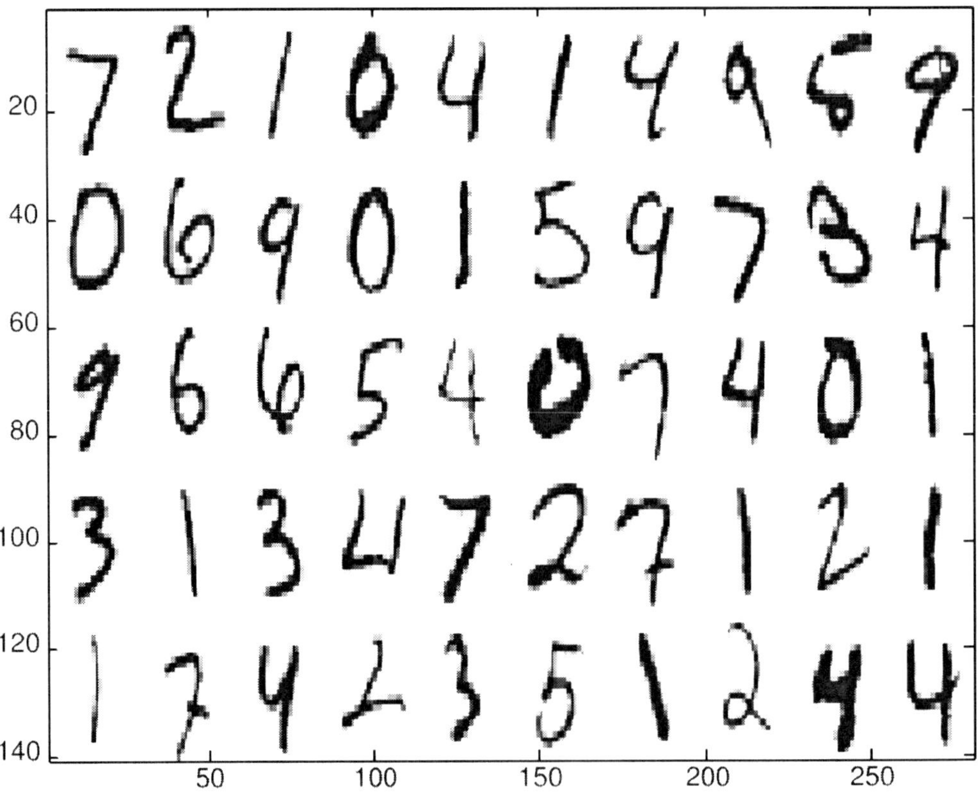

Figure 1: Examples of digits from the MNIST database.

Table 2: Number of examples in the original data

| Digit | 0 | 1 | 2 | 3 | 4 | 5 | 6 | 7 | 8 | 9 | Total |
|---|---|---|---|---|---|---|---|---|---|---|---|
| Training | 5923 | 6742 | 5958 | 6131 | 5842 | 5421 | 5918 | 6265 | 5851 | 5949 | 60000 |
| Test | 980 | 1135 | 1032 | 1010 | 982 | 892 | 958 | 1028 | 974 | 1009 | 10000 |
| Total | 6903 | 7877 | 6990 | 7141 | 6824 | 6313 | 6876 | 7293 | 6825 | 6958 | 70000 |

### 3.2.2. DONOR OF DATABASE

This version of the database was prepared for the "unsupervised and transfer learning challenge" by Isabelle Guyon, 955 Creston Road, Berkeley, CA 94708, USA (`isabelle@clopinet.com`).

### 3.2.3. DATE PREPARED FOR THE CHALLENGE

November 2010.

## 3.3. Past usage

Many methods have been tried on the MNIST database, in its original data split (60,000 training examples, 10,000 test examples, 10 classes.) Table 3 is an abbreviated list from `http://yann.lecun.com/exdb/mnist/`:

This dataset was used in the NIPS 2003 Feature Selection Challenge under the name GISETTE and in the WCCI 2006 Performance Prediction Challenge and the IJCNN 2007 Agnostic Learning vs. Prior Knowledge Challenge under the name GINA.

## References

**Gradient-based learning applied to document recognition.** Y. LeCun, L. Bottou, Y. Bengio, and P. Haffner. In *Proceedings of the IEEE*, 86(11):2278–2324, November 1998.

**Result Analysis of the NIPS 2003 Feature Selection Challenge.** Isabelle Guyon, Asa Ben Hur, Steve Gunn, Gideon Dror, Advances in Neural Information Processing Systems 17, MIT Press, 2004.

**Agnostic Learning vs. Prior Knowledge Challenge.** Isabelle Guyon, Amir Saffari, Gideon Dror, and Gavin Cawley. In *Proceedings IJCNN 2007*, Orlando, Florida, August 2007.

**Analysis of the IJCNN 2007 Agnostic Learning vs. Prior Knowledge Challenge.** Isabelle Guyon, Amir Saffari, Gideon Dror, and Gavin Cawley, Neural Network special anniversary issue, in press. [Earlier draft]

Table 3: Previous results for MNIST (ULE)

| METHOD | TEST ERROR RATE (%) |
|---|---|
| linear classifier (1-layer NN) | 12.0 |
| linear classifier (1-layer NN) [deskewing] | 8.4 |
| pairwise linear classifier | 7.6 |
| K-nearest-neighbors, Euclidean | 5.0 |
| K-nearest-neighbors, Euclidean, deskewed | 2.4 |
| 40 PCA + quadratic classifier | 3.3 |
| 1000 RBF + linear classifier | 3.6 |
| K-NN, Tangent Distance, 16x16 | 1.1 |
| SVM deg 4 polynomial | 1.1 |
| Reduced Set SVM deg 5 polynomial | 1.0 |
| Virtual SVM deg 9 poly [distortions] | 0.8 |
| 2-layer NN, 300 hidden units | 4.7 |
| 2-layer NN, 300 HU, [distortions] | 3.6 |
| 2-layer NN, 300 HU, [deskewing] | 1.6 |
| 2-layer NN, 1000 hidden units | 4.5 |
| 2-layer NN, 1000 HU, [distortions] | 3.8 |
| 3-layer NN, 300+100 hidden units | 3.05 |
| 3-layer NN, 300+100 HU [distortions] | 2.5 |
| 3-layer NN, 500+150 hidden units | 2.95 |
| 3-layer NN, 500+150 HU [distortions] | 2.45 |
| LeNet-1 [with 16x16 input] | 1.7 |
| LeNet-4 | 1.1 |
| LeNet-4 with K-NN instead of last layer | 1.1 |
| LeNet-4 with local learning instead of ll | 1.1 |
| LeNet-5, [no distortions] | 0.95 |
| LeNet-5, [huge distortions] | 0.85 |
| LeNet-5, [distortions] | 0.8 |
| Boosted LeNet-4, [distortions] | 0.7 |
| K-NN, shape context matching | 0.67 |

**Hand on Pattern Recognition, challenges in data representation, model selection, and performance prediction**. Book in preparation. Isabelle Guyon, Gavin Cawley, Gideon Dror, and Amir Saffari Editors.

### 3.4. Experimental design

We used the raw data:

- The feature names are the $(i, j)$ matrix coordinates of the pixels (in a $28 \times 28$ matrix.)

- The data have gray level values between 0 and 255.

- The validation set and the final test set have approximately even numbers of examples for each class.

### 3.5. Number of examples and class distribution

Table 4: Data statistics for ULE

| Dataset | Domain | Feat. num. | Sparsity (%) | Development num. | Transfer num. | Validation num. | Final Eval. num. |
|---------|--------|-----------|--------------|------------------|---------------|-----------------|------------------|
| ULE | Handwriting | 784 | 80.85 | 26808 | 10000 | 4096 | 4096 |

All variables are numeric (no categorical variable). There are no missing values. The target variables are categorical. Here is class label composition of the data subsets:

**Validation set:** X[4096, 784] Y[4096, 1]

```
One:      1370
Three:    1372
Seven:    1354
```

**Final set:** X[4096, 784] Y[4096, 1]

```
Zero:     1376
Two:      1373
Six:      1347
```

**Development set:** X[26808, 784] Y[26808, 1]

```
Zero:     2047
One:      2556
Two:      2089
```

```
Three:    2198
Four:     3426
Five:     3179
Six:      2081
Seven:    2314
Eight:    3470
Nine:     3448
```

**Transfer labels (10000 labels):**

```
Four:     2562
Five:     2301
Eight:    2564
Nine:     2573
```

### 3.6. Type of input variables and variable statistics

The variables in raw data are pixels. We also produced baseline results using as variables Gaussian RBF values with 20 cluster centers generated by the Kmeans clustering algorithm. The algorithm was run on the validation set and the final evaluation set separately. The development set and the transfer labels were not used. The cluster centers are shown in Figure 2.

### 3.7. Baseline results

We used a linear classifier making independence assumptions between variables, similar to Naïve Bayes, to generate baseline learning curves from raw data and preprocessed data. The normalized ALC (score used in the challenge) are shown in Figures 3 and 4 and summarized in Table 5.

Table 5: Baseline results (normalized ALC for 64 training examples).

| ULE | Valid | Final |
|---|---|---|
| Raw | 0.7905 | 0.7169 |
| Preprocessed | 0.8416 | 0.3873 |

## 4.  B – AVICENNA

### 4.1.  Topic

The AVICENNA dataset provides a feature representation of Arabic Historical Manuscripts.

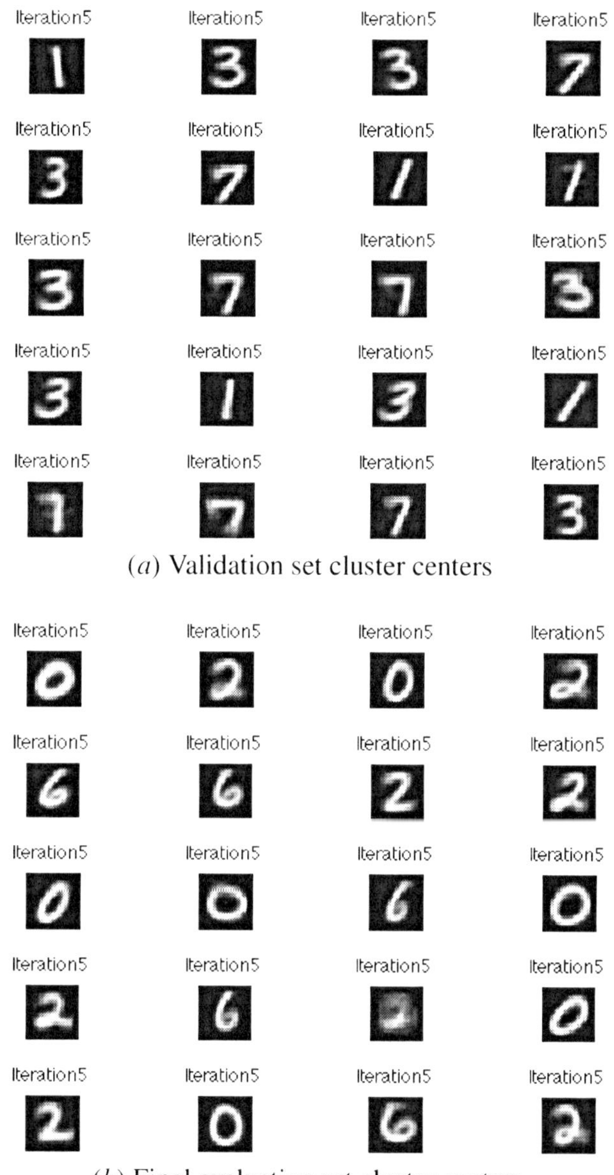

(*a*) Validation set cluster centers

(*b*) Final evaluation set cluster centers

Figure 2: Clusters obtained by Kmeans clustering

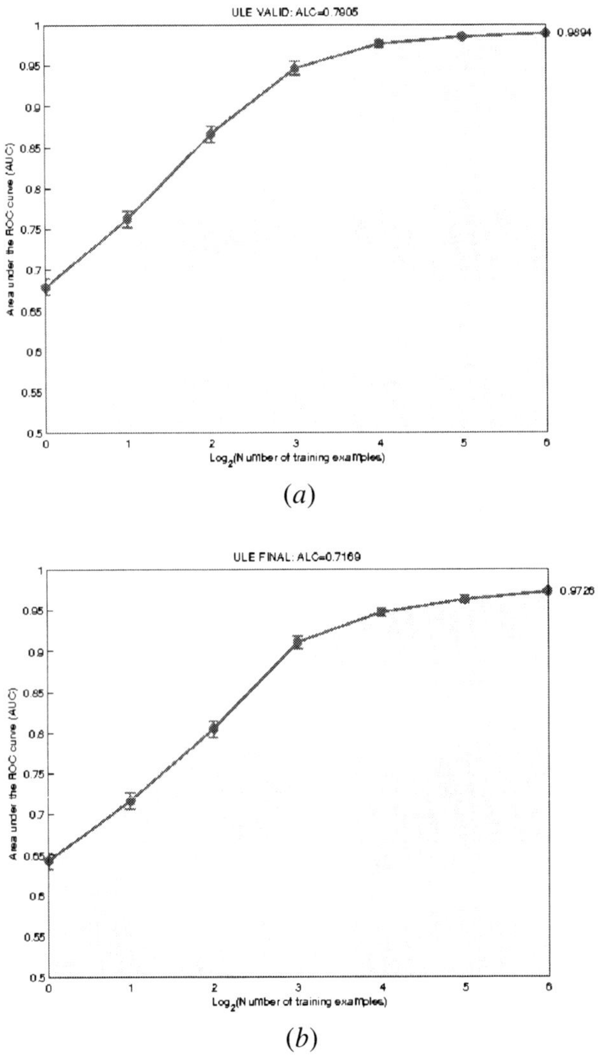

Figure 3: Baseline results on raw ULE data. Top: validation set. Bottom: final evaluation set.

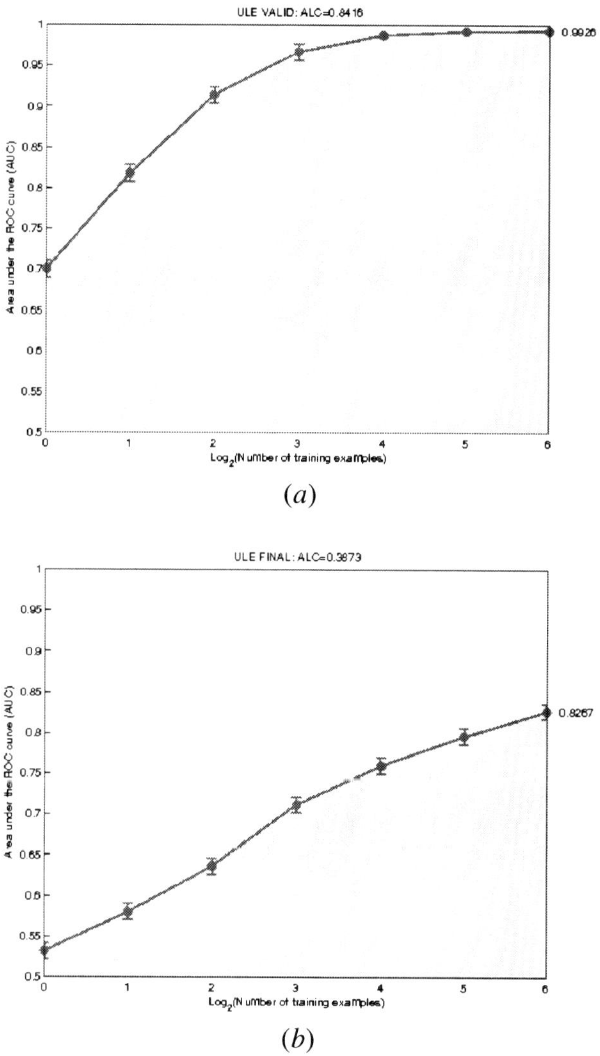



Figure 4:  Baseline results on preprocessed ULE data.  Top:  validation set.  Bottom: final evaluation set.

## 4.2. Sources

4.2.1. ORIGINAL OWNERS

The dataset is prepared on manuscript images provided by The Institute of Islamic Studies (IIS), McGill.

**Manuscript author:** Abu al-Hasan Ali ibn Abi Ali ibn Muhammad al-Amidi (d. 1243 or 1233)

**Manuscript title:** Kitab Kashf al-tamwihat fi sharh al-Tanbīhāt (Commentary on Ibn Sina's al-Isharat wa-al-tanbihat)

**Brief description:** Among the works of Avicenna, his *al-Isharat wa-al-tanbihat* received the attention of the later scholars more than others. The reception of this work is particularly intensive and widespread in the period between the late twelfth century to the first half of the fourteenth century, when more than a dozen comprehensive commentaries on this work were composed. These commentaries were one of the main ways of approaching, understanding and developing Avicenna's philosophy and therefore any study of Post-Avicennian philosophy needs to pay specific attention to this commentary tradition. *Kashf al-tamwihat fi sharh al-Tanbihat* by Abu al-Hasan Ali ibn Abi Ali ibn Muhammad al-Amidi (d. 1243 or 1233), one of the early commentaries written on *al-Isharat wa-al-tanbihat*, is an unpublished commentary which still await scholars' attention.

4.2.2. DONORS OF THE DATABASE

Reza Farrahi Moghaddam, Mathias Adankon, Kostyantyn Filonenko, Robert Wisnovsky, and Mohamed Cheriet.

**Contact:**

> Mohamed Cheriet
> Synchromedia Laboratory
> ETS, Montréal, (QC) Canada H3C 1K3
> `mohamed.cheriet@etsmtl.ca`
> Tel: +1(514)396-8972
> Fax: +1(514)396-8595

4.2.3. DATE RECEIVED:

December 2010

## 4.3. Past usage:

Part of the data was used in the active learning challenge (`http://clopinet.com/al`).

## 4.4. Experimental design

The features were extracted following the procedure described in the JMLR W&CP paper: IBN SINA: A database for handwritten Arabic manuscripts understanding research, by Reza Farrahi Moghaddam, Mathias Adankon, Kostyantyn Filonenko, Robert Wisnovsky, and Mohamed Chériet. The original data includes 92 numeric features. We added 28 distracters then rotated the feature space with a random rotation matrix. Finally, the features were quantized and rescaled between 0 and 999.

## 4.5. Data statistics

Table 6: Data statistics for AVICENNA.

| Dataset | Domain | Feat. num. | Sparsity (%) | Development num. | Transfer num. | Validation num. | Final Eval. num. |
|---|---|---|---|---|---|---|---|
| AVICENNA | Arabic manuscripts | 120 | 0 | 150205 | 50000 | 4096 | 4096 |

Table 7: Original feature statistics

| Name | Type | Min | Max | Num val |
|---|---|---|---|---|
| Aspect_ratio | continuous | 0 | 999 | 395 |
| Horizontal_frequency | ordinal | 1 | 13 | 13 |
| Vertical_CM_ratio | continuous | 0 | 999 | 539 |
| Singular points | continuous | 0 | 238 | 51 |
| Height_ratio | continuous | 0 | 999 | 163 |
| Hole_feature | binary | 0 | 1 | 2 |
| End_points | continuous | 0 | 72 | 43 |
| Dot_feature | binary | 0 | 1 | 2 |
| BP_hole_1 | binary | 0 | 1 | 2 |
| BP_EP_1 | binary | 0 | 1 | 2 |
| BP_BP_1 | binary | 0 | 1 | 2 |
| BP_hole_2 | binary | 0 | 1 | 2 |
| BP_EP_2 | binary | 0 | 1 | 2 |
| BP_BP_2 | binary | 0 | 1 | 2 |
| BP_hole_3 | binary | 0 | 1 | 2 |
| BP_EP_3 | binary | 0 | 1 | 2 |
| BP_BP_3 | binary | 0 | 1 | 2 |
| BP_hole_4 | binary | 0 | 1 | 2 |
| BP_EP_4 | binary | 0 | 1 | 2 |

*Continued overleaf*

*Continued from previous page*

| Name | Type | Min | Max | Num val |
|---|---|---|---|---|
| BP_BP_4 | binary | 0 | 1 | 2 |
| BP_hole_5 | binary | 0 | 1 | 2 |
| BP_EP_5 | binary | 0 | 1 | 2 |
| BP_BP_5 | binary | 0 | 1 | 2 |
| BP_hole_6 | binary | 0 | 1 | 2 |
| BP_EP_6 | binary | 0 | 1 | 2 |
| BP_BP_6 | binary | 0 | 1 | 2 |
| EP_BP_1 | binary | 0 | 1 | 2 |
| EP_EP_1 | binary | 0 | 1 | 2 |
| EP_VCM_1 | ordinal | 0 | 2 | 3 |
| EP_BP_2 | binary | 0 | 1 | 2 |
| EP_EP_2 | binary | 0 | 1 | 2 |
| EP_VCM_2 | ordinal | 0 | 2 | 3 |
| EP_BP_3 | binary | 0 | 1 | 2 |
| EP_EP_3 | binary | 0 | 1 | 2 |
| EP_VCM_3 | ordinal | 0 | 2 | 3 |
| EP_BP_4 | binary | 0 | 1 | 2 |
| EP_EP_4 | binary | 0 | 1 | 2 |
| EP_VCM_4 | ordinal | 0 | 2 | 3 |
| EP_BP_5 | binary | 0 | 1 | 2 |
| EP_EP_5 | binary | 0 | 1 | 2 |
| EP_VCM_5 | ordinal | 0 | 2 | 3 |
| EP_BP_6 | binary | 0 | 1 | 2 |
| EP_EP_6 | binary | 0 | 1 | 2 |
| EP_VCM_6 | ordinal | 0 | 2 | 3 |
| BP_dot_UP_1 | binary | 0 | 1 | 2 |
| BP_dot_DOWN_1 | binary | 0 | 1 | 2 |
| BP_dot_UP_2 | binary | 0 | 1 | 2 |
| BP_dot_DOWN_2 | binary | 0 | 1 | 2 |
| BP_dot_UP_3 | binary | 0 | 1 | 2 |
| BP_dot_DOWN_3 | binary | 0 | 1 | 2 |
| BP_dot_UP_4 | binary | 0 | 1 | 2 |
| BP_dot_DOWN_4 | binary | 0 | 1 | 2 |
| BP_dot_UP_5 | binary | 0 | 1 | 2 |
| BP_dot_DOWN_5 | binary | 0 | 1 | 2 |
| BP_dot_UP_6 | binary | 0 | 1 | 2 |
| BP_dot_DOWN_6 | binary | 0 | 1 | 2 |
| EP_dot_1 | binary | 0 | 1 | 2 |
| EP_dot_2 | binary | 0 | 1 | 2 |

*Continued overleaf*

278

*Continued from previous page*

| Name | Type | Min | Max | Num val |
|---|---|---|---|---|
| EP_dot_3 | binary | 0 | 1 | 2 |
| EP_dot_4 | binary | 0 | 1 | 2 |
| EP_dot_5 | binary | 0 | 1 | 2 |
| EP_dot_6 | binary | 0 | 1 | 2 |
| Dot_dot_1 | binary | 0 | 1 | 2 |
| Dot_dot_2 | binary | 0 | 1 | 2 |
| Dot_dot_3 | binary | 0 | 1 | 2 |
| Dot_dot_4 | binary | 0 | 1 | 2 |
| Dot_dot_5 | binary | 0 | 1 | 2 |
| Dot_dot_6 | binary | 0 | 1 | 2 |
| EP_S_Shape_1 | ordinal | 0 | 2 | 3 |
| EP_clock_1 | ordinal | 0 | 3 | 4 |
| EP_UP_BP_1 | binary | 0 | 1 | 2 |
| EP_DOWN_BP_1 | binary | 0 | 1 | 2 |
| EP_S_Shape_2 | ordinal | 0 | 2 | 3 |
| EP_clock_2 | ordinal | 0 | 3 | 4 |
| EP_UP_BP_2 | binary | 0 | 1 | 2 |
| EP_DOWN_BP_2 | binary | 0 | 1 | 2 |
| EP_S_Shape_3 | ordinal | 0 | 2 | 3 |
| EP_clock_3 | ordinal | 0 | 3 | 4 |
| EP_UP_BP_3 | binary | 0 | 1 | 2 |
| EP_DOWN_BP_3 | binary | 0 | 1 | 2 |
| EP_S_Shape_4 | ordinal | 0 | 2 | 3 |
| EP_clock_4 | ordinal | 0 | 3 | 4 |
| EP_UP_BP_4 | binary | 0 | 1 | 2 |
| EP_DOWN_BP_4 | binary | 0 | 1 | 2 |
| EP_S_Shape_5 | ordinal | 0 | 2 | 3 |
| EP_clock_5 | ordinal | 0 | 3 | 4 |
| EP_UP_BP_5 | binary | 0 | 1 | 2 |
| EP_DOWN_BP_5 | binary | 0 | 1 | 2 |
| EP_S_Shape_6 | ordinal | 0 | 2 | 3 |
| EP_clock_6 | ordinal | 0 | 3 | 4 |
| EP_UP_BP_6 | binary | 0 | 1 | 2 |
| EP_DOWN_BP_6 | binary | 0 | 1 | 2 |

There are no missing values. The data were split as follows:

**Validation set:** `X[4096, 120] Y[4096, 5]`

```
EU:   1113
HU:   875
```

        bL:    1105

        jL:    837

        tL:    1110

**Final set:** X[4096, 120] Y[4096, 5]

        dL:    966

        hL:    1188

        kL:    896

        qL:    982

        sL:    863

**Development set:** X[150205, 120] Y[150205, 52]

        AU:    7

        BU:    2

        CU:    1

        DU:    773

        EU:    4712

        FU:    2

        HU:    506

        IU:    67

        JU:    2

        KU:    552

        LU:    8

        NU:    7

        QU:    182

        RU:    4

        SU:    777

        TU:    372

        VU:    3

        WU:    2

        XU:    161

        YU:    6

        aL:    27219

        bL:    3462

```
cL:    567
dL:    2204
eL:    7
fL:    4225
hL:    6969
iL:    35
jL:    483
kL:    2722
lL:    16345
mL:    9475
nL:    8276
qL:    2270
rL:    4582
sL:    360
tL:    3217
uL:    14
vL:    9750
wL:    468
xL:    557
yL:    9201
zL:    416
```

**Transfer labels (50000 labels):**

```
aL:    25610
lL:    15407
rL:    4301
vL:    9152
yL:    8687
```

## 4.6.  Baseline results

We show first the ridge regression performances obtained by separating one class vs. the rest, training and testing on a balanced subset of examples.

Class 50 -- xL = 619 patterns -- AUC=0.9411
Class 36 -- jL = 1350 patterns -- AUC=0.9168
Class 19 -- SU = 958 patterns -- AUC=0.9135
Class 49 -- wL = 534 patterns -- AUC=0.9134
Class 30 -- dL = 3477 patterns -- AUC=0.9080
Class 20 -- TU = 470 patterns -- AUC=0.9078
Class 4 -- DU = 849 patterns -- AUC=0.9045
Class 45 -- sL = 1274 patterns -- AUC=0.8987
Class 52 -- zL = 537 patterns -- AUC=0.8961
Class 37 -- kL = 3734 patterns -- AUC=0.8861
Class 48 -- vL = 10828 patterns -- AUC=0.8766
Class 34 -- hL = 8677 patterns -- AUC=0.8709
Class 17 -- QU = 194 patterns -- AUC=0.8668
Class 11 -- KU = 597 patterns -- AUC=0.8584
Class 8 -- HU = 1450 patterns -- AUC=0.8555
Class 28 -- bL = 4858 patterns -- AUC=0.8543
Class 5 -- EU = 6103 patterns -- AUC=0.8491
Class 29 -- cL = 677 patterns -- AUC=0.8472
Class 46 -- tL = 4672 patterns -- AUC=0.8434
Class 27 -- aL = 29217 patterns -- AUC=0.8399
Class 43 -- qL = 3437 patterns -- AUC=0.8384
Class 51 -- yL = 10939 patterns -- AUC=0.8342
Class 24 -- XU = 180 patterns -- AUC=0.8270
Class 44 -- rL = 5080 patterns -- AUC=0.8221
Class 40 -- nL = 9209 patterns -- AUC=0.8172
Class 38 -- lL = 18869 patterns -- AUC=0.8138
Class 39 -- mL = 10833 patterns -- AUC=0.7895
Class 32 -- fL = 4709 patterns -- AUC=0.7771
Class 1 -- AU = 10 patterns -- AUC=0.5000
Class 2 -- BU = 2 patterns -- AUC=0.5000
Class 3 -- CU = 1 patterns -- AUC=0.5000
Class 6 -- FU = 3 patterns -- AUC=0.5000
Class 7 -- GU = 0 patterns -- AUC=0.5000
Class 10 -- JU = 2 patterns -- AUC=0.5000
Class 12 -- LU = 8 patterns -- AUC=0.5000
Class 13 -- MU = 1 patterns -- AUC=0.5000
Class 14 -- NU = 8 patterns -- AUC=0.5000
Class 15 -- OU = 0 patterns -- AUC=0.5000
Class 16 -- PU = 0 patterns -- AUC=0.5000
Class 18 -- RU = 6 patterns -- AUC=0.5000
Class 21 -- UU = 0 patterns -- AUC=0.5000
Class 22 -- VU = 5 patterns -- AUC=0.5000
Class 23 -- WU = 2 patterns -- AUC=0.5000

```
Class 25 -- YU = 8 patterns -- AUC=0.5000
Class 26 -- ZU = 0 patterns -- AUC=0.5000
Class 31 -- eL = 7 patterns -- AUC=0.5000
Class 33 -- gL = 0 patterns -- AUC=0.5000
Class 35 -- iL = 41 patterns -- AUC=0.5000
Class 41 -- oL = 0 patterns -- AUC=0.5000
Class 42 -- pL = 0 patterns -- AUC=0.5000
Class 47 -- uL = 16 patterns -- AUC=0.5000
Class 9 -- IU = 79 patterns -- AUC=0.0385
```

The performances of ridge regression are rather good on the classes selected for validation and final testing, when training and testing on a balanced subset of examples (half of the examples ending up in the training set and half in the test set):

**Validation set:**

```
Class 4 -- DU = 837 patterns -- AUC=0.8802
Class 2 -- BU = 875 patterns -- AUC=0.8193
Class 3 -- CU = 1105 patterns -- AUC=0.8172
Class 5 -- EU = 1110 patterns -- AUC=0.7938
Class 1 -- AU = 1113 patterns -- AUC=0.7470
```

**Final evaluation set:**

```
Class 1 -- AU = 966 patterns -- AUC=0.9348
Class 3 -- CU = 896 patterns -- AUC=0.8910
Class 2 -- BU = 1188 patterns -- AUC=0.8663
Class 5 -- EU = 863 patterns -- AUC=0.8336
Class 4 -- DU = 982 patterns -- AUC=0.7712
```

However, when we make learning curves, the classes are not well balanced and the number of training examples is small, so the performances are not as good. We show results on raw data in Figure 5. The baseline results obtained by preprocessing with K-means clustering are even worse. Note that we verified that rotating the space and quantizing does not harm performance. The baseline results indicate that this dataset is much harder than ULE.

Table 8: Baseline results (normalized ALC for 64 training examples).

| AVICENNA | Valid | Final |
| --- | --- | --- |
| Raw | 0.1034 | 0.1501 |
| Preprocessed | 0.0856 | 0.0973 |

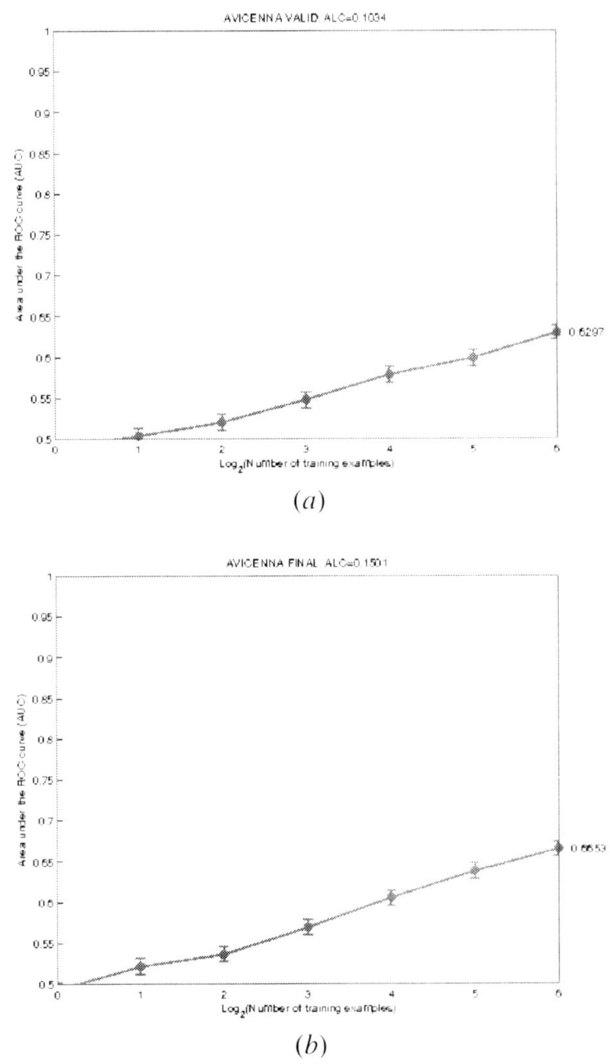

Figure 5: Baseline results on raw data (top valid, bottom final).

# 5. C – HARRY

## 5.1. Topic

The task of HARRY (Human Action Recognition) is action recognition in movies.

Figure 6: Action Recognition in Movies

## 5.2. Sources

### 5.2.1. ORIGINAL OWNERS

Ivan Laptev and Barbara Caputo collected and made publicly available the KTH human action recognition datasets. Marcin Marszałek, Ivan Laptev and Cordelia Schmid collected and made publicly available the Hollywood 2 dataset of human actions and scenes.

We are grateful to Graham Taylor for providing us with the data in preprocessed STIP feature format and for providing Matlab code to read the format and create a bag-of-STIP-features representation.

### 5.2.2. DONOR OF DATABASE

This version of the database was prepared for the "unsupervised and transfer learning challenge" by Isabelle Guyon, 955 Creston Road, Berkeley, CA 94708, USA (`isabelle@clopinet.com`).

### 5.2.3. DATE PREPARED FOR THE CHALLENGE:

November–December 2010.

## 5.3. Past Usage

The original Hollywood-2 dataset contains 12 classes of human actions and 10 classes of scenes distributed over 3669 video clips and approximately 20.1 hours of video in total. The dataset intends to provide a comprehensive benchmark for human action recognition in realistic and challenging settings. The dataset is composed of video clips extracted from 69 movies, it contains approximately 150 samples per action class and 130 samples per scene class in training and test subsets. A part of this dataset was originally used in the paper "Actions in Context", Marszałek et al. in Proc. CVPR'09. Hollywood-2 is an extension of the earlier Hollywood dataset.

The feature representation called STIP on which we based the preprocessing have been successfully used for action recognition in the paper "Learning Realistic Human Actions from Movies", Ivan Laptev, Marcin Marszałek, Cordelia Schmid and Benjamin Rozenfeld; in Proc. CVPR'08. See also the on-line paper description `http://www.irisa.fr/vista/actions/`.

The results on classifying KTH actions reported by the authors are listed in Table 9.

Table 9: Results on classifying KTH actions reported by authors

| Method | Schuldt et al. [icpr04] | Niebles et al. [bmvc06] | Wong et al. [iccv07] | ours |
|---|---|---|---|---|
| Accuracy | 71.7% | 81.5% | 86.7% | 91.8% |

And those from Hollywood movie actions are listed in Table 10.

Table 10: Hollywood movie actions

| | Clean | Automatic | Chance |
|---|---|---|---|
| AnswerPhone | 32.1% | 16.4% | 10.6% |
| GetOutCar | 41.5% | 16.4% | 6.0% |
| HandShake | 32.3% | 9.9% | 8.8% |
| HugPerson | 40.6% | 26.8% | 10.1% |
| Kiss | 53.3% | 45.1% | 23.5% |
| SitDown | 38.6% | 24.8% | 13.8% |
| SitUp | 18.2% | 10.4% | 4.6% |
| StandUp | 50.5% | 33.6% | 22.6% |

The Automatic training set was constructed using automatic action annotation based on movie scripts and contains over 60% correct action labels. The Clean training set was obtained by manually correcting the Automatic set.

## 5.4. Experimental Design

The data were preprocessed into STIP features using the code of Ivan Laptev: `http://www.irisa.fr/vista/Equipe/People/Laptev/download/stip-1.0-winlinux.zip`.

The STIP features are described in:

**"On Space-Time Interest Points"** (2005), I. Laptev; in *International Journal of Computer Vision*, vol 64, number 2/3, pp.107–123.

This yielded both HOG and HOF features for every video frame (in the original format, there are 6 ints followed by 1 float confidence value followed by 162 float HOG/HOF features). The code does not implement scale selection, Instead interest points are detected at multiple spatial and temporal scales. The implemented descriptors HOG (Histograms of Oriented Gradients) and HOF (Histograms of Optical Flow) are computed for 3D video patches in the neighborhood of detected STIPs.

The final representation is a "bag of STIP features". The vectors of HOG/HOF features were clustered into 5000 clusters (we used the KTH data for clustering), using on on-line version of the kmeans algorithm. Each video frame was then assigned to its closest cluster center. We obtained a sparse representation of 5000 features, each feature representing the frequency of presence of a given STIP feature cluster center in a video clip.

To create a large dataset of video examples, the original videos were cut in smaller clips:

Each Hollywood2 movie clip was further split into 40 subsequences and each KTH movie clip was further split into 4 subsequences. Not normalization for sequence length was performed.

## 5.5. Data statistics

Table 11: Data statistics for HARRY

| Dataset | Domain | Feat. num. | Sparsity (%) | Development num. | Transfer num. | Validation num | Final eval. num. |
|---------|--------|------------|--------------|------------------|---------------|----------------|------------------|
| HARRY | Human Action Recognition | 5000 | 98.12 | 69652 | 20000 | 4096 | 4096 |

All variables are numeric (no categorical variable). There are no missing values. The target variables are categorical. The patterns and categories selected for the validation and final evaluation sets are all from the KTH dataset. Here is class label composition of the data subsets:

**Validation set:** X[4096, 5000] Y[4096, 3]

```
    boxing:          1370

    handclapping:    1377

    jogging:         1349
```

**Final set:** X[4096, 5000] Y[4096, 3]

```
    handwaving:      1360

    running:         1369
```

```
walking:        1367
```

**Development set:** `X[69652, 5000] Y[69652, 18]`

```
boxing:          218
handclapping:    207
handwaving:      232
jogging:         251
running:         231
walking:         233
AnswerPhone:     5200
DriveCar:        7480
Eat:             2920
FightPerson:     4960
GetOutCar:       4320
HandShake:       3080
HugPerson:       5200
Kiss:            8680
Run:             11040
SitDown:         8480
SitUp:           2440
StandUp:         11120
```

**Transfer labels (20000 labels):**

```
DriveCar:        5831
Eat:             2213
FightPerson:     3847
Run:             8547
```

## 5.6. Baseline results

The data were preprocessed with kmeans clustering as described in Section 3.

# 6. D – RITA

## 6.1. Topic

The task of RITA (Recognition of Images of Tiny Area) is object recognition.

Table 12: Baseline results (normalized ALC for 64 training examples).

| **HARRY** | Valid | Final |
|---|---|---|
| Raw | **0.6264** | **0.6017** |
| Preprocessed | 0.2230 | 0.2292 |

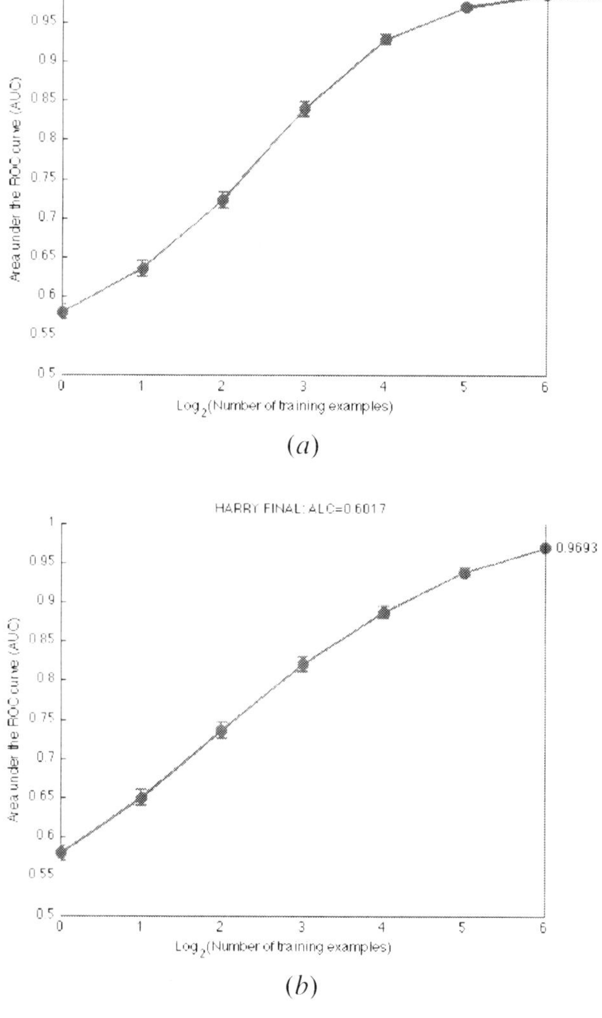



Figure 7: Baseline results on raw data (top valid, bottom final).

Figure 8: Recognition of Images of Tiny Area

## 6.2. Sources

### 6.2.1. ORIGINAL OWNERS

Antonio Torralba, Rob Fergus, and William T. Freeman, collected and made available publicly the 80 million tiny image dataset. Vinod Nair and Geoffrey Hinton collected and made available publicly the CIFAR datasets.

### 6.2.2. DONOR OF DATABASE

This version of the database was prepared for the "unsupervised and transfer learning challenge" by Isabelle Guyon, 955 Creston Road, Berkeley, CA 94708, USA (`isabelle@clopinet.com`).

### 6.2.3. DATE PREPARED FOR THE CHALLENGE:

December 2010.

## 6.3. Past usage

**Learning Multiple Layers of Features from Tiny Images,** by Alex Krizhevsky, Master thesis, Univ. Toronto, 2009.

**Semi-Supervised Learning in Gigantic Image Collections,** Rob Fergus, Yair Weiss and Antonio Torralba, *Advances in Neural Information Processing Systems (NIPS).*

See also many other citations of CIFAR-10 and CIFAR-100 on Google.

## 6.4. Experimental design

We merged the CIFAR-10 and the CIFAR-100 datasets. The CIFAR-10 dataset consists of 60000 $32 \times 32$ colour images in 10 classes, with 6000 images per class. The original categories are:

**airplane**

**automobile**

**bird**

**cat**

**deer**

**dog**

**frog**

**horse**

**ship**

**truck**

The CIFAR-100 dataset is similar to the CIFAR-10, except that it has 100 classes containing 600 images each. The 100 classes in the CIFAR-100 are grouped into 20 superclasses. Each image comes with a "fine" label (the class to which it belongs) and a "coarse" label (the superclass to which it belongs).

Table 13 lists the classes in the CIFAR-100

The raw data came as $32 \times 32$ tiny images coded with 8-bit RGB colors (i.e. $3 \times 32$ features with 256 possible values). We converted RGB to HSV and quantized the results as 8-bit integers. This yielded $30 \times 30 \times 3 = 900 \times 3$ features. We then preprocessed the gray level image to extract edges. This yielded $30 \times 30$ features (1 border pixel was removed). We then cut the images into patches of $10 \times 10$ pixels and ran kmeans clustering (an on-line version) to create 144 cluster centers. We used these cluster centers as a dictionary to create features corresponding to the presence of one the 144 shapes at one of 25 positions on a grid. This created another $144 \times 25 = 3600$ features.

## 6.5. Data statistics

All variables are numeric (no categorical variable). There are no missing values. The target variables are categorical. All the categories of the validation and final evaluation sets are from the CIFAR-10 dataset. Here is class label composition of the data subsets:

**Validation set:** `X[4096, 7200] Y[4096, 3]`

```
automobile:   1330
```

Table 13: Classes in the CIFAR-100

| **Superclass** | **Classes** |
| --- | --- |
| fish | aquarium fish, flatfish, ray, shark, trout |
| flowers | orchids, poppies, roses, sunflowers, tulips |
| food containers | bottles, bowls, cans, cups, plates |
| fruit and vegetables | apples, mushrooms, oranges, pears, sweet peppers |
| household electrical devices | clock, computer keyboard, lamp, telephone, television |
| household furniture | bed, chair, couch, table, wardrobe |
| insects | bee, beetle, butterfly, caterpillar, cockroach |
| large carnivores | bear, leopard, lion, tiger, wolf |
| large man-made outdoor things | bridge, castle, house, road, skyscraper |
| large natural outdoor scenes | cloud, forest, mountain, plain, sea |
| large omnivores and herbivores | camel, cattle, chimpanzee, elephant, kangaroo |
| medium-sized mammals | fox, porcupine, possum, raccoon, skunk |
| non-insect invertebrates | crab, lobster, snail, spider, worm |
| people | baby, boy, girl, man, woman |
| reptiles | crocodile, dinosaur, lizard, snake, turtle |
| small mammals | hamster, mouse, rabbit, shrew, squirrel |
| trees | maple, oak, palm, pine, willow |
| vehicles 1 | bicycle, bus, motorcycle, pickup truck, train |
| vehicles 2 | lawn-mower, rocket, streetcar, tank, tractor |

Table 14: Data statistics for RITA

| Dataset | Domain | Feat. num. | Sparsity (%) | Development num. | Transfer num. | Validation num | Final eval. num. |
| --- | --- | --- | --- | --- | --- | --- | --- |
| RITA | Object recognition | 7200 | 1.19 | 111808 | 24000 | 4096 | 4096 |

Figure 9: 144 cluster centers computed from patches of line images.

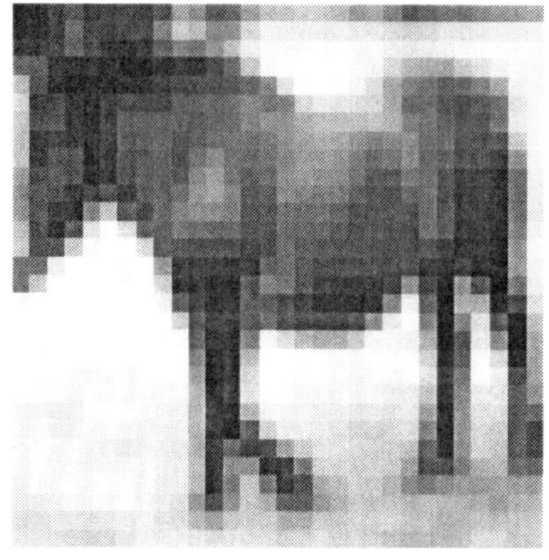

Figure 10: Example of tiny image.

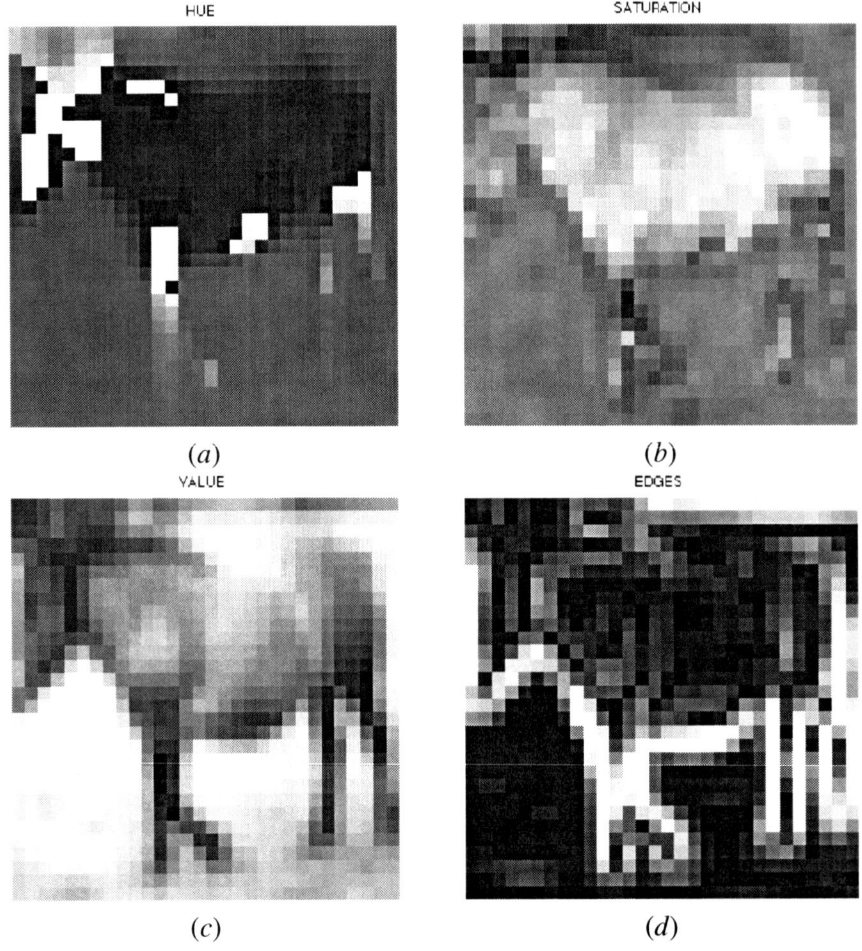



(a)  (b)  (c)  (d)

Figure 11: Image represented by Hue, Saturation, Value, and Edges´ (3600 features). We computed another 3600 features from the edge image using the matched filters computed by clustering.

```
horse:          1377
truck:          1389
```

**Final set:** X[4096, 7200] Y[4096, 3]

```
airplane:   1384
frog:       1370
ship:       1342
```

**Development set:** X[111808, 7200] Y[111808, 110]

```
airplane:                                    4616
automobile:                                  4670
bird:                                        6000
cat:                                         6000
deer:                                        6000
dog:                                         6000
frog:                                        4630
horse:                                       4623
ship:                                        4658
truck:                                       4611
fruit_and_vegetables.apple:                  600
fish.aquarium_fish:                          600
people.baby:                                 600
large_carnivores.bear:                       600
aquatic_mammals.beaver:                      600
household_furniture.bed:                     600
insects.bee:                                 600
insects.beetle:                              600
vehicles_1.bicycle:                          600
food_containers.bottle:                      600
food_containers.bowl:                        600
people.boy:                                  600
large_man-made_outdoor_things.bridge:        600
vehicles_1.bus:                              600
insects.butterfly:                           600
```

large_omnivores_and_herbivores.camel:          600
food_containers.can:                           600
large_man-made_outdoor_things.castle:          600
insects.caterpillar:                           600
large_omnivores_and_herbivores.cattle:         600
household_furniture.chair:                      600
large_omnivores_and_herbivores.chimpanzee:     600
household_electrical_devices.clock:            600
large_natural_outdoor_scenes.cloud:            600
insects.cockroach:                             600
household_furniture.couch:                     600
non-insect_invertebrates.crab:                 600
reptiles.crocodile:                            600
food_containers.cup:                           600
reptiles.dinosaur:                             600
aquatic_mammals.dolphin:                       600
large_omnivores_and_herbivores.elephant:       600
fish.flatfish:                                 600
large_natural_outdoor_scenes.forest:           600
medium_mammals.fox:                            600
people.girl:                                   600
small_mammals.hamster:                         600
large_man-made_outdoor_things.house:           600
large_omnivores_and_herbivores.kangaroo:       600
household_electrical_devices.keyboard:         600
household_electrical_devices.lamp:             600
vehicles_2.lawn_mower:                         600
large_carnivores.leopard:                      600
large_carnivores.lion:                         600
reptiles.lizard:                               600
non-insect_invertebrates.lobster:              600
people.man:                                    600
trees.maple_tree:                              600
vehicles_1.motorcycle:                         600

```
large_natural_outdoor_scenes.mountain:          600
small_mammals.mouse:                            600
fruit_and_vegetables.mushroom:                  600
trees.oak_tree:                                 600
fruit_and_vegetables.orange:                    600
flowers.orchid:                                 600
aquatic_mammals.otter:                          600
trees.palm_tree:                                600
fruit_and_vegetables.pear:                      600
vehicles_1.pickup_truck:                        600
trees.pine_tree:                                600
large_natural_outdoor_scenes.plain:             600
food_containers.plate:                          600
flowers.poppy:                                  600
medium_mammals.porcupine:                       600
medium_mammals.possum:                          600
small_mammals.rabbit:                           600
medium_mammals.raccoon:                         600
fish.ray:                                       600
large_man-made_outdoor_things.road:             600
vehicles_2.rocket:                              600
flowers.rose:                                   600
large_natural_outdoor_scenes.sea:               600
aquatic_mammals.seal:                           600
fish.shark:                                     600
small_mammals.shrew:                            600
medium_mammals.skunk:                           600
large_man-made_outdoor_things.skyscraper:       600
non-insect_invertebrates.snail:                 600
reptiles.snake:                                 600
non-insect_invertebrates.spider:                600
small_mammals.squirrel:                         600
vehicles_2.streetcar:                           600
flowers.sunflower:                              600
```

```
fruit_and_vegetables.sweet_pepper:           600
household_furniture.table:                   600
vehicles_2.tank:                             600
household_electrical_devices.telephone:      600
household_electrical_devices.television:     600
large_carnivores.tiger:                      600
vehicles_2.tractor:                          600
vehicles_1.train:                            600
fish.trout:                                  600
flowers.tulip:                               600
reptiles.turtle:                             600
household_furniture.wardrobe:                600
aquatic_mammals.whale:                       600
trees.willow_tree:                           600
large_carnivores.wolf:                       600
people.woman:                                600
non-insect_invertebrates.worm:               600
```

**Transfer labels (24000 labels):**

```
bird:   6000
cat:    6000
deer:   6000
dog:    6000
```

## 6.6. Baseline results

The data were preprocessed with kmeans clustering as described in Section 3.

Table 15: Baseline results (normalized ALC for 64 training examples).

| **RITA** | Valid | Final |
|---|---|---|
| Raw | **0.2504** | **0.4133** |
| Preprocessed | 0.2417 | 0.3413 |

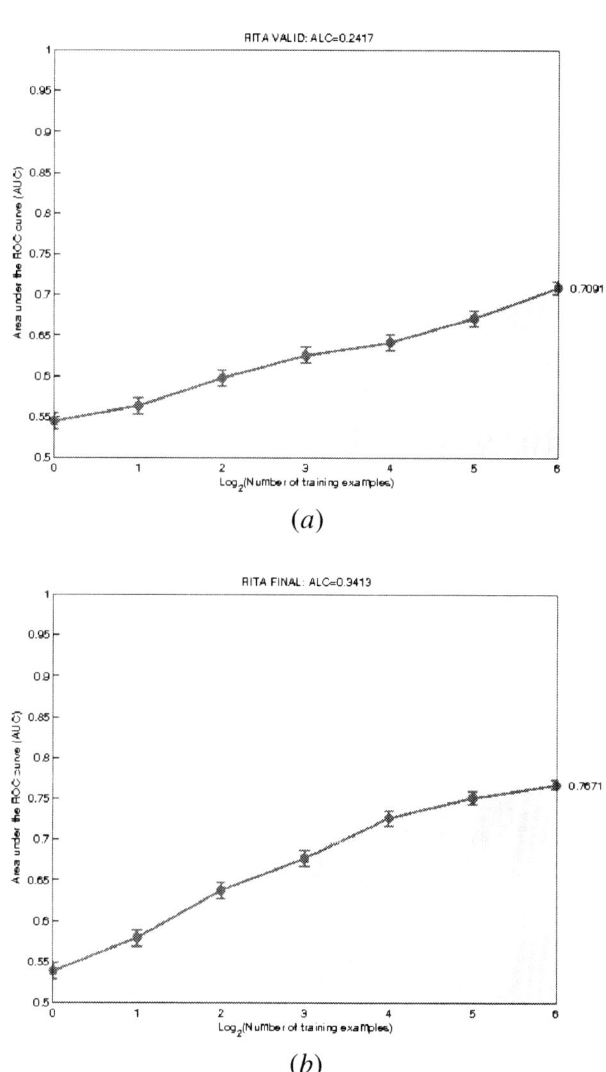

Figure 12: Baseline results on preprocessed data (top valid, bottom final).

# 7. E – SYLVESTER

## 7.1. Topic

The task of SYLVESTER is to classify forest cover types. The task was carved out of data from the US Forest Service (USFS). The data include 7 labels corresponding to forest cover types. We used 2 for transfer learning (training), 2 for validation and 3 for testing.

## 7.2. Sources

### 7.2.1. ORIGINAL OWNERS

Remote Sensing and GIS Program
Department of Forest Sciences
College of Natural Resources
Colorado State University
Fort Collins, CO 80523


(contact Jock A. Blackard, `jblackard/wo_ftcol@fs.fed.us`
or Dr. Denis J. Dean, `denis@cnr.colostate.edu`)
Jock A. Blackard
USDA Forest Service 3825 E. Mulberry
Fort Collins, CO 80524 USA
`jblackard/wo_ftcol@fs.fed.us`


Dr. Denis J. Dean
Associate Professor
Department of Forest Sciences
Colorado State University
Fort Collins, CO 80523 USA
`denis@cnr.colostate.edu`


Dr. Charles W. Anderson
Associate Professor
Department of Computer Science
Colorado State University
Fort Collins, CO 80523 USA
`anderson@cs.colostate.edu`

**Acknowledgements, Copyright Information, and Availability**   Reuse of this database is unlimited with retention of copyright notice for Jock A. Blackard and Colorado State University.

### 7.2.2. Donor of database

This version of the database was prepared for the "unsupervised and transfer learning challenge" by Isabelle Guyon, 955 Creston Road, Berkeley, CA 94708, USA (`isabelle@clopinet.com`).

### 7.2.3. Date received (original data):

August 28, 1998, UCI Machine Learning Repository, under the name Forest Cover Type.

### 7.2.4. Date prepared for the challenge:

September–November 2010.

## 7.3. Past usage

Blackard, Jock A. 1998. "Comparison of Neural Networks and Discriminant Analysis in Predicting Forest Cover Types." Ph.D. dissertation. Department of Forest Sciences. Colorado State University. Fort Collins, Colorado.

Classification performance with first 11,340 records used for training data, next 3,780 records used for validation data, and last 565,892 records used for testing data subset: – 70% backpropagation – 58% Linear Discriminant Analysis.

The subtask SYLVA prepared for the "performance prediction challenge" and the "agnostic learning vs. prior knowledge" (ALvsPK) challenge is a 2-class classification problem (Ponderosa pine vs. others). The best results were obtained with Logitboost by Roman Lutz who obtained 0.4% error in the PK track and 0.6% error in the AL track. See `http://clopinet.com/isabelle/Projects/agnostic/Results.html`. The data were also used in the "active learning challenge" under the name "SYLVA" during the development phase and "F" (for FOREST) during the final test phase. The best entrants (Intel team) obtained a 0.8 area under the learning curve, see `http://www.causality.inf.ethz.ch/activelearning.php?page=results`.

## 7.4. Experimental design

The original data comprises a total of 581012 instances (observations) grouped in 7 classes (forest cover types) and having 54 attributes (features) corresponding to 12 measures (10 quantitative variables, 4 binary wilderness areas and 40 binary soil type variables). The actual forest cover type for a given observation ($30 \times 30$ meter cell) was determined from US Forest Service (USFS) Region 2 Resource Information System (RIS) data. Independent variables were derived from data originally obtained from US Geological Survey (USGS) and USFS data. Data is in raw form (not scaled) and contains binary (0 or 1) columns of data for qualitative independent variables (wilderness areas and soil types).

Given in Table 16 are the variable name, variable type, the measurement unit and a brief description. The forest cover type is the classification problem. The order of this listing corresponds to the order of numerals along the rows of the database.

### Table 16: Variable Information for SYLVESTER

| Name | Data Type | Measurement | Description |
|---|---|---|---|
| Elevation | quantitative | meters | Elevation in meters |
| Aspect | quantitative | azimuth | Aspect in degrees azimuth |
| Slope | quantitative | degrees | Slope in degrees |
| Horizontal_Distance_To_Hydrology | quantitative | meters | Horz Dist to nearest surface water features |
| Vertical_Distance_To_Hydrology | quantitative | meters | Vert Dist to nearest surface water features |
| Horizontal_Distance_To_Roadways | quantitative | meters | Horz Dist to nearest roadway |
| Hillshade_9am | quantitative | 0 to 255 index | Hillshade index at 9am, summer solstice |
| Hillshade_Noon | quantitative | 0 to 255 index | Hillshade index at noon, summer soltice |
| Hillshade_3pm | quantitative | 0 to 255 index | Hillshade index at 3pm, summer solstice |
| Horizontal_Distance_To_Fire_Points | quantitative | meters | Horz Dist to nearest wildfire ignition points |
| Wilderness_Area | (4 binary columns) qualitative | 0 (absence) or 1 (presence) | Wilderness area designation |
| Soil_Type | (40 binary columns) qualitative | 0 (absence) or 1 (presence) | Soil Type designation |
| Cover_Type | (7 types) integer | 1 to 7 | Forest Cover Type designation |

### 7.4.2. Code Designations

**Wilderness Areas:**

1 – Rawah Wilderness Area

2 – Neota Wilderness Area

3 – Comanche Peak Wilderness Area

4 – Cache la Poudre Wilderness Area

**Soil Types:**

**1 to 40** : based on the USFS Ecological Landtype Units for this study area.

**Forest Cover Types:**

1 – Spruce/Fir

2 – Lodgepole Pine

3 – Ponderosa Pine

4 – Cottonwood/Willow

5 – Aspen

6 – Douglas-fir

7 – Krummholz

### 7.4.3. Class Distribution

| | |
|---|---|
| Number of records of Spruce-Fir: | 211840 |
| Number of records of Lodgepole Pine: | 283301 |
| Number of records of Ponderosa Pine: | 35754 |
| Number of records of Cottonwood/Willow: | 2747 |
| Number of records of Aspen: | 9493 |
| Number of records of Douglas-fir: | 17367 |
| Number of records of Krummholz: | 20510 |
| **Total records:** | 581012 |

### 7.4.4. Data preprocessing and data split

We mixed mixed the classes to get approximately the same error rate in baseline results on the validation set and the final evaluation set.

We used the original data encoding from the data donors, transformed by an invertible linear transform (an isometry). To make it even harder to go back to the original data, non-informative features (distractors) were added, corresponding to randomly permuted column values of the original features, before applying the isometry. We then randomized the order of the features and patterns. We quantized the values between 0 and 999.

## 7.5. Number of examples and class distribution

Table 17: Statistics on the SYLVESTER data

| Dataset | Domain | Feat. type | Feat. num. | Sparsity (%) | Label | Development num. | Transfer num. | Validation num | Final eval. num. |
|---|---|---|---|---|---|---|---|---|---|
| SYLVESTER | Ecology | Numeric | 100 | 0 | Binary | 572820 | 10000 | 4096 | 4096 |

There are no missing values. Here is class label composition of the data subsets:

**Validation set:** X[4096, 100] Y[4096, 1]

```
Ponderosa Pine:    2044
Aspen:             2052
```

**Final set:** X[4096, 100] Y[4096, 1]

```
Spruce/Fir:    1319
Douglas-fir:   1404
Krummholz:     1373
```

**Development set:** X[572820, 100] Y[572820, 1]

```
Spruce/Fir:          210521
Lodgepole Pine:      283301
Ponderosa Pine:      33710
Cottonwood/Willow:   2747
Aspen:               7441
Douglas-fir:         15963
Krummholz:           19137
```

**Transfer labels (10000 labels):**

```
Lodgepole Pine:     9891
Cottonwood/Willow:  109
```

## 7.6. Type of input variables and variable statistics

100 numeric variables transformed via a random isometry from the raw input variables to which 46 distractors were added. The distractors were obtained by picking real variables and randomizing the order of the values. The final variables were quantized between 0 and 999.

## 7.7. Baseline results

We show results using our baseline classifier shown in appendix. The prepreprocessing in kmeans clustering (20 clusters).

Table 18: Baseline results (normalized ALC for 64 training examples).

| **SYLVESTER** | Valid | Final |
|---|---|---|
| Raw | **0.2167** | **0.3095** |
| Preprocessed | 0.1670 | 0.2362 |

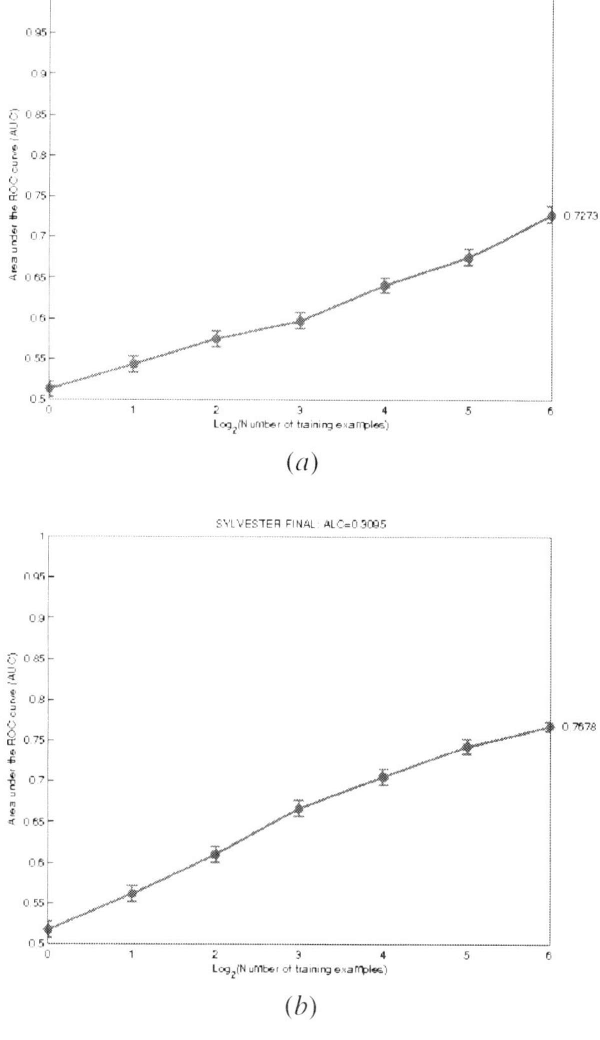

Figure 13: Baseline results on raw data (top valid, bottom final).

## 8. F – TERRY

### 8.1. Topic

The task of TERRY is the Text Recognition dataset.

### 8.2. Sources

#### 8.2.1. ORIGINAL OWNERS

The data were donated by Reuters and downloaded from: Lewis, D. D. RCV1-v2/ LYRL2004: The LYRL2004 Distribution of the RCV1-v2 Text Categorization Test Collection (12-Apr-2004 Version). `http://www.jmlr.org/papers/volume5/ lewis04a/lyrl2004_rcv1v2_README.htm`.

#### 8.2.2. DONOR OF DATABASE

This version of the database was prepared for the "unsupervised and transfer learning challenge" by Isabelle Guyon, 955 Creston Road, Berkeley, CA 94708, USA (`isabelle@clopinet.com`).

#### 8.2.3. DATE PREPARED FOR THE CHALLENGE:

November–December 2010.

### 8.3. Past usage

Lewis, D. D.; Yang, Y.; Rose, T.; and Li, F. RCV1: A New Benchmark Collection for Text Categorization Research. Journal of Machine Learning Research, 5:361-397, 2004. `http://www.jmlr.org/papers/volume5/lewis04a/lewis04a. pdf`.

### 8.4. Experimental design

We used a subset of the 800,000 documents of the RCV1-v2 data collection, formatted in a bag-of-words representation. The representation uses 47,236 unique stemmed tokens. The representation was obtained from on-line appendix B.13. The list of stems was found in on-line appendix B14. We used as target values the topic categories (on-line appendices 3 and 8). We considered all levels of the hierarchy to select the most promising categories.

The features were obfuscated by making a non-linear transformation of the values then quantizing them between 0 and 999. Further, the raws and lines of the data matrix were permuted.

### 8.5. Data statistics

All variables are numeric (no categorical variable). There are no missing values. The target variables are categorical. The data are very sparse, so they were stored in a sparse matrix. Here is class label composition of the data subsets:

Table 19: Data statistics for TERRY

| Dataset | Domain | Feat. num. | Sparsity (%) | Development num. | Transfer num. | Validation num | Final eval. num. |
|---|---|---|---|---|---|---|---|
| TERRY | Text recognition | 47236 | 99.84 | 217034 | 40000 | 4096 | 4096 |

**Validation set:** X[4096, 47236] Y[4096, 5]

```
ENERGY MARKETS:                        808
EUROPEAN COMMUNITY:                    886
PRIVATISATIONS:                        817
MANAGEMENT:                            863
ENVIRONMENT AND NATURAL WORLD:         826
```

**Final set:** X[4096, 47236] Y[4096, 5]

```
SPORTS:                       797
CREDIT RATINGS:               804
DISASTERS AND ACCIDENTS:      829
ELECTIONS:                    856
LABOUR ISSUES:                829
```

**Development set:** X[217034, 47236] Y[217034, 103]

```
STRATEGY/PLANS:                   6944
LEGAL/JUDICIAL:                   2898
REGULATION/POLICY:               10279
SHARE LISTINGS:                   2166
PERFORMANCE:                     42290
ACCOUNTS/EARNINGS:               21832
ANNUAL RESULTS:                   2243
COMMENT/FORECASTS:               21315
INSOLVENCY/LIQUIDITY:             494
FUNDING/CAPITAL:                 11885
SHARE CAPITAL:                    5378
BONDS/DEBT ISSUES:                3147
LOANS/CREDITS:                     705
```

CREDIT RATINGS:                     1453
OWNERSHIP CHANGES:                  13853
MERGERS/ACQUISITIONS:               11739
ASSET TRANSFERS:                    1312
PRIVATISATIONS:                     1370
PRODUCTION/SERVICES:                7749
NEW PRODUCTS/SERVICES:              1967
RESEARCH/DEVELOPMENT:               751
CAPACITY/FACILITIES:                8895
MARKETS/MARKETING:                  11832
DOMESTIC MARKETS:                   1199
EXTERNAL MARKETS:                   1999
MARKET SHARE:                       282
ADVERTISING/PROMOTION:              513
CONTRACTS/ORDERS:                   4360
DEFENCE CONTRACTS:                  339
MONOPOLIES/COMPETITION:             1264
MANAGEMENT:                         2245
MANAGEMENT MOVES:                   2044
LABOUR:                             2971
CORPORATE/INDUSTRIAL:               105241
ECONOMIC PERFORMANCE:               2462
MONETARY/ECONOMIC:                  7044
MONEY SUPPLY:                       632
INFLATION/PRICES:                   1924
CONSUMER PRICES:                    1642
WHOLESALE PRICES:                   288
CONSUMER FINANCE:                   615
PERSONAL INCOME:                    84
CONSUMER CREDIT:                    63
RETAIL SALES:                       365
GOVERNMENT FINANCE:                 12008
EXPENDITURE/REVENUE:                4066
GOVERNMENT BORROWING:               8052

```
OUTPUT/CAPACITY:                   679
INDUSTRIAL PRODUCTION:             482
CAPACITY UTILIZATION:              13
INVENTORIES:                       30
EMPLOYMENT/LABOUR:                 4087
UNEMPLOYMENT:                      484
TRADE/RESERVES:                    6412
BALANCE OF PAYMENTS:               933
MERCHANDISE TRADE:                 3994
RESERVES:                          546
HOUSING STARTS:                    104
LEADING INDICATORS:                1556
ECONOMICS:                         33239
EUROPEAN COMMUNITY:                5554
EC INTERNAL MARKET:                945
EC CORPORATE POLICY:               559
EC AGRICULTURE POLICY:             620
EC MONETARY/ECONOMIC:              2219
EC INSTITUTIONS:                   561
EC ENVIRONMENT ISSUES:             50
EC COMPETITION/SUBSIDY:            524
EC EXTERNAL RELATIONS:             1142
EC GENERAL:                        18
GOVERNMENT/SOCIAL:                 63881
CRIME, LAW ENFORCEMENT:            8380
DEFENCE:                           2506
INTERNATIONAL RELATIONS:           11105
DISASTERS AND ACCIDENTS:           1488
ARTS, CULTURE, ENTERTAINMENT:      1078
ENVIRONMENT AND NATURAL WORLD:     790
FASHION:                           76
HEALTH:                            1744
LABOUR ISSUES:                     4161
OBITUARIES:                        184
```

```
HUMAN INTEREST:                            667
DOMESTIC POLITICS:                         15654
BIOGRAPHIES, PERSONALITIES, PEOPLE:        1668
RELIGION:                                  804
SCIENCE AND TECHNOLOGY:                    638
SPORTS:                                    8671
TRAVEL AND TOURISM:                        223
WAR, CIVIL WAR:                            9323
ELECTIONS:                                 3539
WEATHER:                                   821
WELFARE, SOCIAL SERVICES:                  484
EQUITY MARKETS:                            12424
BOND MARKETS:                              6179
MONEY MARKETS:                             13574
INTERBANK MARKETS:                         7279
FOREX MARKETS:                             6599
COMMODITY MARKETS:                         21557
SOFT COMMODITIES:                          12155
METALS TRADING:                            3092
ENERGY MARKETS:                            5162
MARKETS:                                   51279
```

**Transfer labels (40000 labels):**

```
DOMESTIC POLITICS:     12865
MONEY MARKETS:         11322
REGULATION/POLICY:     8508
GOVERNMENT FINANCE:    9900
```

## 8.6. Baseline results

The data were preprocessed with kmeans clustering as described in Section 3.

We see in Table 20 and Figure 14 that the performances in preprocessed data in the final evaluation set are not good. This is another example of preprocessing overfiting: we used the clusters found with the validation set to preprocess the test set.

Table 20: Baseline results (normalized ALC for 64 training examples).

| TERRY | Valid | Final |
|---|---|---|
| Raw | **0.6969** | **0.7550** |
| Preprocessed | 0.6602 | 0.3440 |

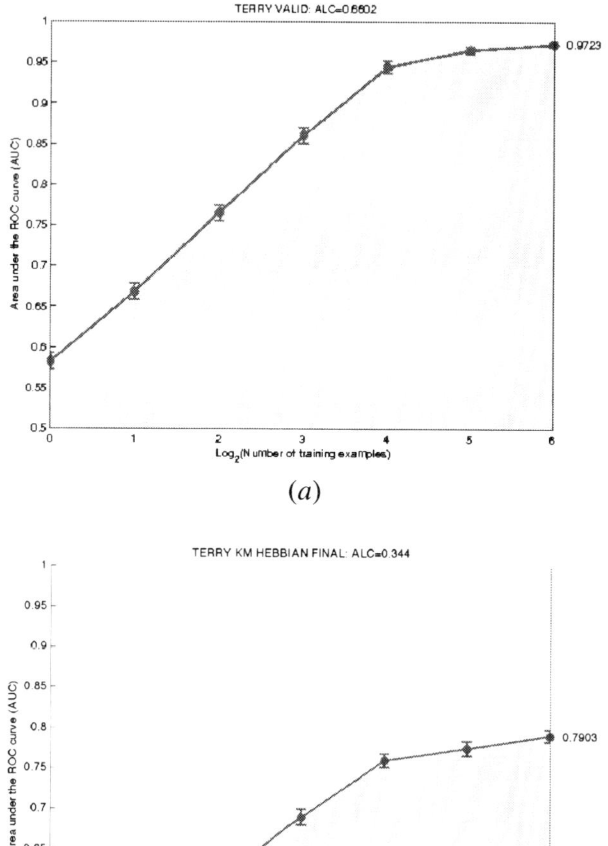



(a)

(b)

Figure 14: Baseline results on preprocessed data (top valid, bottom final).

## Appendix

Code for the linear classifier

```matlab
function [data, model]=train(model, data)
%[data, model]=train(model, data)
% Simple linear classifier with Hebbian-style learning.
% Inputs:
% model     -- A hebbian learning object.
% data      -- A data object.
% Returns:
% model     -- The trained model.
% data      -- A new data structure containing the results.
% Usually works best with standardized data.
% Standardization is not performed here for computational
% reasons (we put it outside the CV loop).


% Isabelle Guyon -- isabelle@clopinet.com -- November 2010


if model.verbosity>0
    fprintf('==> Training Hebbian classifier ...  ');
end


Posidx=find(data.Y>0);
Negidx=find(data.Y<0);


if pd_check(data)
    % Kernelized version
    model.W=zeros(1, length(data.Y));
    model.W(Posidx)=1/(length(Posidx)+eps);
    model.W(Negidx)=-1/(length(Negidx)+eps);
else
    n=size(data.X, 2);
    Mu1=zeros(1, n); Mu2=zeros(1, n);
    if ~isempty(Posidx)
        Mu1=mean(data.X(Posidx,:), 1);
    end
    if ~isempty(Negidx)
        Mu2=mean(data.X(Negidx,:), 1);
    end
```

```matlab
    model.W=Mu1-Mu2;
    B=(Mu1+Mu2)/2;
    model.b0=-model.W*B';
end


% Test the model
if model.test_on_training_data
    data=test(model, data);
end


if model.verbosity>0, fprintf('done\n'); end
```

CPSIA information can be obtained at www.ICGtesting.com
Printed in the USA
LVOW09*2333300713

345489LV00010B/78/P